# Small Cities

*Small Cities* explores the economic, political, socio-spatial and cultural practices and processes relating to small cities. While intense attention from a range of disciplines and by policy makers has been paid over the past twenty years to the large metropolitan centre, there is a growing awareness that cities further down the urban hierarchy have as much to tell us about contemporary urban change. Drawing together a team of international contributors, a small-cities research and policy agenda is discussed in a single volume for the first time.

Among the topics addressed in *Small Cities* are:

- What roles do small cities occupy in the urban hierarchy?
- What are the economic, political, socio-spatial and cultural practices and processes that specifically relate to small cities?
- What kinds of local and regional development strategies have been used by small cities?
- How are small cities produced and consumed? How is 'smallness' commodified?
- How are issues of identity, lifestyle, social exclusion, multiculturalism and tourism mediated by and experienced in small cities?

With case studies from around the world, *Small Cities* raises important conceptual issues surrounding urban change and the urban hierarchy. Illustrated throughout, this comprehensive and integrated book will be essential reading for all those seeking to understand urban practices and processes outside the metropolis.

**David Bell** is Senior Lecturer in Human Geography at the University of Leeds.

**Mark Jayne** is Lecturer in Human Geography at the University of Manchester.

## Questioning Cities

Edited by Gary Bridge, *University of Bristol, UK* and
Sophie Watson, *The Open University, UK*

The 'Questioning Cities' series brings together an unusual mix of urban scholars under the title. Rather than taking a broadly economic approach, planning approach or more socio-cultural approach, it aims to include titles from a multidisciplinary field of those interested in critical urban analysis. The series thus includes authors who draw on contemporary social, urban and critical theory to explore different aspects of the city. It is not therefore a series made up of books which are largely case studies of different cities and predominantly descriptive. It seeks instead to extend current debates, through in most cases, excellent empirical work, and to develop sophisticated understandings of the city from a number of disciplines including geography, sociology, politics, planning, cultural studies, philosophy and literature. The series also aims to be thoroughly international where possible, to be innovative, to surprise, and to challenge received wisdom in urban studies. Overall it will encourage a multidisciplinary and international dialogue always bearing in mind that simple description or empirical observation which is not located within a broader theoretical framework would not – for this series at least – be enough.

**Global Metropolitan**
Globalizing cities in a capitalist world
*John Rennie Short*

**Reason in the City of Difference**
Pragmatism, communicative action and contemporary urbanism
*Gary Bridge*

**In the Nature of Cities**
Urban political ecology and the politics of urban metabolism
*Edited by Nik Heynen, Maria Kaika and Erik Swyngedouw*

**Ordinary Cities**
Between modernity and development
*Jennifer Robinson*

**Urban Space and Cityscapes**
Perspectives from modern and contemporary culture
*Edited by Christoph Lindner*

**City Publics**
The (dis)enchantments of urban encounters
*Sophie Watson*

**Small Cities**
Urban experience beyond the metropolis
*Edited by David Bell and Mark Jayne*

**Cities and Race**
America's new black ghetto
*David Wilson*

**Cities in Globalization**
Practices, policies and theories
*Edited by Peter J. Taylor, Ben Derudder, Piet Saey and Frank Witlox*

# Small Cities

## Urban experience beyond the metropolis

**Edited by David Bell and Mark Jayne**

First published 2006 by Routledge
2 Park Square, Milton Park, Abingdon, Oxon OX14 4RN

Simultaneously published in the USA and Canada by Routledge
270 Madison Avenue, New York, NY 10016

*Routledge is an imprint of the Taylor & Francis Group, an informa business*



Typeset in Times and Bauhaus by
Bookcraft Ltd, Stroud, Gloucestershire
Printed and bound in Great Britain by
MPG Books Ltd, Bodmin



*British Library Cataloguing in Publication Data*
A catalogue record for this book is available from the British Library

*Library of Congress Cataloging-in-Publication Data*
Small cities : urban experience beyond the metropolis / edited by David Bell and Mark Jayne.
p. cm. — (Questioning cities series)
Includes bibliographical references and index.
1. Cities and towns. 2. Urban policy.
I. Bell, David, 1965 Feb. 12– II. Jayne, Mark, 1970– III. Series.
HT151.S523 2006
307.76—dc22 2005037581

ISBN10: 0-415-36657-7 (hbk)
ISBN10: 0-415-36658-5 (pbk)
ISBN10: 0-203-01926-1 (ebk)

ISBN13: 978-0-415-36657-1 (hbk)
ISBN13: 978-0-415-36658-8 (pbk)
ISBN13: 978-0-203-01926-9 (ebk)

# Contents

# Illustrations

## Figures



## Tables

# Contributors

**David Bell** teaches Human Geography at the University of Leeds, UK. His recent publications include *Ordinary Lifestyles* (edited with Joanne Hollows; Open University Press 2005), *Science, Technology and Culture* (Open University Press 2005) and *Cyberculture Theorists* (Routledge 2006).

**Andrew Bradley** is Lecturer in Sports Tourism and Events Management in Gloucestershire Business School, UK. His PhD, from the School of Environment, University of Gloucestershire, looked at the relationship between festivals and events, place promotion and local economic development. He has also researched and published across a range of pedagogic research areas.

**T. C. Chang** is Associate Professor at the Department of Geography at National University of Singapore. He received his PhD at McGill University (Canada) in 1997 and his research interests include tourism in South-East Asia, urban redevelopment and conservation, and arts and culture. He is the co-editor of *Interconnected Worlds: Tourism in Southeast Asia* (Pergamon 2001).

**Frank Eckardt** is Professor of Urban Sociology at the Institute for European Urban Studies at the Bauhaus-University Weimar, Germany. He holds a PhD in Political Science and has published many articles and books on related subjects. He is currently the coordinator of the *Future of Urban Research* training programme of the European Commission (2005–9). His most recent book is *Soziologie der Stadt* (Transcript Publishers 2004).

**Tim Edensor** teaches Human Geography at Manchester Metropolitan University, UK. He is the author of *Tourists at the Taj* (Routledge 1998), *National Identity, Popular Culture and Everyday Life* (Berg 2002) and *Industrial Ruins: Space, Aesthetics and Materiality* (Berg 2005). He has written widely on tourism, automobility, walking and Mauritius.

**Graeme Evans** is Director of the Cities Institute at London Metropolitan University, UK. He is a principal investigator on two EPSRC-funded research projects under the Sustainable Urban Environments programme, *VivaCity* and *AUNT-SUE*, studying urban design, inclusion, compact city living and transport. He leads an international comparative project, *Creative Spaces*, for the London Development Agency and City of Toronto that is reviewing creative industry strategies worldwide. Graeme's research interests include urban design, city planning and urban cultures. He has written widely in academic and professional publications, including the book *Cultural Planning: An Urban Renaissance?* (Routledge 2001) and led the review of *Evidence on the Role of Culture in Regeneration* for the Department for Culture, Media and Sport (2004).

**Tom Fleming** is Director of TF Creative Consultancy, UK, and specializes in research and support for the cultural sector and creative industries. Key areas of specialism include local and regional cultural and creative industries strategies, cultural planning and culture-led regeneration, diversity and creativity, and the establishment of targeted support mechanisms that include approaches to finance, business support, network and cluster development. Tom has developed a portfolio of work that focuses on the key issues for the economies, cultures and creativity of cities, regions and nations. Recent clients include the UK's National Endowment for Science, Technology and the Arts, The Council of Europe, Arts Council England, The British Council, The London Development Agency and The European Commission.

**Jo Foord** is Principal Research Fellow in the Cities Institute at London Metropolitan University, UK. She is currently working on an international study of *Creative Spaces* for the London Development Agency; a comparative project of urban sustainability in Canadian and British cities; and *VivaCity* researching the quality of life in mixed-use inner-city neighbourhoods. Jo has a research and teaching background in Human Geography and has worked in local government. She has researched and published on local cultural policy and creative industries, consumption and retail change, and gendered employment, all within the context of the uneven development of city spaces.

**Lia Ghilardi** is the founder and director of Noema Research and Planning Ltd, an international cultural planning consultancy. As well as assisting public institutions in the implementation of cultural planning strategies across Europe, she has worked regularly as a member of the special committee of advisers for the selection of projects concerning the regeneration of cities through cultural initiatives for DG XVI of the European Commission. Her academic work involves lecturing on issues of urban cultural development in universities in the UK, Sweden and Italy. She writes on issues of cultural policies, identity, heritage and diversity and her work has featured in *Cultural Policy and Cultural Diversity* (Tony Bennett; Council of Europe 2001), *Culture and Neighbourhood: A Comparative Report* (with Franco Bianchini), *City of Quarters* (Bell and Jayne; Ashgate 2004) and *ŒCultural Planning: Thinking Culturally about Diversity in A Manual of Cultural Diversity* (Arts Council England, forthcoming).

**Tim Hall** is Lecturer in the School of Environment, University of Gloucestershire, UK. His research interests include urban and cultural geography. He is the author of *Everyday Geography* (2006), *Urban Geography* (Routledge 1998; 2002; 2006) and co-editor of *The Entrepreneurial City* (Wiley 1998); *The City Cultures Reader* (Routledge 2000; 2003) and *Urban Futures* (Routledge 2003).

**Tim Hewitt** was an undergraduate in the School of Earth and Environmental Sciences at the University of Wollongong, Australia.

**Mark Jayne** is Lecturer in Human Geography at the University of Manchester, UK. His research interests include consumption, urban and regional regeneration, cultural economy and the creative industries. Mark has published numerous articles in journals such as *Progress in Human Geography, Environment and Planning A, The International Journal of Cultural Policy, Environment and Planning C: Government and Policy* and *Capital and Class*. His books include *Cities and Consumption* and *City of Quarters: Urban Villages in the Contemporary City* (edited with David Bell).

**Ellen Kraly** (Ellen Percy) is Professor of Geography (1984) at Colgate University, Hamilton, New York State, USA. From 2000 to 2004 she held the position of Director, Division of Social Sciences, Colgate University. Ellen has a wide range of interests in demography and population geography including population measures, statistical systems, international migration, ethnicity, and the relationships between environment, population and immigration policy. She has written numerous articles and book chapters and her work has featured in books such as *New York's Immigrants* (Foner 2001) and *Immigration Today: Pastoral and Research Challenges* (Powers 2000). Ellen's professional experience includes her roles as a panellist, National Academy of Sciences; and consultant to the National Science Foundation, United Nations Statistical Office, US Commission on Immigration Reform, US Immigration and Naturalization Service, US General Accounting Office, and US Bureau of the Census.

**Loretta Lees** is Reader in Geography at King's College London, UK. She has published numerous articles on North American and British gentrification and has recently edited *The Emancipatory City? Paradoxes and Possibilities* (Sage 2004).

**Joseph Leibovitz** is Lecturer in Urban and Regional Studies at the Institute of Geography, the University of Edinburgh, UK. His research and teaching interests are in comparative urban and regional political economy, the impact of European integration on urban and regional transformation, and planning theory. Joseph has conducted research on American and Canadian urban politics, the restructuring of local government in different national settings, and economic change in less-favoured regions. He is currently undertaking research on urban citizenship and ethnic mobilization in Israeli cities.

**Greg Lloyd** is Head of the School of Town and Regional Planning at the University of Dundee, UK. He has served as an Adviser to the Scottish Affairs Committee, the Welsh Assembly Government and currently serves on the Tayside Economic Forum. He has published widely with respect to spatial planning, community planning and regeneration. He has undertaken numerous funded research studies and is currently involved in research into modernizing planning practice, community planning and local vulnerability, and 'what works'. He has published a number of books, articles in journals and research reports.

**John McCarthy** is Senior Lecturer at the School of Town and Regional Planning at the University of Dundee, UK. He worked formerly in local government and sits on the Dundee Partnership. He has published widely in the fields of urban regeneration and spatial planning. He has undertaken funded research studies for the Economic and Social Research Council, Scottish Executive and Carnegie Trust for Scottish Universities, and is currently involved in research into culture-led regeneration and waterfront regeneration in historic port cities.

**Malcolm Miles** is Reader in Cultural Theory at the University of Plymouth, UK, author of *Arts, Space & the City* (Routledge 1997) and *Urban Avant-Gardes* (Routledge 2004), and co-editor of *The City Cultures Reader* (Routledge 2003, 2nd edn).

**Steven Miles** is Senior Lecturer in Sociology at the University of Liverpool, UK, where he is also part of the management team of the Liverpool Model, a longitudinal research project looking at the impacts of European Capital of Culture 2008 conducted by the University of Liverpool and Liverpool John Moores University. He is author of *Consuming Cities* (with Malcolm Miles) and recently co-edited a review issue of *Urban Studies* on Culture-Led Regeneration (with Ronan Paddison). Steven was formerly Head of Research at the Centre for Cultural Policy and Management at Northumbria University.

**Nancy K. Napier** is Professor of International Business and Executive Director of the Global Business Consortium at Boise State University, USA. She was a former Associate Dean of the College of Business and Economics and Chairman of the Management Department. She also managed Boise State's nine-year involvement in an $A8.5 Swedish- and USAID-funded capacity-building project at the National Economics University in Hanoi, Vietnam. Her most recent book is *Managing Relationships in Transition Economies* (with D. Thomas), published by Praeger. Her articles appear in such journals as *Journal of Management Inquiry*, *Human Resource Management Journal*, *Human Resource Planning*, *Organisation*, *Academy of Management Review*, *Journal of Management Studies*, and *Journal of International Business Studies*.

**Deborah Peel** is Lecturer at the School of Town and Regional Planning at the University of Dundee, UK. She is a Member of the Royal Town Planning Institute and sits on the RTPI Scottish Executive Committee and Planning Summer School Council. Formerly she worked in local government. She has published widely with respect to spatial planning, community planning and regeneration. She has undertaken funded research studies for the Scottish Executive, Carnegie Trust for Scottish Universities, Calouste Gulbenkian and the Development Trusts Association Foundation and is currently involved in research into modernizing planning practice, marine spatial planning and community planning.

**John Joe Schlichtman** is a recent graduate of the New York University Department of Sociology, and Visiting Assistant Professor at Queens College, City University of New York, Flushing, New York, USA. His current research interests are in the areas of urban political economy, economic change, development strategies and homelessness.

**AbdouMaliq Simone** is an urbanist in the broad sense that his work focuses on various communities, powers, cultural expressions, governance and planning discourses, spaces and times in cities across the world. He is Professor of Sociology at Goldsmiths College, UK, and he has taught at New School University, the University of Witwatersrand, the University of Khartoum, University of Ghana, University of the Western Cape and the City University of New York, as well as working for several African non-governmental organizations and regional institutions. Key publications include *In Whose Image?: Political Islam and Urban Practices in Sudan* (University of Chicago Press 1994) and *For the City Yet to Come: Changing Urban Life in Four African Cities* (Duke University Press 2004).

**Gordon Waitt** is Associate Professor in the School of Earth and Environmental Sciences at the University of Wollongong, New South Wales, Australia. His research interests include social and cultural geographies, particularly tourism geographies and the geographies of sexuality. Gordon is co-author of *Introducing Human Geography* (Pearson 2000) and *Gay Tourism: Culture and Context* (Haworth 2006) and numerous articles in various journals including the *Australian Geographer*, *Annals of the Association of American Geographers*, *Environment and Planning D* and *Social and Cultural Geography*. Gordon is actively involved in the professional activities of the Cultural Study Group of the Institute of Australian Geographers and the Geographical Society of New South Wales.

# Acknowledgements

The editors would like to thank the contributors for ensuring that the compilation of this book was an enjoyable process. We would also like to thank the people of Stoke-on-Trent, the small city where both editors lived and worked, for giving us the impetus to produce this book. There are many people to whom we are grateful for their help and support during the editing of this book. We would like to thank our new colleagues and students at the University of Manchester and the University of Leeds for making us very welcome, and old friends who also left Stoke-on-Trent and Staffordshire University, in particular Tim Edensor, Ruth Holliday, Derek Longhurst, Maggie O'Neill and Tracey Potts. We would also like to thank other colleagues, including Jon Binnie, Sarah Holloway, Lottie Kenten, Charlotte Knell and Gill Valentine, and the numerous others who have been helpful and supportive in a number of different ways. We would like to thank the series editors Gary Bridge and Sophie Watson for giving us the chance to contribute to the series, as well as Andrew Mould and Zoe Kruze at Routledge. Special thanks to Ruth and Daisy.

The editors, contributors and publisher thank the following for granting permission to reproduce the following material in this work:

The *Strait Times* (Singapore Press Holdings) for Figure 5.1 'Thinking beyond Singapore's borders' and Figure 5.2 'Media depictions of Singapore's regionalization programme'.

Singapore Tourist Board for Figure 5.3 '"New Asia-Singapore": an Asian city blending tradition and modernity'.

# 1 Conceptualizing small cities

David Bell and Mark Jayne

> There are a large number of cities around the world which do not register on intellectual maps that chart the rise and fall of global and world cities. They don't fall into either of these categories, and they probably never will – but many managers of these cities would like them to.
>
> (Robinson 2002: 531)

The idea for this book was conceived during the time we both lived and worked in the city of Stoke-on-Trent, UK. We were both busy spending a lot of our time delivering our employer's mission for 'third stream' activity, by undertaking consultancy on behalf of the university for public-sector clients in the locality. Although broadly focused on regeneration and economic development, these projects were mostly concerned with culture-led regeneration and creative industries development. This work reflected what were at the time (and still are) significant policy agendas unfolding in the majority (if not all) of Western cities. What we observed 'on the ground' in our role as consultants was a desire by local actors, such as council officers or employees of assorted local and regional development bodies, for the adoption and adaptation of policies and practices proven elsewhere to the local context of that city and its sub-region. We were often engaged in acts of translation and what we might call 're-scaling', attempting to make a fit between policy agendas and the scale, scope and reach of Stoke-on-Trent.

At one level this reflects the commonplace 'me-too-ism' produced by conditions of inter-urban competitiveness. It also reflects the obsession with 'trading up' that is part of the same competitiveness: the desire for upward mobility on the part of middling cities – those places Gray and Markusen (1999: 313) refer to as 'would-be cities' – a kind of emulation mixed with jealousy mixed, contrarily, with often obsessive parochialism. This kind of love/hate relationship between the small and large city, we want to argue here, poses all kinds of problems. We will remark on two now and others will crop up as we progress. The first concerns the adaptation of big-city policies and ideas in small-city contexts: a process labelled 'mundanization' (Bell 2006) in which lofty ideals and policy promises are translated into ineffectual practical outcomes as a result of a variety of local cultural factors such as staunch localism, conservatism, risk aversion, traditionalism and lack of ambition (Jayne 2003). 'Would-be cities' may have a vision of future greatness, but that vision is often stymied by on-the-ground resistances and reluctances which, we suggest, are all about the smallness of small cities.

The second problem is also a question of translation and relation: how are small cities to find a place for themselves, find their 'Unique Selling Point' (USP), tap into tradable capital, given the emphasis on the *bigness* of cities as their defining feature? In an urban hierarchy topped by so-called global cities, followed by so-called second-tier cities – places with national importance but moving towards global reach (Markusen *et al.* 1999) – can small, third-tier cities find a meaningful and valuable use of their third-tierness, their localness, their smallness (Jayne 2004b)? Caught between the bigness of the global metropolis dominating global flows of capital, culture and people, and the openness of the rural, small cities are faced with a problem of definition and redefinition, caught between bulking up and staying small.

Added to this, the woeful neglect of the small city in the literature on urban studies means that we don't yet have to hand wholly appropriate ways to understand what small cities are, what smallness and bigness mean, how small cities fit or don't fit into the 'new urban order', or what their fortunes and fates might be. Generalized accounts of 'the city' always imagine something big: 'too often, single cities – most recently, Los Angeles – are wheeled out as paradigmatic cases, alleged conveniently to encompass all urban trends everywhere' (Amin and Graham 1997: 411). Yet a quick bit of scoping and counting shows that small cities are, numerically speaking, the typical size of urban form the world over (Jayne 2004b). It was very clear to us, moreover, busily consulting away, that Stoke-on-Trent was *not* LA. It wasn't London or Manchester, either. Stoke-on-Trent had its own mode of 'cityness' – it may be 'a city in name only' (see Jayne 2003), but it was made to embody cityness, to behave in a city-like way, to aspire to heightened 'cityfication'.

With a population of around 250,000, Stoke-on-Trent (aka The Potteries) sits between Manchester and Birmingham in the English Midlands and has a global reputation for its most famous export, ceramics (aside, that is, from pop singer Robbie Williams). However, due to the nature of the ceramics process – capital was tied up in the production process and there was little need for significant administrative, financial and banking institutions or retail infrastructure because goods were exported around the world, but people did not come in significant numbers to the city to purchase ceramic products – the city has not had a large representation of middle-class or ethnically diverse residents (see Phillips 1993; Edensor 2000). Moreover, the spatial configuration of the city, made up of six towns (seven if you include the nearby but separate political entity of Newcastle-under-Lyme, with its population of around 110,000), has ensured that there has never been a dominant city centre. Prior to unification in 1911, for example, each of the Potteries towns had developed its own cinemas, shopping streets and squares, markets, town halls, political structures and infrastructure for utilities provision. While Hanley eventually edged ahead of the other towns and was

*Figure 1.1* Stoke-on-Trent's city centre. (Source: courtesy of Mark Jayne.)

officially designated as the 'city centre', the combination of the dominant working-class production and consumption cultures combined with the dispersed spatial configuration has ensured that the city has failed to 'punch its weight' for a conurbation of its size (see Figure 1.1). Although relatively slow to respond to the dramatic economic and social change associated with de-industrialization, the city has, over the past decade, embarked on a range of attempts to undertake regeneration and instigate economic development strategies. These have tended to reproduce, rather than transcend, the historic failings of the city, further exacerbating the quantitative and qualitative gap between Stoke-on-Trent and its more successful neighbours in an urban hierarchy dominated by intense competition (Bell and Jayne 2003a and b; Jayne 1999, 2000, 2004b, 2005).

Spurred by teaching, researching and consulting about cities *and* about Stoke-on-Trent, we resolved to commission essays for an edited volume with a focus on small cities around the world, where smallness was a qualitative thing as much as anything. Cities where smallness was a state of mind, an attitude, a disposition. Cities like Stoke-on-Trent, where small meant small-minded more than small-sized. We started to ask questions about the cultures of smallness, and the cultures of small cities. Was smallness something that could be alchemically made into gold for such places? Were these places seemingly small because of the big-city boosterist discourse that made small cities seem inadequate, underdeveloped and retarded? Was there a stubborn, perverse pride in smallness that these cities evoked and invoked? Or were such places confined to the dustbin of post-industrial urbanism, leftovers from a bygone age, still trading on worn-out ideas of cityness that others read only as smallness, as lacking in the kind of cosmopolitan urbanism and economic dynamism associated with real (large) cities?

To a certain degree, someone beat us to it by publishing a book called *Crap Towns* (Jordison and Kieran 2003, 2004), a sort of anti-guidebook listing the UK's least desirable places; places too crap even to be cities. (For those who are now curious, the place 'topping' the Crap Towns league is Hull; in the book's sequel, that dubious honour goes to Luton). Undeterred by having been beaten to it, we stuck with the idea which became *Small Cities: urban experience beyond the metropolis*, which we hope to be a more critically informed approach to the thorny problem of *smallness*.

## SIZE MATTERS

How small is small? And what ways of measuring size are useful? Given our allergy to generalization, we decided against any minimal or maximal requirements. Just as Thrift (2000) reminds us that 'one size does not fit all' in ways of thinking about cities, so we didn't want size to be an absolute. A quick scan across the scant previous work confirms this: in studies from the USA, a small city is often defined as having less than 50,000 inhabitants (Brennan and Hoene 2003); in studies from 'developing countries', a small urban centre might be classed as one which has 5000 to 20,000 inhabitants, although here national and regional variation make such a definition unsustainable (Hardoy and Satterthwaite 1986b). There is great difficulty in defining smallness through population size or economic growth, in the context of very different urban hierarchies in countries in Europe, Asia and North America (Prakesh 1982; Markussen *et al.* 1999; Bagnasco and Le Gales 2000). Attempts to schematize cities into a pecking order also tend to be inadequate due to this discursive and differential construction of cities and urban hierarchies around the world. Peter Hall's (2004: 36) classification of a global urban hierarchy based on population size, for example, develops and adds to a number of previous schemas, all of which (beyond the global and sub-global level) fail to accommodate or account for geographically specific urban hierarchies around the world:

1. *Global cities* – typically with 5 million people within their administrative boundaries and up to 20 million within their hinterlands, but effectively serving very large global territories.
2. *Sub-global cities* – typically with 1–5 million people and up to perhaps 10 million in their hinterlands, performing global service functions for certain specialized services (banking, fashion, culture, media).
3. *Regional cities* – with populations of 250,000 to 1 million.
4. *Provincial cities* – with populations of 100,000 to 250,000.

Smallness, we would argue, needs to be assessed in other ways which are at least complementary to more standard numerical measures. Smallness is as much about reach and influence as it is about population size, density or growth. At what level in the global urban hierarchy does a small city 'trade'? To which other cities (and nonurban places) does it link and what forms do those linkages take? It's not size, it's what you do with it. In a global urban order characterized at once by dense networks of interconnection and by intense inter-urban competition, absolute size is less important; 'the value of small urban centers is not so much in their ... sizes as in their functional characteristics' (Rondinelli 1983: 385). So smallness is in the urban habitus; it's about ways of acting, self-image, the sedimented structures of feeling, sense of place and aspiration. You are only as small as you think you are – or as other cities make you feel.

Now, there is a further dimension to the sizing of cities and that is a kind of 'sizism' that marks urban theory and urban policy: 'small and intermediate urban centres remain the least studied and perhaps the least understood elements within national and regional urban systems' (Hardoy and Satterthwaite 1986b: 6). But 'small cities' is not oxymoronic; small cities are typical in a quantitative sense. Too many theorists have been wooed and wowed by spectacular urbanism to notice them, however. Small cities are just not a topic that urban studies has engaged with, that theorists have seen as deserving of concerted, coherent and focused research and writing. So the discourses of cities – the ways they are talked about and thought about by different people, including academics, planners, managers, inhabitants – have tended to follow the logic that cities should be big things, either amazing or terrifying in their bigness, but big nonetheless. The very idea of cities is to be big and to get bigger: shrinkage, even stasis, is a sign of failure. Small cities, therefore, are not as good as big cities by the very fact of their size, even when those small cities (or spaces and places within them) punch above their weight. Metropolises and mega-cities are the acme of urbanism and small cities are therefore something of an embarrassment: as already noted, they are a strange in-between category, neither one thing (rural) nor the other (properly urban). Small towns and villages can stay small, where smallness is quaint, whimsical, old-fashioned; but *small cities*? What is the point of them?

## SMALL ACTS

Jennifer Robinson's (2002) important discussion of cities 'off the map' of the global cities research agenda calls for an urban theory that accounts for a wider variety of cities and, crucially, for a bridging between research on global or world cities and that focused on small cities in 'less developed' countries. Arguing against the view of 'third world cities' as 'not [yet] cities' and as 'irrelevant' to world cities theorizing, Robinson calls for alternative analytical approaches that might think of these 'irrelevant' cities differently. For example, in the literature on small cities in the global South, a very clear (if sometimes contested) role for small cities is often highlighted: spreading more equitably across space the 'benefits of development' by producing a more evenly distributed and better developed urban network (Hardoy and Satterthwaite 1986b). There is therefore a pressing research agenda here to understand the role of small cities in social and economic development though, as Robinson (2002: 540) again reminds us, this 'developmentalist' view of 'third world cities' is in itself problematic, assuming that 'all poor cities are infrastructurally poor and economically stagnant yet (perversely?) expanding in size'. This assumption stereotypes small cities in the global South on the basis of ideas about the defining function of economic globalization in the fortunes of cities. Yet Hardoy and Satterthwaite (1986a) note that these cities perform other key functions too:

> one aspect of small and intermediate urban centres which has gone virtually unmentioned [in work to date] is the fact that they so often serve as the focus for social life and social contacts in their area. … Such centres may be the place where young people in the area socialize, where there are important opportunities to meet people of the opposite sex and where there are opportunities for sport, recreation and for attending religious services and festivals. Certainly it is within many such centres that the culture of the area has its most concentrated expression.
>
> (Hardoy and Satterthwaite 1986a: 410)

The social and cultural functions of small cities have also been flagged as important, maybe even unique, in cities in the global North, as we shall see. Moreover, the stress on economic globalization neglects what Harris (quoted in Robinson 2002: 541) called the 'real urban economy'. This differently articulated economic function (which is also at once both social and cultural) is exemplified by Rondinelli (1983), who cites a study from the 1970s of Oaxaca, a small Mexican city (population at the time of just over 100,000) with a central market role. Rondinelli summarizes the functions of Oaxaca for its locality:

> The market in Oaxaca provides outlets for agricultural produce, livestock, nonagricultural goods like fibers and firewood, and artisanal products such as pottery, baskets, mats, and household and agricultural implements. An impressive array of people find employment directly or indirectly through market activities – carpenters, stonecutters, healers and curers, butchers, blacksmiths, small-parts sellers, marriage arrangers, mechanics, and vendors of seeds and equipment. The market offers opportunities for farmers to sell their goods and for a large number of intermediaries to engage in trade. Oaxaca supports traders who buy and resell goods in the market, traders who travel to small rural markets to collect goods for resale in Oaxaca, and traders who buy goods in the market and resell them door-to-door in the city. Rural visitors have the opportunity to shop in stores along the periphery of the market and to call on doctors, dentists, lawyers, and lenders. Wholesalers collect small quantities of local products from the Oaxaca market and sell them in bulk to retailers in larger cities and bring small lots of goods back to Oaxaca. The employment network of the market is thus extended to include field buyers, agents, truckers, and small-load haulers.
>
> (Rondinelli 1983: 387)

So here we see ways in which small cities can and do 'work' in the urban hierarchy; they are much more than fillers, not (yet) cities or would-be cities – they are important nodes in the networks between places of different scales, and they are seen to mediate between the rural and the urban, as well as between the local and the global. Yet in studies of small cities from the global North, they are more often figured as always facing problems, even as being problems themselves, as a result of the restructuring of economies and geographies in recent decades. Now, while we are mindful that in categorizing cities in terms of their political-economic geography, we are going against Robinson's (2002: 549) call for a non-categorizing 'cosmopolitan project of understanding ordinary cities', we are interested in sketching both similarities and differences in the shapes and fortunes of small cities in different places, in their *specificity*. We wholly agree with Robinson that 'simply mobilizing evidence of difference and possibly deviation ... is not enough' (ibid.); at the same time, however, we are very interested in the stories of small cities and the lessons learnt from those stories.

In post-socialist states, for example, many small cities have been twice ravaged, by de-industrialization and by 'liberalization'. They thus face some particular challenges and enact some particular responses. Jörg Dürrschmidt's (2005) work on the twin city Guben/Gubin, on the German–Polish border, captures these problems vividly. Guben/Gubin, one of many twin cities built on the border corridor between socialist states, now finds itself shrinking: it is suffering a haemorrhage of population, especially on the German side (Guben), leaving behind an ageing population marked, moreover, by what Dürrschmidt refers to as 'mental shrinkage' – a collective fatalism, a resignation to the city's fate and a turning away from civic culture. Those people left behind, he says, have already mentally left the place, retreating into the family and nostalgia. Efforts to tackle these problems, such as the rebranding of the city as *Apfelstadt* (Appletown) with a cider festival, have failed to gain popular support in a city with a 'modern counter-modern milieu', in which its people reflexively negotiate new mobilities (due to liberalization and EU accession) and old traditions (the revival of hobbies such as shooting). Beyond that, the future of cities like Guben/Gubin is very uncertain. At a workshop Dürrschmidt attended with officials from these border cities, one town planner said the only solution left was to evacuate the city altogether, build a wall around it and 'let nature take its course'. As Dürrschmidt summarizes, Guben/Gubin, a dying city on the European periphery, tells a particular story about the global and the local, a more 'ordinary story' than that told by the global city. So, while we agree with Robinson's (2002: 533) criticism of the 'geography of urban theory' and its role in neglecting those cities off its map, we must also note that there is a geography to small cities, too.

## Smallsville, USA

One country where small cities have had a relatively higher profile in policy (if not in academic terms) is the United States of America. For example, work has explored the changing fate of small cities under conditions of de-industrialization and (sometimes) subsequent regeneration, but under conditions and embodying responses that are marked by geography, at global, regional, national and local levels. So, demographers have spotted a trend towards migration from metropolitan US centres to small towns and cities, suggesting that this move 'may be becoming a more pervasive, middle-class phenomenon that could constitute a viable alternative to further suburbanization' (Frey and Liaw 1998: 216). It is, of course, a selective, elective movement of particular groups, repatterning population stratification across new lines, with middle-class families opting out of big city life for both economic and 'lifestyle' reasons. As we shall see, these migrations provide ambivalent opportunities and provoke ambivalent responses in small cities.

There has been a notable focus on downtown districts in work on US small cities, based on the argument that 'Downtown is the heart of small cities. Downtown reflects the character and image of the city and provides the first impression of its overall quality of life' (Haque 2001: 278). Downtown becomes an asset which may, in Gray and Markusen's (1999) nice term, be 'parlayed' as small cities try to re-capitalize in the face of economic misfortune. This misfortune, and the search for responses to it, arguably takes shape in particular ways in small US cities. While Burayidi (2001: 3) writes that 'at this point in time we know very little about the pattern in economic activity shifts in free standing small urban areas', he concludes from the skimpy evidence that 'there is a distinctive difference in the forces that shape these small cities versus those of the large urban centers or their suburbs' (ibid.). What happens in and to small cities is different – but, of course, much more typical, more ordinary, than the experiences of and in global cities.

Robertson has done a lot of work on small-city downtowns in the US, with an emphasis on highlighting success stories: 'for every Seattle there are scores of smaller cities that can boast equally impressive downtown success stories' (Robertson 2001: 9). His work emphasizes the strong sense of place and the 'human scale' of small cities as their USPs; he lists eight key differences between large- and small-city downtowns. Small-city downtowns:

1 are more human scale, less busy, more walkable
2 do not exhibit the problems of big cities – congestion, crime, etc
3 aren't dominated by corporate presence
4 lack large-scale flagship or signature projects
5 have retailing distinguished by independents
6 aren't subdivided into monofunctional districts
7 are closely linked to nearby residential neighbourhoods
8 possess higher numbers of intact historic buildings

All of these, he argues, are tradable assets.

These differences, all spun positively for the borderline-messianic Robertson, produce a set of principles for successful small-city downtown redevelopment, including public/private partnership, clear visioning, multifunctionality – including city-centre living, visitor attractions and what we might call the '24-hour small city' – preservation of heritage, use of waterfronts as an amenity, walkability, urban design and planning regulation, and not placing too much emphasis on car parking (according to Robertson a common small-city ailment). Notably, Robertson concludes by emphasizing the appropriate scale of ambition and pace of change: don't rush, and stay small: '"Big fix" solutions rarely work ... in smaller cities. Rather, a continuous series of small-scale organizational, aesthetic/design and economic improvements that makes downtown distinctive from other settings – a strong sense of place – is the foundation for successful downtown development in small cities' (Robertson 2001: 20).

Haque (2001) also looks at success stories in US small-city downtowns, comparing strategies of diversification with those of specialization. She also notes the lack of vision in some small cities: 'Although small cities and towns typically possess a significant amount of determination, energy and spirit, studies suggest that small cities lack proper understanding of their strengths and weaknesses and how to undertake economic redevelopment planning' (Haque 2001: 277). She concludes that uniqueness is the key and that visioning, again, can produce revitalization by trading up the small city's USP. This is illustrated in one of her case studies, Dothan, Alabama, the 'mural city' (also self-proclaimed 'Peanut capital of the world'), where tourism based around historic and contemporary murals – and peanuts – has given the downtown the necessary boost.

Paradis has also studied tourism-based small-city downtown revitalization in the US, similarly exploring how small-city sense of place is used in place promotion but also may be affected positively or negatively by the influx of tourists that the successful promotion produces. His work on Roswell, New Mexico, for example (Paradis 2002), shows tensions between residents and visitors over the rebranding of the city around UFO themes because of its cultural association with a flying saucer crash landing in the 1940s. Fifty years on, Roswell has a clear, if contested, theming around the crash and UFO mythologies which attracts large numbers of visitors. The conversion of ailing small-city downtowns into 'theme towns' is a more widespread phenomenon across the US and beyond, as small cities concur with Haque's assessment that going all out for uniqueness is the key to survival. Paradis is also interested in growth coalitions that form in small cities to drive forward downtown redevelopment, a feature of his study of Roswell and of previous work in Galena, Illinois (Paradis 2000a, 2000b). In Galena, he tracks the development of the 'tourist business district' (TBD), and the consequent tensions between the 'old-timer' view of downtown as a community centre and the 'newcomer' and visitor's view of downtown as a theme park. The preservation or heritization of downtown epitomizes this ambivalence, in maintaining the unique character of the area but turning it into a museum, making its residents more like the 'cast members' of a Disney park. At the same time, incoming new residents, seduced by the sense of place, caused a hike in property prices: the place becomes too tourist-focused and not 'local' enough, while the very people attracted by its history and amenities have a negative impact on the sense of place and structure of feeling as they become the focus of resentment. While the picture is in fact more complex than an 'us versus them' divide, the lesson is clear: small-city downtown redevelopment treads a fine line in trading on uniqueness – the line between maintaining the sense of place while turning it into a commodity. While this is common to many tourist destinations of different scales, Paradis argues that small-city downtowns, in the US at least, are particularly prone to this approach to revitalization, given the shared features listed by Robertson (2001).

In his study of Roswell, Paradis (2002) makes another important comment, in terms of the distinctive ways in which the global and the local come together in small cities. He phrases this point in a concluding question which has broader resonance for analyses of small cities: 'in what ways can Roswell be placed into the conceptual framework of the global–local nexus, and what can be learned from studying a small city that has to some degree successfully "plugged in" to the global economy?' (Paradis 2002: 40). By tapping a global cultural phenomenon – ufology – and utilizing global technologies such as the Internet, Roswell re-marks its place on the map as a (maybe *the*) global centre of UFO tourism. At the same time, through its theming, 'Roswell is attempting to assert itself as a distinctive local place in the midst of a globalizing society' (ibid.). As Paradis summarizes, at present Roswell is trading on this local–global uniqueness at the expense of other, previously strong local place identities. The long-term outcome of this contest over the meaning of Roswell has yet to be decided.

Stepping back from particular cases, Burayidi (2001b) highlights common features and issues shaping downtown regeneration in small-city USA. These include, importantly, understanding the local political culture and that also means understanding that this culture is local: locally focused, sensitive to the electorate and so on (Jayne 2003). Quick fixes may be seen as more attractive, depending in part on the stage in the political cycle into which they come – how close are the next elections? Local state politics must also be matched by considerations of grassroots support: without the local community on their side, local politicians are unable (indeed unwilling) to press for change. Local money is also preferred together with local control of design and aesthetic considerations. As he concludes, downtown revitalization must be a community effort – local apathy and entrenched historic identities can be an insurmountable obstacle. Notice the repetition of the key word 'local' here: the small city works at the level of the local. Smallness is about reach and scope, but this means it can also be about smallness of ambition and vision. The off-the-peg solutions offered by Burayidi and Robertson, for example, threaten to produce a series of monocultural 'unique' small-city downtowns that are consumed, not by locals but by cosmopolitan visitors who want to consume localness and smallness. Burayidi (2001b) ends with the question of sustainability, raised also in Paradis's work. If being small means doing only one thing, even if that thing has global reach, as in Roswell's case, then how sustainable is that? What kind of competitive advantage does being the world's number one UFO tourism destination give a city once grounded in the local economy and community? How is localness mobilized in this kind of intervention and how might it be mobilized otherwise? Pile (1999) offers heterogeneity as a defining feature of the cityness of cities, having already junked size; 'cities are huge juxtapositions', his colleague Massey (1999: 165) writes in the same volume.

While there are interesting, even jolting juxtapositions in the US small-city downtowns referred to here, such as aliens in 'cow town' or peanuts on murals, we are not sure these are the kinds of juxtaposition of which Massey was thinking. But that's the problem: what are we thinking about when we're thinking about the (small) city?

So, there is a real need to look closely at small cities and their interrelationships. Urban theory needs to account for how different kinds and 'sizes' of cities operate, cooperate and compete – processes which also occur at different scales, local, regional, national and global. We need to understand both the *smallness* of small cities and the *cityness* of small cities: both terms have specificity (on these two characteristics in general, see Pile 1999). These are cities we are studying, not 'the city'. Writing about downtowns generally, MacLeod *et al.* (2003) reveal the problem of reconciling 'small' with 'city' and the trouble with generalization:

> On entering the *typical* city centre, one is *invariably* dazzled by an imposing array of gleaming towers, which overlook a kaleidoscopic geography of corporate glamour, conspicuous consumption, café culture and street-life chic, all in close proximity to some elegantly restored town houses and gentrified terraces.
>
> (MacLeod *et al.* 2003: 1660, our emphasis)

One can only guess at the 'typical' cities these authors 'invariably' hang out in; certainly not Roswell, New Mexico; or Dothan, Alabama; there are no UFOs, murals or peanuts in this vista. Given that 97 per cent of US cities have populations below 50,000 and 87 per cent below 10,000, the typical city centre is, as Brennan and Hoene (2003) write, actually to be found in places like Watertown, South Dakota; or Whiteville, North Carolina. And, of course, on a global scale, small cities are best typified by places like Oaxaca, Mexico, servicing the surrounding rural hinterland and other cities at other levels of the urban hierarchy in complex ways.

## Selling smallness

Other current vogues in urban discourses are ambivalent about smallness. 'New urbanism' values human scale and community, making cities more livable by taking away the bad cityness and buffing up the good cityness (Talen 1999). Planning and building new small cities under the mantra of new urbanism marks one clear attempt to re-capitalize smallness – even villageness (see Bell and Jayne 2004) – emphasizing features of neighbourliness and the cosiness of human-scaled living. Perceived as a solution to the much-talked-up loss of social capital in cities (Putnam 2000), critics nonetheless warn that new urbanism is being turned into 'little more than a marketing strategy' in which urban design is supposed to catalyse socialities (Talen 1999). Designing cities to be small is one thing; finding ways to market already existing smallness is something else, as we have seen.

The still resounding 'creative cities' agenda, one of the most prominent packaged solutions to de-industrialization currently being hawked by its many proponents, has less favourable things to say about the creative pull of small cities, in spite of evidence of urban out-migration among at least some creative workers, attracted by small-city and country amenities as sites for 'lifestyle businesses' (Florida 2002; Nelson 2005). Yet, as Erickcek and McKinney (2004: 2) note, small cities are more often seen as lacking the particular 'growth-facilitating amenities' attractive to young professionals – the café culture highlighted by MacLeod *et al.* (2003) as 'typifying' cities. Nevertheless, in their statistical analysis of small-city growth in the US, Erickcek and McKinney did spot some positive correlations with creative-city-like features, such as enterprise culture, learning and so on (even if other variables proved equally influential to growth, such as the weather). But the creative and culture-led regeneration agendas have tended to emphasize big-city features, leaving small cities with the choice between emulation and differentiation. In a rare big book about one small city, Kamloops, British Columbia, Canada, Garrett-Petts and Dubinsky (2005: 1) start out from this very problem: 'if not by definition, then certainly by default, "culture" is associated with big city life: big cities are commonly equated with "big culture"; small cities with something less'. Part of the mission of *The Small Cities Book* is corrective; to map the culture of Kamloops and to foster it through strategic interventions that aim to address the core issues facing this city and, by association, other similar small cities. Those problems are listed as:

> Accommodating to new migration patterns, establishing growth-oriented social and economic networks, linking local planning to community functions and identity, recognizing the challenges of integrating 'newcomers,' preserving a viable downtown, animating local history, resisting the forces of purely commercial gentrification, promoting the multiple faces and facets of the city, generating a strong sense of place, and taking advantage of scale to promote community involvement.
>
> (Garrett-Petts and Dubinsky 2005: 2)

Resonating with the designs and desires of new urbanism, this formulation of the challenges and promises of the old urbanism of small cities like Kamloops sets the agenda of a large multidisciplinary research and engagement programme in the city, detailed in *The Small Cities Book*. This programme includes a look at the 'creative cities' agenda, noting how Florida's metro-centric analysis misses the creativity-nurturing assets of small cities, dismissing them as 'hopeless places' (quoted in Nelson 2005: 92). The distinct attractions of small cities are shown to be varied – in some cases the presence of higher education institutions is significant, in others the natural amenities are shown to appeal to creatives – yet some research in the Kamloops project suggests that the social traditionalism of small cities might work against other assets, placing greater emphasis on nurturing indigenous creativity than on attracting it from elsewhere (Bratton and Garrett-Petts 2005). This is, we think, a crucial point, though we are less convinced by the response provided here: the small city is marked by a cultural smallness, a conservatism or traditionalism that can be a keen source of conflict over change, as Paradis's work shows. But some studies suggest that even this parochialism can be re-capitalized, packaged and used as a marketing strategy.

A particularly vivid example is provided by Gray and Markusen (1999) in a case study of Colorado Springs, USA. Once a small city heavily dominated by military bases, Colorado Springs has had to repackage its assets in the wake of the downscaling of military-related activities in the post-cold war period. Drawing on networks established through its military past, the city has been doubly rebranded as a sports city and, perhaps more surprisingly, as a node in Christian Right evangelism. As Gray and Markusen write, the city 'parlayed' its local culture – 'conservative, pro-military, religious and antiunion' – and its lack of cultural diversity as a USP to draw in fundamentalist Christian Right organizations such as Focus on the Family as well as more niche bodies such as the Fellowship of Christian Cowboys or Mission to Mormons. If creative cities, in Richard Florida's ubiquitous formulation, are hot spots for gays, geeks and the creative class, then places like Colorado Springs are going to struggle to compete. Instead, this parlaying of provincialism and parochialism marks a fascinating contextual response to the economic downturn faced by the city. Provincialization and anti-cosmopolitanism may well be appropriate situated strategies in the face of pressures to 'go global' and they can be witnessed playing out elsewhere, too (Varma 2004).

More romanticized versions of the same logic are, of course, at the heart of new urbanism and in the celebration of small-city vernaculars and traditions. Aritha van Herk (2005: 143) hymns the 'different corpus of locality' found in the small city, with its 'small mythologies' and its 'modesty'. While much of this discourse retreads old ideas, of *Gemeinschaft* and *Gesellschaft*, of urban anomie and other anti-urban sentiments, the small city is not strictly anti-urban. It may be anti-metropolitan and anti-cosmopolitan, but it nonetheless trades on its cityness. Erickcek and McKinney's (2004) survey research generated a typology of US small cities which is worth sketching here and which needs a symptomatic reading beyond the scope of this introduction (but note all the sprawling, leaking and dependency). The eight types they identify are:

1 old economy places in slow decline
2 private sector-dependent places
3 sprawling places
4 company towns left behind
5 college places leaking graduates
6 company towns left behind but still socially viable
7 growing new economy places
8 growing university/government/business complexes

(Erickcek and McKinney 2004: 20–21)

While they were unable to crunch any strong conclusions from their data, this typology shows at a glance the fates awaiting small US cities and the likely shape of future growth opportunities. Note, too, that growth is still figured as important;

even in *The Small Cities Book* growth is discussed, as stagnation or shrinkage are unviable alternative futures. So it might be worth commenting briefly on one last size matter: the fact that cities *do* continue to grow – small cities perhaps especially so (Brennan and Hoene 2003) – despite recent premonitions of their death. A couple of decades ago, futurologists were telling us that cities would, by now, be a thing of the past, shrunken and shrivelled as new technologies and new life patterns made city living obsolete. 'Networkers' hard-wired into cyberspace simply wouldn't want or need cities and cities would be locked into a tech-noir downward spiral into dystopia. Instead, cities have grown and diversified, even as they have become wired (Graham 2003). The world's population is becoming increasingly urban. Cities are still here, even small ones. This book will help us understand why.

## Reading small cities

The chapters in this book focus on small cities from around the world and when read together a particular story, or set of stories, can be seen to emerge. While small cities have been woefully neglected in urban studies, as this introduction has shown, there are nonetheless (if you're looking for the small cities needle in the urban studies haystack) a small number of useful books, articles and chapters that together raise a variety of interesting theoretical issues and outline rich empirical material.

There has not, however, been a concerted attempt to synthesize small cities research in a focused way that avoids abstraction or overcomes an over-simplification of the complexity and nuances of individual small cities. A key failing of this de facto and/or fragmented research agenda is that work tends to based on a case study approach that ultimately depicts specific small cities in an abstract or disembedded way that does not enable generalizations to be made (see Castree 2005). This book seeks to overcome this weakness, and chapters have been commissioned in order to pursue the theoretical and empirical concern to map the political, economic, social, cultural and spatial proliferation of Western neo-liberal governance. Each section and each chapter is oriented towards addressing the same questions – how is the uneven proliferation of political, economic, social, spatial and cultural practices and processes bound up with the discursive and differential construction of neo-liberalism unfolding in small cities? How, and in what ways, does smallness enable or constrain the ways in which small cities pursue neo-liberal policy? Is there something about size and scale that delineates the ways in which broad urban agendas are interpreted in, or articulated from, the fabric of small cities?

We argue that urban life beyond the metropolis must be theorized in terms of the uneven, differential and discursively constructed proliferation of neo-liberalism in cities throughout urban hierarchies throughout the world. We see this volume as a starting point, a rallying call for small cities to be taken seriously, a catalyst for sustained and coherent theoretical and empirical debate and research. We hope that others will share our and the contributors' interests, to join in, run with and expand on the ideas and debates signposted in the following pages.

To kick off this agenda, *Small Cities* is made up of four parts that reflect on broad urban practices and processes – political economy, the urban hierarchy and competitive advantage, cultural economy, and identity, lifestyle and forms of sociability. While each chapter touches on all of these issues to varying degrees, each has a particular weighting. Part one is a series of theoretical and empirical essays that address the interpenetration between political and economic agendas and trajectories of small cities. Chapters consider issues of urban productivity and the up-scaling of small cities as they seek regional, national and global significance, but also offer cautionary tales of fragile economic specificities. Part two builds on this work to look in more detail at how small cities seek to obtain competitive advantage in an urban hierarchy and attract flows of capital, culture and people. Chapters focus on festivalization, culture and entertainment-led regeneration, the agglomeration and clustering of post-industrial economic activity, and the role of myth and image in helping small cities to overcome perceived limitations of size and scale. Part three considers the interpenetration between culture and economy, and chapters focus on cultural and creative-led economic development strategies as well as exploring ideas of conviviality and human scale. The final substantive section considers the ways in which everyday life in small cities impacts upon identity, lifestyle and forms of sociability. Chapters consider waterfront developments, youth identities and the impact of urban regeneration on historic local identities.

Part one begins with AbdouMaliq Simone, who unpacks the economic productivity and competitiveness of the city of Port Louis, Mauritius. The chapter focuses on the attempts of city authorities over the past three decades to overcome Port Louis's relative geographical isolation and to develop links with Europe, Africa, South-East Asia and the Middle East. AbdouMaliq Simone depicts a process of reflection on the city's character that has been undertaken over this time – specifically in terms of ethnic division and unrest. The chapter critically assesses the attempts of policy makers to link structural economic restructuring and attempts to 'globalize' economic development with physical redevelopment of the city on the one hand and the celebration of a cosmopolitan urban and national identity on the other.

In chapter three, John Joe Schlichtman invites us to visit Highpoint, North Carolina, the two-hundred-and-eighty-ninth largest city in the USA. Schlichtman shows that Highpoint has sacrificed the cultural and economic diversity of its downtown in order to cement its position as 'Temp Town' – a city dominated by the biannual International Home Furniture Market which draws 80,000 visitors from 110 countries. Furniture is championed both politically and in economic development strategies as the key basis for Highpoint's competitive advantage over larger cities. Schlichtman interestingly traces the development of Highpoint from the 1970s in its fight with Chicago for the title of furniture capital of the world. This role was fortified during the early 1990s, via the growth in the number of showrooms in the downtown in order to compete with Dallas, which had built its own furniture showroom. Schlichtman ends the chapter by introducing a new round of growth as city authorities re-assert pro-growth strategies as Las Vegas becomes a new competitor.

In chapter four, Joseph Leibovitz investigates the economic cooperation between four cities in Canada's Technology Triangle – Cambridge, Guelph, Kitchener and Waterloo – which in close proximity with Toronto have sought to work together to 'jump scale' and to build an imagined urban–regional identity. Leibovitz shows the ways in which each of the four cities appreciates the combined advantage of size, economic linkage, proximity and agglomeration, and he unpacks the fragility of the alliance underpinned by political barriers to collaboration due to entrenched local identities, resentment towards regionalization and fears over a loss of political and economic control. In the last chapter in this section, T.C. Chang describes the central role of tourism in the strategies of Singapore's goal to overcome the constraints of size and resources to become a global player. This trajectory is shown to be pursued via the dual elements of developing Singapore as an international business hub and a cosmopolitan tourist centre. Chang describes politics of enlargement based around metaphors of size and scale and associated images of the city as being 'larger than itself'.

Part two begins with a review of the role of place marketing in urban development strategies through the medium of festivals. Andrew Bradley and Tim Hall discuss the ways in which small cities are marketed as great places to live as well as do business. They argue that there is little empirical evidence assessing the impact of place promotion, and represent findings from research from the city of Cheltenham, UK. Cheltenham's festivals represent a balance between culture and sport, and Bradley and Hall discuss their benefits in giving the city a brand image and hence an advantage over its rivals.

In chapter seven, Loretta Lees describes the complex relationship between small provincial cities and nearby larger metropolitan centres. Lees's focus is Portland, Maine, USA, located 115 miles from Boston. Portland has successfully renovated its port area and experienced an influx of out-of-state professionals and an associated arts- and entertainment-led urban renaissance. Lees argues that Portland has not simply 'kept up' with its neighbour Boston and global cities such as New York, but that the city has been ahead of the game. However, she expresses caution about the way in which Portland is now known as a best-practice model for other small US cities. Lees points to the specificity of local contingent factors – the city's local entrepreneurial base, its success in the regional service economy and a healthy arts and creative economy – as specific features of Portland's success, conditions that are not necessarily replicable in other small cities.

In chapter eight, Greg Lloyd, John McCarthy and Deborah Peel reflect on the fortunes of a small Scottish city. They show how local authorities in Dundee have focused on retail and cultural facilities, coupled with attempts to develop economic clusters around biotechnology, medical science and multimedia software (as well as the historic presence of the comic, *The Beano*). Key to this strategy has been the designation of a cultural quarter and initiatives to promote the development of a knowledge-based economy. Lloyd, McCarthy and Peel contextualize these initiatives in terms of Dundee's position, as Scotland's fourth city, and in terms of regional/national competition with Edinburgh, Glasgow and Aberdeen.

In chapter nine, Frank Eckardt investigates the role of myth and 'reputation effect' in the symbolic sizing of the city of Weimar, Germany. Eckardt unpacks the city's historic trajectory and associations as the national home to the arts and avant-garde. Eckardt traces the fortunes of the city from its early history through Nazi and socialist periods as well as through the reunification of Germany. He then focuses on the impact of the city's designation as European Capital of Culture in 1999 and the continued central role that Weimar plays in terms of German national identity and imagination.

Part three, which focuses on the cultural economy of cities, opens with Malcolm Miles's chapter, which revisits the idea of the Garden City. Miles discusses issues of sentimentality relating to smallness and uses the examples of Letchworth Garden City and the planned rural community of Poundbury, both in the UK. Miles contrasts these examples with squatter gardens in New York's Lower East Side and critically investigates issues of human scale and community.

In chapter twelve, Gordon Waitt considers the role of creative industries in the re-imaging of Wollongong, Australia. Waitt describes certain narratives that constitute the repositioning of Wollongong as a 'city of innovation' where science and technology drive change and, in order to offset the decline of traditional industries, policy makers seek to fold art and creativity into the economic realm. Waitt reflects on the ability of small cities to compete with larger urban centres and generate creative economies. Tom Fleming, Lia Ghilardi and Nancy K. Napier also investigate culture-led regeneration in Malmö, Sweden, Boise, USA, and Tuzla, Bosnia-Herzegovina. In chapter thirteen, they identify common challenges facing all three cities and their ubiquitous engagement with developing a 'creative city' agenda. The authors assess the relative merits of each city's attempts to halt decline or stagnation and to facilitate socially and culturally inclusive policy.

Part four investigates the relationship between identity, lifestyle and forms of sociability in small cities. In chapter fourteen, Tim Edensor returns to the city that is the focus of chapter two, Port Louis, Mauritius. Complementing and building on the arguments of AbdouMaliq Simone, Edensor's focus is on one part of that small city – the redevelopment of the waterfront area. Edensor argues that, rather than being a homogeneous urban space produced to serve the interests of global capital, the redevelopment has generated a locally specific social and cultural milieu. Unlike other waterfront developments throughout the world, with their focus on global retail brands and outlets, Edensor describes how Caudan Waterfront Centre is characterized by small Mauritian retail businesses and is a space in which inhabitants can find some release from dense urban living. Edensor argues that Caudan further generates inclusive social interaction that is in stark contrast to rigid ethnic and religious divisions found elsewhere in the city.

In chapter fifteen, Gordon Waitt, Tim Hewitt and Ellen Kraly describe the experience of being young in an anonymized US small city, Townsville. The authors show the ways in which smallness heightens young people's sense for and of place. This emerges as frustrations and tensions due to 'living on the margins' and a perception of restricted opportunities in comparison with peers in larger cities. In contrast Steven Miles, in chapter sixteen, describes how local identity has been central to the regeneration of NewcastleGateshead in the north-east of England.

Miles describes how two cities, which face each other on opposite banks of the River Tyne, have sought to overcome traditional rivalry. Newcastle-upon-Tyne and Gateshead have combined resources in order to instigate culture-led regeneration, with a focus on rejuvenating the dilapidated riverside. Miles describes the ways in which the two cities, while embarking on ambitious, spectacular projects, have been careful to ensure that historic identities and associations with the docks area and river have been maintained in order to gain popular support for the re-imagining of the allied cities.

In chapter seventeen, we review the key themes and issues raised in each of the sections and chapters, reflecting on the political, economic, social, cultural and spatial features of small cities as they seek to compete in urban hierarchies dominated by neo-liberal agendas and trajectories. We also highlight the need for further work and point towards a future research agenda.

# Part 1

# The political economy of small cities

# 2 On the dynamics of ambivalent urbanization and urban productivity in Port Louis, Mauritius

AbdouMaliq Simone

For the past two decades notions of urban productivity have constituted a major emphasis in reflections on cities. This has been centred on attempts to identify a city's actual or potential comparative advantage in an increasingly regionalized and/or globalized domain of economic transactions and reference. Here, the initial question is how specific cities can most effectively maximize geographic position, history, infrastructural resources, institutional capacity and social composition to make their current economic activities more competitive. Concomitantly, what investments – in infrastructure, governance reform and social capital – seem necessary to incubate or attract economic activities potentially well suited to the city's position within a specific regional constellation?

The terms of assessment usually centre on gross city product, as well as the city's fiscal health calculated in terms of its capital base, revenue generation, budgetary surplus, credit rating and more ephemeral characteristics like its 'buzz' and status as a destination. Urban productivity has also entailed (although to a much lesser extent) a focus on how cities can mobilize and deploy the city's own fundamental differentiations, exteriorities and incessantly recombinant inclinations and practices to render the city a more mobile and resilient participant in proliferating intersections with external territories, actors and institutions. In other words, how can the city use its own 'cityness' – its own capacity to assemble and re-assemble people, activities, practices, infrastructure and objects in new and often unpredictable ways so as to make its inhabitants more vital and dynamic actors?

This focus does not necessarily entail the search for governance and economic formulas that would attempt to calibrate what each resident does in terms of all the others. Nor does it mean that particular ways of life would be continuously reproduced or valorized in the same ways. But what such notions of urban productivity do entail is incessantly finding means of putting various dimensions and actors of the city 'in face of' each other – to promote ongoing intersections of economic activities, resources, capacities, institutions and places and to adapt flexibly to the outcome of such intersections in terms of elaborating guiding policies, setting economic and investment priorities and even in decisions relating to budgeting and the processes of governance themselves.

My focus in this discussion is how these two different notions of urban productivity play out in Port Louis, Mauritius. I believe that approaching the city with this problematic is particularly salient given the preoccupation of Mauritius over the past three decades with attempting to overcome its relative geographical isolation and small size through urbanizing its position within a larger economic arena. In other words, by attempting to exceed its historical position as an entrepôt, an island economy producing sugar and nothing else, finding ways of thickening and diversifying its articulations to Europe, Africa, South and South-East Asia as well as the Middle East, Mauritius urbanizes its position within the world. As Port Louis comes increasingly to manage these multiple and diverse articulations – as it attempts to manage this 'urbanization process' – it thus must continuously reassess its own character as a city.

## HOW THE SMALL BECOMES LARGE: PORT LOUIS

What is particularly interesting about this process is that Port Louis, a small city of 250,000, must attempt to manage an urbanizing process that exceeds what would otherwise be expected for a city of this size. It is true that Mauritius as a whole has one of the world's highest population densities and also true that if one could conceptualize those municipalities proximate to Port Louis – Rose Hill, Quatre Bornes, Triolet, Roche Terre, Beau Bassin, Curepipe, Vacoas and Phoenix – as one metropolitan area, it would consist of a population close to 750,000. Nevertheless, Port Louis remains a discrete governmental entity in which are located the managerial functions related to its economic and political affiliations with the larger world.

Whereas Port Louis has received a great deal of well-deserved recognition for its past and ongoing efforts to reconstitute itself as an economically dynamic city, my concern is how these efforts to act with the capacity usually attributable to a large city end up reiterating certain dimensions of its 'smallness' that need not necessarily be the case if other notions of urban productivity were in play. By pursuing particular strategies of urban development, the question is whether Port Louis really cultivates the productivity of the resource which largely constitutes the platform on which it bases its urbanizing positions with the rest of the world – a cosmopolitan human population and its relatively successful integration.

Of course, the character of Port Louis – its development trajectories, its capacity to determine a particular urban future – is wrapped up in the overall economic direction undertaken by the Mauritian state. Port Louis is the nation's capital, housing the vast majority of its administrative operations, and the municipality itself therefore has limited scope to exercise autonomous powers. The decisions for Port Louis are basically decisions for and by Mauritius as a whole, and so the aspirations of Port Louis to act as a large city really reflect the aspirations of Mauritius to constitute itself as a viable and growing economic entity.[1]

For the past 25 years, Mauritius has been considered an economic and political success story, particularly in the area of poverty reduction. It has gone from an economy solely dependent upon the production of sugar to a much more diversified profile, including textile and clothing production (primarily within export processing zones), tourism geared toward high-end markets, financial services, transshipment and, more recently, an emphasis on information and communication technology (for the moment geared toward providing back-office services for major Indian companies seeking to penetrate a francophone market). The country's ability to sustain consistently high growth rates for over two decades, coupled with well-targeted public investments in human welfare and housing, have lifted the majority of the population out of relative impoverishment into the ranks of some semblance of middle-class status (African Development Bank 2003).

The diversification of economic activities has also been accompanied by a comprehensive overhaul of investment codes, banking procedures, corporate ownership capital repatriation schemes and double taxation formulas that have greatly boosted investor confidence, maximized transparencies and situated the country as a particularly adept platform in the financial services field. Like other cities, the capital-intensive character of financial services generates specific configurations of the built environment, which include the construction of high-density office space equipped with premium services – construction that itself constitutes a speculative financial activity hedged through various financial instruments (United Nations Development Program 2003).

Given the range of externalities associated with any specific financial services firm – networks of computing, information and communication technologies, accounting, advertising, media, banking, public relations and legal services – advances in this sector require urban settings where discrete service components can have physical proximity to each other, even when the client base, transactional targets and operating markets are widely dispersed. As firms are increasingly specialized, various webs of interdependency and complementarity are established on a case- or time-delimited basis among these firms. Similarly, as large firms seek to expand and diversify their financial products, this expansion will also be reflected within the built environment.

This dynamic has certainly been the case in Port Louis, where a significant remaking of the city centre has taken place over the past two decades, with the construction of several large office towers and the rehabilitation of the waterfront area contiguous to the centre. The most substantial addition has been the construction of the Cauldan Waterfront Centre – a series of hotels, shopping centres and leisure zones, largely modelled on a waterfront project in Cape Town (see Edensor, this volume). Up until ten years ago, Port Louis was considered to be a highly unattractive city for the most part, with few investments made in the built environment. Even governmental buildings were largely dilapidated or subject to fairly egregious attempts at architectural modernism completely disjointed from the prevailing ambiance of a slightly seedy colonial town.[2] However, beginning in the mid-1980s, key historical governmental buildings, such as the Prime Minister's office and the Supreme Court Complex, have been rehabilitated.

Given this trajectory of an increasingly 'corporatized' urban centre, much of the city's historical commercial district is experiencing an almost suspended state of transition, which makes it difficult to anticipate the shape of what will inevitably be a key part of the emerging city. This area, just north of the major government and corporate complexes, is the location for the country's largest produce market, the historic Chinatown area, and approximately 2200 small retail shops selling clothes, textiles, hardware, home supplies and dry goods. The number of businesses in this area has been reduced by nearly one-third in the past 15 years, giving way to the encroachment of several new mini-malls, fast-food restaurants and cafés and, most particularly, to the building of parking garages to accommodate the substantial increase in the volume of vehicular traffic to the city each day.[3]

Whereas entrepreneurs and their families customarily would reside above their businesses, this is no longer the case within this central commercial district. Part of the reason has been that these entrepreneurs have moved to nearby residential areas of the city, such as Plaine Verte, Cité Martial and Vallee Pitot, as well as to Pailles and Vallijee to the south. With the exception of the remaining mosques, the encroachment of gentrification from the centre and its concomitant initial pressure on the social institutions of the district undermined the area's residential viability.[4]

Additionally, whereas the often complicated leasehold arrangements through which entrepreneurs have secured commercial space – often paying a relatively high proportion of their monthly earnings in return for a measure of long-term security – ensured some stability to particular businesses within the centre of the city, these arrangements were not usually available to residential space.[5] Flats became too expensive. Owners, many of them living outside of Mauritius, stopped providing services and were able to reclassify residential space as commercial and take tax write-offs for their under-utilization. As fire damage is one of the only legal recourses available to property owners to renege on leasehold arrangements for commercial property, arson has been frequent in this area as a means of clearing out buildings.[6]

The continued viability of this historic area is largely dependent upon the very factors which make it vulnerable – its proximity to the government and administrative centres. These generate a large pool of lunchtime consumers who still maintain some attachment to consumption based on a network of social ties and, more particularly, to speciality needs, such as clothing for the many different religious festivals. Additionally, in order to maintain some competitiveness with the hypermarkets in the suburban areas where more and more Mauritians are doing their shopping, there are still significant levels of corruption at the port – particularly as it attempts to capture larger shares of the transshipment market – which are able to divert a wide range of goods into these shops and are thus sold at cut-rate prices.[7] Perhaps more important, however, is the persistence of the interweaving of family, ethnic and religious ties into strong entrepreneurial networks that continue to exert significant political influence in maintaining a particular shape to trade, much of it becoming increasingly informalized.

For example, it was believed that when the produce market burned down in 1999 and was rebuilt by the municipality, the power of the few families dominating the market would subsequently be diluted. This was not the case and their position has even been strengthened as they control groups of hawkers that fan out across the city. Yet, in general, this commercial district appears to be in terminal decline.[8] The interpenetration of familial ties, residential proximity and social and religious life has been more fully elaborated, particularly for the district's Muslim majority in the neighbouring community of Plaine Verte. In Chinatown, while many shopkeepers remain, the social clubs and nightlife that made this a dynamic place at all hours have largely moved outside the city, although there has been the addition of some major real estate developments that implicitly reiterate some sense of Chinese control.[9]

While Cassis, Plaine Verte and Cité Martial do provide an ambience of dynamic residential life, Port Louis continues to lose population and after people leave work, the city appears lifeless. This has its antecedents in malaria epidemics of 1866–8 where nearly 40,000 people lost their lives. There was a major exodus of inhabitants from Port Louis to the cooler elevated cities of Curepipe, Rose Hill, Vacoas, Floreal and Quatre Bornes. As Port Louis had been a highly congested and dirty city, with a physical layout in the residential sections that reflected the segmentations of slavery with broad streets for whites alternating with narrow lanes to which the slaves were confined, the enforced opportunity to resettle in new towns institutionalized a form of domesticity largely antithetical to urban life and one to which the economic growth of Mauritius has provided the means of reinforcement.[10]

## URBAN SOCIALITY

The contrast between night and day in Port Louis is striking. The closest the entire country comes to a crowded intersection of diversely identified bodies is the lunch hour between noon and two pm. These hours host one of the few visibly aggregate demonstrations of the country's cosmopolitan composition. French, Hindu Indian, Muslim Indian, coloured Creole, black Creole, Malagasy and Chinese – to use the most banal of the ascriptions deployed to account for Mauritian diversity – can be viewed as having something publicly to do with each other during this time. This is not to say that every lunch table or grouping navigating the street takes on a multicultural composition nor that such multicultural compositions are even primarily the purview of the lunch hour, rarely to appear elsewhere in Mauritian everyday life. But as this 'convocation' is the most regularized setting for the dense proximity of diverse Mauritians, most of whom will board buses at four pm to return to clapboard houses on overgrown lanes, there is the sense that this kind of urban intensity holds out certain dangers.

In important ways this quotidian vacating of the city is of little significance, since Beau Bassin, Rose Hill and Quatre Bornes to the South and Triolet, Terre Rouge and Pamplemousses to the North are all less than a thirty-minute express bus ride from the centre of the city. In a small island, well covered with a functional bus transportation system, the dearth of a residential base and of nocturnal activities in Port Louis may not be a critical factor in either discerning or determining the country's urban 'abilities'. On the other hand, Port Louis is situated in the coastal recess of surrounding mountains, which constitute a physical barrier for growth west-to-east or centre-to-south. Thus the city has limited access to and from the rest of the country, a situation that can be remedied only by massive engineering projects. All significant transport corridors north to south in the country pass through the capital city. In addition, the liberalization of tariff regimes applied to automobiles, new and used, has resulted in a substantial increase in the number of cars in use. Access to the city has become more difficult, with daily traffic jams and long commute times.

So while transportation possibilities – both public and private – would suggest that Port Louis could be extensively integrated into its residential hinterland, the concrete difficulties of accessing and navigating the city, coupled with the relative dearth of non-work activities taking place within the city, reiterate Port Louis primarily as a place of administration. But as cities also, in various ways and scales, both explicit and implicit, attempt to administer, or at least 'minister to' cultural diversities and social intersections of all kinds, the question then becomes, what does Port Louis have to work with?

This is a country of intense ethnic, racial and cultural mixtures, but one where the bulk of the miscegenation transpired well before Port Louis was thoroughly vacated in 1868. Miscegenation was largely driven by necessity, given the vastly disproportionate ratio between men and women that existed prior to the advent of family-level settlement. The strength of religious sentiments characterizing both Indian Hindu and Muslim residents – the vast majority of Mauritians – is capable of reinforcing communalism within certain circumstances. Even the designation 'Creole' attempts to constitute some semblance of clarity out of what, in most respects, is a kind of 'refuse' identity – refuse both in the sense of what is left over, with accompanying connotations of what is not useful, and refuse in the sense of a refuge for those who have refused the identity to which they are otherwise inscribed. For black Afro-Mauritians, 'Creole' has been a particularly tricky identity through which to navigate, since they share this designation with those whose 'mongrel' racial composition has made them outcasts from clearly demarcated ethnic or religious identities and those few who seek to intentionally shake them off.[11]

Whereas a multicultural history is the critical dimension through which Mauritians understand their national identity, as well as past and future capacities, its concrete social manifestations can often be full of ambivalence. In a small island, there are no readily available opportunities for escape. People are compelled to try and deal effectively with each other if their own interests are to be secured.[12] If Indian Hindus do constitute at least 50 per cent of the population, sufficient regional and historical differences exist among them to preclude them from being a monolithic bloc in every area of national life. The only exception is that of believing they are entitled to control national politics. For example, the designation of Hindu in Mauritius does not include Tamil migrants who have preferred to encapsulate themselves in a religious-based separateness. Minority groups such as the Telegus and Marathis emphasize regional cohesion rather than caste differences (Bates 2000).

As with the logics at the fore of the nation's effort to urbanize its relationship with the larger world – rationalization of the public sector, diversification of economic activities, export-led economic growth, trade liberalization and governance reform –

the management of cross-ethnic intersections approaches what could be called a 'corporate cosmopolitanism'. Here, such intersections are valued and practised as long as they are subject to a discourse that can keep the concrete messiness of such transactions from compromising the reproduction of valued communal orientations. In public discourse, and in everyday transactions, Mauritians seem to display a proficiency of interchange that on the surface far exceeds that of other multiracial and multicultural societies. But the continued emphasis on family domesticity, as well as the systematic underplaying of Port Louis's social urbanity in the state's efforts to make a city capable of 'acting big', constitutes a firewall that ends up precluding any real thickening of hybrid ties.

## PORT LOUIS'S URBAN MAJORITY: MUSLIM AND CREOLE

A century after the exodus from the city in 1868, the 1968 riots that tore apart Port Louis's Creole and Muslim communities remain a traumatic and often-cited event for the city's residents. The concerns of multiculturality often centre on the status of the women of a community. To control the trajectory of a young woman's sexuality has in many contexts come to stand in for the integrity of the community or society as a whole. Urban life has always raised the prospects of unbridled sexuality; in the city individuals are made strangers to each other, their desires are freed from the strictures of past obligations and socialization and thus individuals come to navigate a more unencumbered field of desire.[13] The riots of 1968, the year of national independence, were riots about many things. Muslims in all twelve wards of the city had largely voted against independence, worried that the imminent political dominance of Hindus would upend their dominance over mercantile affairs. Port Louis was, and continues to be, a predominantly Muslim city and then secondarily a black city. Fears were rampant as to how a newly independent national government would exercise its power over the capital, particularly as it was the home of minority populations.[14]

Yet, the actual catalyst of the riots was about women and about sex. There was strong competition at the time between Creole and Muslim prostitution syndicates. Thus it was considered a major coup when the Creole syndicates secured the services of a young Muslim prostitute. Although these syndicates had clashed violently in the past, this event was enough in this particularly turbulent political context to run rampant.[15] Roche Bois and Briquetterie were primarily Muslim neighbourhoods and Trou Fanfaron, Cité Martial and Plaine Verte were highly mixed Creole and Muslim areas. For several weeks, there was an ethnic cleansing of neighbourhoods, with houses burned and families driven out. The result is a stolid ethnic homogeneity that remains to this day, reflected primarily in the Creole district of Roche Bois and the Muslim district of Plaine Verte. While the Cassis district at the southern edge of the city remains a mixed neighbourhood, the strict segregation predominating in the northern districts sets the tone for the city as a whole.

As for the Creole, the socio-economic and political situation remains grim. Although the level of impoverishment is nowhere near as stark as it is in many

African cities, there remains a profound disenfranchisement of real participation in Mauritian society. The so-called Creole 'malaise', pointing to a combination of systematic institutional blockages and an internally generated inertia, at times approaches a silent decimation. Without the local political clout of Muslims, the Creole have seen their primary residential base in the city bifurcated by a new highway and subject to the proximity of hazardous industrial sites – the expansion of the port, chemical plants, solid waste dumping and petroleum processing.

The malaise has been compounded by an educational policy that required students completing the basic six years of primary school to pass a skills test in order to proceed with their education. Nearly 70 per cent of the Creole student population consistently failed to pass this examination, thus ending their school careers. Given that the age of legal employment is 15, there is a considerable lag time between the termination of schooling and the advent of legal employment. Entire generations of Creole children were condemned to an enforced idleness, the ramifications of which meant it was even more difficult for them to find work and a niche in Mauritian society. This education policy was finally changed in 2004 with the guarantee of basic education until the age of 15.

The Creole malaise also points to a more ambivalently insidious dynamic that sometimes stands in sharp contrast to the implosive character of the Creole district of Roche Bois. Cassis is the other major district in Port Louis with a large Creole population. Unlike Roche Bois, Cassis is not an ethnically homogeneous area; neither is the Creole community exclusively poor. While the problems of disenfranchisement, criminality, drug use and unemployment are certainly prevalent here, as in Roche Bois, these problems are wrapped into a veneer of seeming self-sufficiency – a community apart from the self-consciousness that characterizes Roche Bois. It is as if the Creole residents here do not need to feel a part of the larger city nor the larger country and that the quotidian rhythms of conversation, neighbourhood dramas, continued strong relations with the Catholic Church, football games and chatter in the bars is a world unto itself.

## DEPLOYING THE COSMOPOLITAN

In part, this capacity exemplifies what several observers have pointed out concerning how the banal character of everyday interactions is the most incisive locus through which viable interpenetrations of distinct cultural practices take place (Hills 2002). Given the actual blurring of discernible orientations in the context of managing the contingencies of everyday relations and events, it could also be construed that the invocation of clarity in any such orientations is a largely retrospective manoeuvre possible because of the active and concrete reciprocities that take place within neighbourhoods.

Additionally, the multiplicity of discrete cultural references also acts as a tool, not only in offering diverse contexts of refuge to which those at the margins of any one cultural orientation or legacy can have some even tentative affiliation, but also in elaborating a terrain in which any discrete set of practices can attain a sense of their own vitality. A broad range of sensibilities, locatable and attributable to long-standing cultural histories, as opposed to individual characters, can provide neighbourhoods with various inclinations and explanations that mark out highly flexible oscillations of proximity and distance among neighbours. Neighbours neither have to be too close nor too far, too obligated nor too irresponsible. These incessant and minor recalibrations offer a locus of cooperation without requiring some major testimonial of overarching commitment.

This quotidian resilience can also sometimes be seen in the strategic practices elaborated by economic policy makers. For much of the period of structural adjustment, the country was able to maintain a highly open economy premised on a wide range of preferential trade agreements – for example the guaranteed European Union price for sugar set at three times the world price and the duty-free access of locally produced garments to the European Union and the United States of America. As the primary trade agreements upon which Mauritian economic growth was actualized have recently come to an end, the country has been intensifying its efforts to broker various modalities of regional cooperation across Southern Africa and the Indian Ocean.

The country actively seeks to use its bilingual capacity and its geographical location between South Asia and Africa within a cross-section of domains – from political regionalization, financial architectures and transportation flows – to construct itself as an important interlocutor.[16] Particularly in the area of financial intermediation, Mauritius has proven highly adept at steering its tax codes, investment structures and institutional frameworks to give it maximum leverage in economic articulations across a diverse range of regions and national economic orientations. A strategic mix of offshore and onshore banking facilities, coupled with the emergence of an information and communication technology service economy and reorganization underway in the export processing zones as well as in the free port transshipment facilities, reflects an ongoing effort to engineer symmetries across sectors.[17]

Yet resiliency has its limitations. There is only so far it can go to compensate for relative geographical isolation, as well as the lack of resources and small size. While the country has gone a long way towards raising its population out of poverty, it requires substantial investments in the upgrading of its human resource base in order to continue to ply the same kinds of global circuits on which its past economic success was based (Bunwaree 2002). The key danger is posed in what some Mauritians label the 'frog theory'.[18] This refers to the relative ease Mauritians feel in different environments – their resilience in adapting to diverse situations. As indicated before, such resilience is largely honed in the arena of everyday life, in the largely neighbourhood settings in which the majority of the population resides. But, like frogs able to adapt to incremental elevations of temperature, a threshold is eventually crossed which is catastrophic. The frog, able to adapt so quickly, never senses this threshold coming, never experiences a sense of imminent danger, as the new climatic conditions don't serve as warnings.

City life, with all of its anonymity and possibilities of neighbourliness, may promote resilience but it does not necessarily reward or accommodate resilience in all cases. While there are certainly long and varied histories of making do in the most inhospitable of urban environments, adaptation is rarely seamless. Cities exude multiple sources of harm and sites of contestation; interests and agendas are often incompatible no matter how they are articulated and frequently, differences are unable to be accommodated. Cities are sites of disjuncture, and no number of improvisations or amount of flexibility can completely ward off erasures, violence and remaking of all kinds. Many Mauritians believe that keeping a brake on the intensity of urbanization is critical to their ability to act resiliently across many different dimensions of life. At the same time they will also tell you that, in the end, this very resilience defers to a reckoning with critical asymmetries in entitlements and opportunities that, in turn, keep the country from getting all that it could from its human capital base.

In fact, the limitations on urbanization could be playing a major role in obscuring the extensiveness of labour market segmentation, deficiencies in manpower training and fracturing in the social fabric.[19] These limitations are also seen by some as an impediment to dealing with these issues. In response to several politically influential architects, the government initiated an international ideas competition in April 2004 for the design of a master plan for Port Louis. In part, the impetus for this competition was the desperation over the protracted problems of traffic circulation which plague the capital, sometimes making what is otherwise a twenty-minute drive from the major suburb of Quatre Bornes a two-hour commute. But the agenda was also to use major infrastructural renovations – the rebuilding of the two major bus terminals servicing the southern and northern routes out of the city – as a context to reignite a more sustained interest in the long-term development of Port Louis.[20]

The drivers behind this project, most particularly Gaetan Siew, the country's most renowned architect, are clearly operating against the grain of conventional sentiment. Siew believes that the city must attract a new generation of young, middle-class, urban dwellers and that subsidizing their implantation in the city will lead to a broader swathe of up-market development for the city, making it a more attractive and vital place in which to reside. While what Siew and others advocating a major remaking of the capital have in mind tends largely to be an encapsulated urbanization – well-serviced residential enclaves close to a growing financial and services sector with access to a broad range of cultural and leisure accoutrements – the push for a more urban mentality clearly seeks to exceed the conventionally familial and neighbourly domestic orientations of much of Mauritian social life.

Despite the spectral dimensions of Port Louis – the memories of abandonment and the insalubrious, as well as the presently engineered vacancies of much of the city's centre – it embodies a responsibility to forge some kind of working coherence to what is a metropolitan region, albeit with ruralized gaps, that runs through Curepipe, Floreal, Vacoas, Pheonix, Quatre Bornes, Rose Hill, Beau Bassin, Port Louis, Terre Rouge, Pamplemousses, Troilet and Grand Baie. To the west of this corridor lie the sugar fields, many of which may be seeing their penultimate harvest. To the east, there exists a nearly continuous strip of high-end tourist resorts which worry about occupancy rates. The economic future of the country may rest in engineering a more proficient consonance between the earning power of urban residents, the provision and servicing of urban infrastructure and a greater level of densification and infilling that could promote more coherent spatial planning.[21] The most recent long-range development planning framework emphasizes the creation of high-density nodes across this conurbation, intersecting commercial activities, schools, cultural and recreation centres, with a broad range of synergies to special use zones, and all serviced by new forms of public transportation (Government of Mauritius 2003).

The question is how Port Louis rearticulates these emergent developments across the conurbation back to itself, how it comes to resonate with highly particularized residential communities and offers itself as a locus of intersection, beyond simply being a place where people come to work. Given the high cost of land (400–500 square metres averaging $US18,000) and the anticipated need to build seventy thousand units of housing in the next decade (almost half of the current housing stock), Port Louis perhaps cannot avoid becoming an important site of both densification and diversity. While the years of relative residential abandonment have given rise to a spatially and culturally highly cohesive urban Muslim population, there is no reason why the city can't rise to other equally cohesive forms of cultural life. Finding ways to actively recoup a long urban memory of intercommunal collaboration, elaborate upon the obvious fascination with the potentials of urban mixture exemplified by the lunchtime meanderings of workers, and to extend and concretize public spaces of intersection among what are often highly liveable neighbourhoods already constitutes modalities for making Port Louis larger than it is. Instead of relying upon the conventional financial and infrastructural interventions that seek to further implicate particular urban systems in a wider range of economic and cultural networks, there seems much within Mauritian urban history to build upon – forms of life capable of more fully urbanizing the everyday capacities of Mauritians from various walks of life to find other ways of occupying, spatializing and institutionalizing the obvious excitement and long-honed confidence of being amongst Mauritians other than themselves and their kind.

## NOTES

1 Interview with Ms. V. Saha, Town Planning Department, Ministry of Housing and Lands, August 2004.
2 Interview with Cader Kaller, Mauritius Museums Council, July 2004.
3 Interviews with the Municipal Town Planning Council, Municipality of Port Louis; Ms Fong-Weng Poorun; Ramon Cundasawmy, Ministry of Environment.
4 Interview with R. Jahangeer-Choychoo, historian, Mahatma Gandhi Institute and the Ministry of Environment.
5 Interviews with Rafik Gurti, Militant Socialist Movement, Port Louis; T. Raha, planner, Municipal Council of Port Louis, August 2004.

6 Interviews with R.S. Sonea, Local Government Service Commission, July 2004.
7 Interviews with Prakash Maunthrooa, ex-President, Mauritius Port Authority and V. Kallee, Senior Manager, Mauritius Port Authority, July 2004.
8 Interview with Indranee Gopaulo, Director of the Center for Mauritian Studies, August 2004.
9 Interview with Chris Lee Sin Cheong, Head of the Chinese Chamber of Commerce.
10 Interviews with Jocelyn Chan Low, Director, Mauritius Cultural Center and Cader Kalla, August 2004.
11 Interviews with Jocelyn Chan Low; Jean-Claude Augustave, Director, Nelson Mandela Cultural Center; Father Phillippe Fanchette, Paroisse Notre Dame de l'Assomption, and Mario Flore, Le Muvman Mobilisasyon Kreol Afrikain.
12 Interview with Hemsing Hurrynag, Director, Development Indian Ocean Network.
13 Interviews with Cader Kalla, Jocelyn Chan Low; Vinesh Hookoomsing, Dean of the Faculty of Arts and Sciences, University of Mauritius, August 2004; Sheila Bunwaree, Department of Sociology, University of Mauritius, August 2004.
14 Interviews with Muslim Jumeer, Mauritius Heritage Trust and the Mauritius Institute of Education.
15 Interviews with Cader Kalla and Jocelyn Chan Low.
16 Interview with Yvan Martial, journalist.
17 Interviews with D. Dusiruth, Indian Rim Association, July 2004; Assad Bughla, Trade Specialist, Ministry of Finance, July 2004; M.D. Phokeer, Assistant Director, Department of Regional Cooperation, Ministry of Foreign Affairs, August 2004.
18 Interview with Samad Ramolay.
19 Interview with Raj Makoond, Executive Director of the Joint Economic Council, August 2004.
20 Interviews with Gaetan Siew, Lampotang and Siew Architects.
21 Interviews with G. Heemoo, Director, National Development Strategy, Ministry of Housing and Lands.

# 3 Temp Town

## Temporality as a place promotion niche in the world's furniture capital

John Joe Schlichtman

> High Point has one of the few downtowns in North Carolina that does not have a considerable aura of emptiness to it. Many visitors remark how pleasant it is to see a heavy majority of storefronts occupied ... [They] are astonished to learn that a majority of those buildings are used basically only four to six weeks a year for International Home Furnishings Markets.
>
> Tom Blount, Editor
> *High Point Enterprise*, 1999

'The whole face of High Point has changed', stated Judy Mendenhall, the current president of the International Home Furnishings Market Authority and a former High Point mayor, councilwoman and chamber of commerce president (Maheras 2001). 'I don't know that the downtown is ever going to come back the way it was, but I don't think that we're that different from most cities our size across the land' (Maheras 2001). High Point, North Carolina, the two-hundred-and-eighty-ninth largest city in the United States of America with a population of just over 90,000 is indeed similar to its small urban peers. The loss of retail establishments, office space and community meeting places from its downtown over the past 40 years is familiar to those who examine de-industrializing, decentralizing cities such as High Point.

However, High Point is also different from other small cities that have lost their downtown base. For one, the downtown is one of the newest and most well-maintained in the state of North Carolina. It boasts of an extremely low vacancy rate. It is also remarkably international in both its architectural influence and its real estate ownership. Finally, and most obviously, the entire downtown evokes furniture. From the furniture buildings such as the National Furniture Mart, Home Furnishings Centre, and Furniture Design Centre to amenities such as the World's Largest Chest of Drawers, the Furniture Walk of Fame, the Furniture Discovery Museum and the Bernice Bienenstock Furniture Library, downtown High Point is designed to provide a themed landscape of home furnishings.

However, downtown High Point is different in a more profound way. The use of the city centre is *temporary*. Downtown High Point is the location of the biannual

*Figure 3.1* Showplace is one of the Market's 190 buildings. (Source: courtesy of John Joe Schlichtman.)

International Home Furnishings Market (hereafter, I will use its local name, 'Market'), which currently draws over 80,000 visitors from more than 110 countries, effectively doubling the population each April and October. Furniture buyers, executives, designers and media come to the downtown to view just under 12 million square feet of displays by about 3000 manufacturers in 190 buildings. It is at Market that furniture retailers make the purchase decisions that will influence the next year's trends. While exchanging billions of dollars in furniture transactions, they generate just under one billion dollars annually state-wide and about one-third of a billion dollars annually region-wide in other non-furniture revenue such as accommodation, food, entertainment, transportation, retail and tax dollars.

However, all of these activities populate the downtown only during the months of April and October. For this reason, I term downtown High Point 'Temp Town'. Yet, despite its temporariness, the temporality of downtown High Point is not akin to an event site such as a fairground or to a seasonal place such as a summer resort town. Make no mistake; the structure of Temp Town's real estate market is as permanent as that of Wall Street's in New York. While Temp Town is *inhabited* only temporarily, the leases and ownership of its tenants are active all year round.

In fact, High Point's competitive advantage over larger cities that have sought the revenue that the Market brings has been precisely that it allows manufacturers' preparation of their showrooms to take the full six months between Markets. As a result, Temp Town is always in flux. Judy Mendenhall, the sixty-something president of the International Home Furnishings Market Authority, explained this six-month cycle from her large downtown office. During the first month, she began to explain, furniture manufacturers' staff 'are going to look at the fabric that they have chosen to put on upholstered pieces, they are going to look at the designs, the finish, the – whatever – for all of the casegoods, and they are going to start working with the folks at [the factory] to determine how they want those showrooms laid out'. The designs begin coming to fruition in approximately the second and third months. 'At the factory', Mendenhall continued, 'they start laying out the design of how those showrooms are going to look – and they think in their own mind and work with their internal creative people on the best vignettes, the best way to show the new product, the best way to show any nuances in the existing lines'.

What begins as brainstorming in the minds of designers and engineers all over the world takes physical form in Temp Town during months four and five. It is then that manufacturers typically 'get their contractors, the local contractors, who repaint, reposition walls, put new carpet down, put new flooring down, do whatever basics need to be done in that space, in that building'. During the fifth and sixth months, 'the goods will come in ... the product will come in, new will be added, the designers will come – they'll rearrange, they'll configure it, they'll get the people up there, they'll start moving the stuff around'. Then, 'right before Market, they get the fresh flowers in and adjust the lighting. The weekend before Market starts, the reps come in so they can get their training, then the caterer that is going to serve in the showroom moves in right before Market and gets all that ready'.

'And they are ready to go', Mendenhall concluded. 'My point', stated the president, 'is that there is a long planning process'. Although the buildings of High Point's downtown are stages that are spotlighted only twice each year when the curtains rise, they are also sets that are in constant preparation during the six-month periods that precede each production. Perhaps no place better illustrates Zukin's (1995: 9) claim that 'the symbolic economy recycles real estate as it does designer clothes'. Temp Town's totalizing alignment with the temporality of the furniture industry is the city's economic niche:

> Renting exhibition space permanently to be used for only a handful of days a year may seem counterintuitive, but such an arrangement has advantages for furniture manufacturers who find it cost-effective to maintain a single showroom, rather than to bear the cost of shipping, erecting and dismantling a trade show booth every year, explained the *New York Times* (Newman 2005).

This 'arrangement' is the totality of downtown High Point's political, economic and social life.

This chapter will follow the actions of local growth proponents in the four decades between 1963 and the present day in an effort to illuminate how Temp Town's transformation unfolded. I will examine the story of Temp Town decade by decade, observing how, as the city lost its furniture manufacturing base, it rebounded by gradually opening its entire 30-block downtown for showroom investment. As the downtown's showroom space grew from just over 500,000 square feet to just under 12 million square feet, High Point's furniture exposition would rival cities anywhere with five to thirty times its population.

## FACTORY SHOWROOM GROWTH AS VERTICAL INTEGRATION (1963–73): FIGHTING CHICAGO

In the early 1960s, American furniture manufacturers, whose consumers were mostly small, family-owned furniture stores, tended to be large firms that had a tremendous amount of leverage regarding the location of furniture expositions. Many of these companies were based within a two-hundred-mile radius of High Point, the centre of the most dominant furniture production complex in the United States. Manufacturers in this production complex were showing their products in the major furniture exposition in Chicago. During this period, however, they began to integrate showrooms into their factories, enabling them to save money in rent and transportation and to have more control in the preparation of their showrooms.

The North Carolina-based Market grew to usurp the prominent Chicago market when these dominant firms began showing back home. As the North Carolina showroom presence grew, firms outside the south-eastern United States and smaller firms within this region sought exhibition space in order to benefit from the traffic that the larger companies attracted. High Point's booming three-block showroom complex would fill this niche. Between 1963 and 1973, downtown High Point's showroom area almost tripled to 1.5 million square feet. The boom began with the building of the 85,000-square-foot National Furniture Mart, which was erected in 1964. The building was financed by a group of investors led by an original board member of the city's first and largest showroom building, which I will call the Home Furnishings Centre. At the building's opening, he argued that 'anything which is put in High Point helps High Point', adding that he wanted to do his part to 'help High Point keep the market' (Hawkins 1964).

During this period, High Point leaders regularly argued that showroom development was High Point's ticket to national and international prominence. 'The market has made the difference between High Point being just another manufacturing town and the internationally known furniture exhibition centre it has become', expressed a local (now) billionaire and showroom owner in 1970 (Clontz 1977). High Point leaders began rallying the community around the Market, arguing that High Point need not be second to any city. The city's daily newspaper, the *High Point Enterprise*, would report this position as front page news: 'At market time, High Point is warm, bubbling, and charming, and manages to take on an urbane, cosmopolitan air along with courtesy … Those who fear that little towns lack the glitter, excitement and convenience of major metropolitan centers often go away convinced that High Point has advantages over Chicago' (Shires 1963).

While High Point leaders pronounced their city the Market hub, it is imperative to understand that the Market was also hosted in showrooms spread throughout five other towns in North Carolina that spanned 300 miles of road known in the industry as 'The Furniture Highway'. The building of factory showrooms and the sprawl it caused had obvious adverse effects on the efficiency of the Market. In the late 1960s, it was estimated that the average buyer would put about 1200 miles on his car during Market. Manufacturers were often left in the dark as to whether their comparison shoppers would even return to their end of the state. 'The market is too fragmented', said the executive vice-president of a major manufacturer whose firm contributed to the sprawl. 'Personally, I liked the spring and fall shows in Chicago, when everyone was together', reminisced the executive in 1967 (Marks 1967). 'A buyer could look at one line, then go down the hall or to the other building, check another line and make up his mind within 15 minutes which one he wanted to buy. Now the buyer can't see it all at one market. He can't remember how a line looked from one showroom to the next'. Manufacturers began to discuss the need to consolidate the Market.

Meanwhile, High Point boosters continued to articulate that the Market conferred big city status upon the town. 'It's Paris for hemlines. It's Detroit for autos. It's High Point for furniture', stated the *High Point Enterprise* in 1970 (Haislip 1970). Despite all the talk about Market, the city's showroom presence was still only one component of High Point's lively downtown. The downtown was still a vibrant centre of department stores, restaurants, bars, clubs and offices. Its three temporary showroom buildings were built exclusively to be furniture showrooms and they occupied only three blocks of the centre of the city.

## PLANNED SHOWROOM GROWTH (1973–83): GROWTH AT THE EXPENSE OF RESIDENTS

From 1973 to 1983, furniture showroom space in downtown High Point grew sixty per cent to 2.5 million square feet as the demand for showroom space required greater amounts of downtown real estate. In the early 1970s, the city council inquired about four unused adjacent historic properties that were owned by the city of High Point: the former firehouse, city hall, police department and the Paramount Theater, a city landmark built in 1923 in the luxurious vaudeville fashion typical of the era. The city council appointed a special citizens' council chaired by a local real estate leader and comprising members representing local theatre interests to devise a plan for the buildings. The committee developed a consensus that the three buildings be incorporated into the Paramount to create a theatre and exhibition complex of a calibre unparalleled in the region. Early in 1972, voters enthusiastically issued bonds to create the historic complex.

As plans were unfolding, the director of the Home Furnishings Centre retired. The new director moved from his position as the executive vice-president of the National Home Furnishings Association in Chicago to take the post. His first action in his new stewardship of the building was to propose that the Home Furnishings Centre expand to the end of the block over which it had grown during the previous forty years. In 1973, the city leased the Paramount project land to the Home Furnishings Centre. The city, however, had a conundrum: it had just committed to the people by referendum to build the proposed theatre complex on the same land. The city and the exhibition complex determined that the referendum and the lease could both be honoured if the historic buildings were razed and the eastern-most side of the first floor of the massive addition – where the buildings stood – was made into a theatre. And so by 1975, on the site of the Paramount, the drab 1,000-seat High Point Theater stood awaiting its first event. On opening day, the Home Furnishings Centre announced that it would foot the bill for the theatre, its director stating that 'Presenting this theatre to High Point is our way of thanking everyone in the city for showing our market visitors that we are glad to see them and appreciate their business' (Marks 1975). The mayor of High Point announced that the theatre ushered in a 'new era'. Indeed, it did.

While the growth of this period was similar in that the bulk of square footage was built on the Market's existing three blocks, it was also notably different in that the philosophy of showroom growth among the city's real estate interests began to change. Showroom use was taking precedence over more resident-centred land usage. Showrooms began springing up outside the Market's core three blocks and furniture manufacturers and local real estate investors openly supported the renovation of buildings for showroom purposes. During this time, the city celebrated the opening of its first major shopping centre, Westchester Mall, which featured two anchor department stores and adjoining shops. As we will see, this pull of retailers, restaurants and offices out of the downtown also had a significant push. Downtown landowners began to seek the higher revenue and lower maintenance of Market leases.

The retired manager of the city's second largest furniture building had owned 154 South Main Street his entire lifetime. In the early 1970s, this address was fast becoming globally-central real estate in the furniture world. The retiree-turned-real-estate-investor began 'urging furniture producers to take long-term leases' on his and other downtown properties that had been used for retail, offices and other traditional downtown uses (Hawkins 1972). The plan sparked a one-block transformation in which a former department store, café, record store, photography shop, clothing store, shoe store and hat and jewellery store all became showrooms by the end of 1972. The rationale he gave for doing this would be used by the city's leaders to the current day. First, he stated that furniture manufacturers that have long-term leases on their showrooms can 'get known for their location and amortize the costs' (Hawkins 1972). Second, manufacturers with long-term control could 'invest in their spaces so that the areas are fixed to suit their needs'. In other words, manufacturers could renovate and rearrange their spaces on their own schedule and at their own discretion. Finally, utilizing the buildings was good for High Point. Having buildings in use, he suggested to High Pointers 'looks better than having a lost tooth' in the city's increasingly bright architectural smile (Hawkins 1972). He failed to mention a fourth reason for his excitement and this, of course, was that manufacturers' demand was driving up the value of downtown real estate at a time when it would otherwise have been depreciating.

Showrooms, he concluded, 'represent the best current use for otherwise vacant buildings'. His strategy and the city's subsequent acceptance of it would lead to a new commodification of downtown real estate (Hawkins 1972). Without zoning or planning to deter or delay it, this new revaluation of the downtown real estate suddenly had much higher stakes; stakes that would warrant efforts to sculpt zoning resolutions, political decisions, and infrastructure placement. Showrooms began to consume the downtown. In 1973, a manufacturer's representative leased

*Figure 3.2* The Main Street storefronts of yesterday are today's showrooms. (Source: courtesy of John Joe Schlichtman.)

a 17,000-square-foot former department store to show off an international array of furniture lines that he represented at the time. His other furniture-related real estate was in Florence, Italy; Barcelona, Spain; and London, UK. Later that same year, a local group of businessmen led by the brother of a downtown real estate leader joined with the president of a local advertising firm to announce the purchase of a large – and occupied – furniture factory. Although a committee had just been formed to determine how traditional businesses might be kept in the downtown, the announcement of the building's transition from an active factory to the city's second largest temporary showroom space was hailed by city officials as a major factor in the 'revitalization' of the centre of the city.

## FIGHTING OFF DALLAS WITH CONSOLIDATION

In the early 1970s, Dallas, Texas, was constructing a 1.4-million-square-foot showroom complex for home furnishings, gifts and decorative accessories. In 1976, the Dallas Market Center campus became the world's largest with an expansion that brought the campus to 5.3 million square feet. High Point was facing a challenge on a scale that it had never faced before. Although the need for the consolidation of the Furniture Highway was already known, the threat from Dallas motivated local furniture interests to action. Downtown growth was supervised by the newly established Downtown Development Board, composed of a downtown real estate owner, a banker and the Chief Executive Officer of the Home Furnishings Centre, who chaired the committee. The board established a low-interest loan programme by which properties in the downtown could be renovated using federal assistance. Building owners were able to buy, sell or rent their property at whatever price the showroom real estate market allowed. Owners worked to manipulate property to its highest and best use using market criteria.

A shoe store proprietor decried the influence of furniture industry stakeholders on the city council and stated that critical downtown decisions were being made to serve the furniture Market at the expense of other uses. He stated that construction plans to remove downtown parking spaces, for example, were revealed to the merchants at a breakfast meeting only after the project had been finalized. 'The plan stinks', stated the owner of a downtown jewellery store. 'It was not presented to merchants until it was finalized ... Merchants were not consulted on their needs' (Johnson 1978). 'It's just killing downtown', echoed a sporting goods store operator, who stated the project 'put the icing on the cake' for his plans to go out of business (Johnson 1978). High Point leaders continued to encourage residents that supporting the changing downtown was a part of their civic duty. After all, argued the *Enterprise* editor, the Market transforms High Point from 'a medium-sized Southern industrial town' into 'a teeming metropolis' (Brown 1978).

## UNFETTERED SHOWROOM GROWTH (1983–93): HIGH POINT BECOMES THE HUB

While small malls were developing outside the downtown and showroom investment was increasing in the downtown, factory closures around the city were making it clear that High Point's position as a production centre was in question. Automation was simply not an option for some High Point furniture plants because of their age or their lack of investment capital and the building of new furniture factories was slowing. Increasingly, the decision not to invest in factories in North Carolina was sealed by the hefty savings of producing abroad. As United States furniture production deteriorated, 'Asia, Taiwan, Singapore and even the Philippines grew' explained a leading industry analyst (Epperson 2003).

In 1983, there was 2.5 million square feet of furniture showroom space in downtown High Point. In fact, by 1985, after three of the Market's largest manufacturers came to High Point from neighbouring cities and others followed in their wake, 1470 of the 1500 manufacturers showing in the Market had consolidated in downtown High Point. By 1993, the city's showroom space had almost tripled to 7 million square feet. A local construction mogul was head of the Chamber of Commerce, and the city's leading estate agent served as chair of the Economic Development Corporation. Downtown development boomed with unchecked free market growth. Ironically, the once car-based Market, which had been spread across hundreds of miles, was becoming equally hard to navigate on foot as it spread out over 30 blocks. 'It's so spread out people get tired', said a manufacturer from Alabama in 1986, when the square footage reached six million. 'If it doesn't change, we're going to quit showing at this market' (Stephens 1986).

Despite the downtown's unwieldiness, the expansion continued as manufacturers saw the folly in renting space in a large exposition building for more than they could build, purchase or lease their own facility. Increasingly, manufacturers viewed their showrooms as important aesthetic statements. Free-standing showrooms not only had an interior aesthetic presence, but they made an exterior statement as well. They became a symbol of the line represented inside. A local manufacturer overhauled High Point's classically styled legal building with the purpose of making such a statement. 'It's a very distinctive building,' he explained. 'We can use this building in advertising and be recognized … Free standing showrooms are a key to keeping the furniture market entrenched in this area' (Cashwell 1987a). An executive of another multinational manufacturer offered a similar description of their new building: 'We can now present the product in an environment that can be reviewed, seen, and understood' (Cashwell 1987a). The showroom, he explained, is 'a symbol of the direction we're heading in the home furnishings industry'.

Buildings of vastly different styles were erected to accentuate the goods that they would hold inside with little thought of their juxtapositioning with their neighbours. At times the building seemed to reach a frenzied pace, as in 1987, when the same air rights were promised to two new multi-tenant exposition buildings: Market Square and Commerce and Design. Due to the political leverage of Market Square's managing partner, the Commerce and Design coalition was required to have a skywalk built *through* their building, although it would allow no access to it. 'It's ludicrous', said one tenant, noting the managing partner's clout with the city. At least one councilman, an estate agent, seemed rather candid in his acknowledgement of the lunacy: 'We should have made the rules, but we were afraid to' (Cashwell 1987b).

Mayor Judy Mendenhall began to voice concerns that the downtown was becoming a temporary use bubble utilized only by the furniture world. Her views became widely known in 1986 when the front page of the *Enterprise* announced that 'battle lines' over the downtown's future were drawn between, 'in the near corner, standing proud and resting on the weight of societal trends and High Point's furniture orientation is a determined group of downtown property owners' and 'in the far corner, looming extremely tall, and with the weight of state redevelopment statutes behind it, is the city of High Point' (Reis 1986). Never before had the city taken a stance that could be construed as anti-Market. Clearly, however, the small downtown was at a critical crossroads: while it was profitable and occupied, six million square feet of its space was utilized only twice each year.

Showroom building owners were in an uproar. 'The best benefit to the people of High Point is to protect the furniture industry', noted a showroom owner's lawyer. 'I don't understand fooling with that', said a property owner across the street. 'Anything that hurts exhibitors hurts High Point', she added, reciting the common rhetoric of Market leaders (Reis 1986). 'To the mayor', explained the *Enterprise*, 'the targeted blocks are a largely unattractive and abandoned looking cluster of buildings that contribute little to the greater community and dress up only twice a year – during the [Market]' (Reis 1986). In 1988, a signature of the downtown closed after almost 70 years. 'Downtown is becoming all furniture market', explained the owner of High Point Cleaners and Hatters who, as the founder's son, grew up in the shop (Ingram 1988). The mayor's office warned the city that the loss of such businesses was significant. Mendenhall's plan to exert more control over the downtown's transformation would fail and increasing amounts of real estate were aggregated for Market use. In fact, while the mayor was questioning Market development, the Home Furnishings Centre was undergoing another expansion. Previous expansions covered the entire city block except for part of one side where the county courthouse stood. 'We're just filling in the "U"', explained the 'Big Building's' president before the courthouse was demolished (Inman 1989a).

Owners of the city's second largest building were also planning for expansion, as the investment group of Market Square (the former factory discussed in the last section) announced that they would add a fourteen-storey tower to their historic site. The group billed it as the 'first mixed use high-rise in the state' (Inman 1989b). Announcing that it would feature five floors of showrooms and nineteen condominium units, city leaders used a sleight of hand still in use today. The 'residential' developments that would be touted in 'mixed use' projects over the next fifteen years were not resident-owned as one might expect, but were purchased by furniture companies in lieu of renting hotel rooms. High Point's downtown condominiums are another facet of the temporal arrangement of the downtown that is centred on the furniture industry.

Despite the critical review of Market development by the mayor's office, pro-growth leaders continued to promote Market as High Point's road to big city status. 'The prominence of this market adds to this area and North Carolina and is something to be very proud of', conveyed a local radio personality and columnist. 'Information is now being placed in every U.S. embassy and consul in the world. It's printed in five different languages … We are a known commodity throughout the world. Y'all should feel proud' (Martin 1988).

## THE MARKET TODAY (1993–2005): OVERWEIGHT, OUTDATED, BUT STILL ON TOP FOR NOW

Furniture production grew even more scattered throughout the globe during this period as 'Malaysia, then Thailand, then China and now Vietnam burst on the scene' (Epperson 2003). While the furniture production complex once centred on High Point continued to disperse, the city became firmly entrenched as the world's wholesale furniture node. By 2005, Temp Town's showroom presence had risen to almost 12 million square feet. 'There are buildings all over the place that were just parking lots when I first started coming here', I overheard a man remark at a recent Market. Now the International Home Furnishings Market Authority President, Judy Mendenhall, spearheads the coordination and promotion of the massive Market. Although 'everybody doesn't agree on everything', she explained in speaking of the Authority, 'they do agree on the Market', echoing Logan and Molotch's (1987) observation that growth leaders need only agree on growth.

Mendenhall's question in the 1980s of 'how much showroom growth is too much?', was readdressed by the president of the High Point Economic Development Corporation as the 1990s drew to a close. 'The market [lowercase 'm'] determines that. Those of us in the business development area can't control how well an industry is doing. Obviously, the Market [uppercase 'M'] is doing well and I'm

*Figure 3.3* The Natuzzi building is a result of the latest building boom. (Source: courtesy of John Joe Schlichtman.)

happy to see it. It stabilizes our downtown area and promotes the reuse of our facilities' (Whittington 1998). Bringing traditional uses back to the downtown was out of the question. 'They can't afford the dirt', said the chair of the Downtown Improvement Committee. The dearth of restaurants, retail and hotels in the downtown makes Temp Town the butt of jokes by residents and, ironically, the subject of grumblings by marketgoers wishing to rest and refuel.

Meanwhile, the unceasing development of showroom space continues to frustrate visitors. 'If I was coming to market for the first time, [and] saw all the showroom space out there', said a buyer from Rhode Island, 'I'd probably throw my hands up and get back on the plane' (Craver 2000). Temp Town now faces its most serious attack from a larger rival as its leaders address an unprecedented threat on the horizon from Las Vegas, Nevada. Developers from New York and Los Angeles have plans to build a single 12-million-square-foot complex in Las Vegas called the World Market Center. Construction of phase one of six phases was completed in July 2005 and its first market in the same month drew 53,000 marketgoers. Las Vegas, approximately the thirtieth most populated city in the United States today, is five to six times the size of High Point. However, due to its status as a vacation capital, it has the visitor infrastructure of a much larger city. Las Vegas' advantage would seem apparent. Whereas Las Vegas offers 140,000 hotel rooms, High Point claims just under 1100, requiring many marketgoers to rent rooms or entire homes. Whereas the Las Vegas Market will offer 12 million square feet in eight conjoined buildings built specifically for furniture display,

High Point's 12 million square feet is spread across 190 buildings – most originally constructed for other uses.

At critical moments in the past 40 years, especially during times of intense interurban competition, High Point leaders have chosen to give their downtown over to the free market development of the International Home Furnishings Market. Even the downtown's most prominent office building, built in the early 1970s to promote non-showroom development, is expected to be converted into temporary showrooms. A conceptual sketch of the building shows the gutted office tower connected by skywalk to the Showplace furniture exhibition building across the street, its illuminated 'GE Capital' sign replaced with 'Showplace West'. In so doing, the Market has become difficult to administer, unwieldy to navigate and extremely complex for a small city to accommodate. The success of the first Las Vegas Market 'has presented an opportunity for High Point to rally', said Mayor Becky Smothers, one of the High Point leaders engaged in a lively repartee with Las Vegas leaders over the past few years (Johnson 2005). But who is there to rally? A coordinated response to Las Vegas is virtually impossible because of Temp Town's thousands of stakeholders, various lease arrangements and myriad national and international ownerships with varying interests in the city. Why would such a diverse group rally when it is easier to leave? Furthermore, the powerful furniture manufacturers that once favoured the High Point region now have interests scattered all over the world. And the small, local, family-owned furniture retailers that once accounted for the majority of American furniture sales have been replaced by influential national and multinational corporations.

It is this national and global capital that has served as a catalyst for Temp Town's growth, but it may also have raised the stakes to a level on which High Point can no longer compete. The real estate of the entire 12-million-square-foot, thirty-block area of Temp Town has a value of $US500–600 million. Seeking to replace it with the brand new 12-million-square-foot, 57-acre World Market Center, the investment group in Las Vegas will ultimately invest $US2 billion dollars. High Point's hold on the Market seems as fragile as ever. Forty years ago, marketgoers came to High Point's showrooms because it was the world's furniture manufacturing capital. Today, High Point is the world's furniture showroom capital only because marketgoers come.

## Acknowledgements

This chapter is based upon work supported by the US National Science Foundation under Grant No. 0327474. Special thanks to Harvey Molotch, Neil Brenner, Jason Patch and Monique Bobb for their comments on earlier drafts.

# 4 Jumping scale: from small town politics to a 'regional presence'?

## Re-doing economic governance in Canada's technology triangle

Joseph Leibovitz

For some time now, economic sociologists and geographers have been arguing that the economic success of cities, regions and nations is not merely a product of 'hard' economic factors, as important as they are. Rather, in pointing out the organizational and institutional features of economic life, it has been suggested that collective action, interpersonal and institutional ties, trust, collaboration, social behaviour, and norms and routines are all elements which underpin and shape the expectations and conduct of economic actors.

A particular strand of this institutional and social perspective on the economy focuses on the governance structures through which economies are managed. In relation to the challenges of urban and regional economic performance, it has been argued that associative forms of governance can enhance the competitiveness of territorial economies (Amin 1999; Amin and Thrift 1995; Cooke and Morgan 1998; Gertler 2001). Associative governance involves policy-making processes and institutional infrastructures that facilitate the participation of and interaction between a wide range of actors from public agencies, private enterprises, trade unions and the voluntary sector. Such processes are said to increase the capacity of communities to develop inclusive, responsive and strategic frameworks of economic governance. In short, collaborative economic governance is an important institutional vehicle for the promotion of 'progressive competitiveness'.

This chapter offers an empirical analysis of the forces driving and limiting the emergence of 'economic regionalism' in the context of small-city politics in Canada. Economic regionalism signifies an attempt to build regional economic identity, so that actors are mobilized towards the promotion of economic development and are less concerned with inter-(small) city competition and more preoccupied with the potential advantages of regional size and diversity. The focus of the chapter is on a process of institutional development and change which has been unfolding in one Canadian city-region since the late 1980s. An important element of this process has been the attempts made by various actors to 'jump scale', that is to build regional identity and supportive collaborative regional institutions in order to protect and enhance the political presence and economic performance of what is, in effect, an 'imagined region'. The empirical evidence on which this paper is based was gathered in research conducted since the late 1990s.[1]

## THE ASSOCIATIVE FOUNDATIONS OF THE NEW REGIONALISM

It has become almost axiomatic to assume that economic competition between cities, whether on a national or international scale, has become the order of the day. As the welfare state model of centralized, national, state-led regional policy and equalization programmes continues to unravel in most advanced industrialized countries, cities are called upon to fend for themselves. The result has been intensive competition between cities for the retention and attraction of investment, skilled workers and economic activity. All of this has been occurring within a policy discourse, backed by some theoretical reasoning, which emphasizes the imperative of unlocking endogenous capabilities of capital, skills and enterprise.

However, little attention has been given in the literature to the extent to which small cities and mid-sized localities, situated within close proximity to each other, may be reacting to the supposed transformation in urban political economy. Given the emphasis on the potential economic resilience of *regions,* this chapter asks, specifically, whether small and mid-sized cities can hope to create the organizational and institutional capabilities that may enable them to 'jump scale', that is to become embroiled in collaborative regionalism in terms of economic governance. It does so by analysing the balance between incentives and disincentives for associative regional governance in one Canadian case study, using an institutionalist framework.

Much of the recent discussion on the governance of territorial development has referred, either explicitly or implicitly, to a 'third way' of understanding the 'new competition' – based on knowledge, innovation, flexibility and networking (Best 1990) – and has identified the policy implications derived of this understanding (Langendijk 1999). While the 'third way' of economic governance remains a rather vague notion, it is widely believed to constitute an eclectic policy paradigm, seeking to bridge the state and market, the private and the public. In this view, often termed institutionalist, associative, or the networking paradigm, institutions of the state and civil society have an important role to play in providing the mechanism for trust, collaboration and collective learning which are said to be elemental in supporting learning, innovation and 'untraded interdependencies' (Cooke and Morgan 1998; Henry and Pinch 2000; Lawson and Lorenz 1999; Scott 1996; Storper 1995). The value of common conventions and trust as significant elements in economic regulation and governance, and ultimately in economic performance, has been captured by Putnam's (1993) conception of social capital: that is features such as engagement, dialogue, collaboration and networks which work to create mutual benefit. Drawing on the contrasting performance of the prosperous Italian north compared with the economically disadvantaged south, Putnam has concluded that social capital goes a long way in explaining the propensity of some social systems to generate assets through collaboration, active citizenry and trust.

Subsequently, these ideas have been taken a step further by studies of economic systems, knowledge transfer and innovation. As innovation is largely seen as a product of intense interaction, collective learning and long-term trust-based relations, social capital is deemed an essential component in cementing shared expectations, values and tacit understanding (Lundvall 1992; Nelson 1993). There are important geographical and regional governance implications to these arguments, although these implications have not always been made as explicit as they should have been in the literature on urban and regional politics. In particular, economic geographers have argued that innovative and technological development capacities are necessarily spatially differentiated by the ability of certain actors and institutions, within particular locales, to engage in collaborative learning and collective strategic anticipation of market opportunities. These capabilities, in turn, enable certain regions to keep to the 'high road' of economic competitiveness based on the ability to respond rapidly to volatile market conditions (Cooke 1995). Furthermore, flows of knowledge and the nurturing of networking, trust and collaboration are enhanced by territorial proximity where face-to-face interaction and even cultural attributes are said to play a persistently significant role in the successful development and use of advanced technologies (Gertler and Wolfe 1998; Storper 1997). Such attributes vary geographically, a result of complex and path-dependent combinations of economic, social and political factors, which may render some superficial efforts at cross-regional/cross-national imitations rather futile (Massey *et al.* 1992).

From a regional and local governance point of view, supportive institutional frameworks to 'pin-down' collaborative practices are increasingly important in providing sub-national territories with the opportunity to engage a wide sector of the community in economic governance. The implications of this are both practical and normative. From a practical point of view, the continued strength of geographical factors in underpinning economies places a premium on regional or city-region scales of economic and political organization (Parr 2005). For small cities situated in close proximity, the potential of sharing resources, knowledge, capabilities and communities, while also avoiding wasteful competition, is too promising to overlook. From a normative point of view, the development of associative forms of regional politics suggests (although it may not always be the case) the potential for more widely engaging polity (Hirst 1994). For such small cities which are situated in a liberal political economy not endowed with a tradition of corporatist or collaborative engagement, associative regional governance experiments offer a way of democratizing politics and of creating a deeper sense of attachment and belonging between various sectors of the economic community (labour, business, social activists, community organizations and government).

In a related fashion, Amin and Thrift (1995) have suggested the concept of 'institutional thickness' as encapsulating the qualities which are said to enable city-regions to prosper in a global economy. Essentially, this concept builds on the assertion that the success of certain local economies to embed within them the forces of globalization cannot be explained by a narrow focus on economic factors. Instead, they argue that 'social and cultural factors also lie at the heart of success and that those factors are best summed up by the phrase "institutional thickness"' (Amin and Thrift 1995: 101).

In particular, Amin and Thrift indicate four elements that constitute institutional thickness. These are: (1) a variety of institutions – public, quasi-public and private – which can provide for the development of particular local practices and representation of varied interests; (2) a high degree of interaction between the institutions involved en route to the forging of collective identity; (3) the establishment of formal and informal coalitions and hierarchies around which power and discipline could be exercised; and, as an outcome of the successful operation of the previous three elements, (4) the formation of a shared vision or strategy which is pursued by the various participants in the process of institutionalization.

At the same time, however, the appearance of such institutions and governance frameworks cannot come 'out of nowhere'. It is deeply embedded in the political economy in which actors are situated, in local and regional histories and in certain chance conditions. In fact, critical approaches to studies of institutions and urban and regional economic governance have recently drawn attention to the wider political economy in which these are situated, pointing to the centrality of state strategy, power relations and ideological stances in shaping particular institutional forms (Amin 2004; Keating 2001; Leitner and Sheppard 2002; Lovering 1999; MacLeod 2001). Furthermore, institutional 'presence' does not always translate itself into social capital and associative principles of trust and reciprocity for a range of complex factors, including the perceptions and behaviour of local actors, but also the wider political-economic settings in which these practices are situated (Gertler 1997; Phelps and Tewdwr-Jones 2000). It is therefore through empirical case studies that further understanding of the interplay between small-city politics versus the 'regional imperative' could be gauged.

## Charting the Triangle: from small cities to regional formation

Situated in relative proximity to the Greater Toronto Area in Canada, the city-region now popularly known as 'Canada's Technology Triangle' (CTT) is often portrayed as a story of successful local economic adjustment, of a dynamic system of university research commercialization and academic spin-outs and of innovative collaborative governance frameworks (Smith, N. 1996; Walker 1987). The city-region consists of four neighbouring cities: Cambridge, Guelph, Kitchener and Waterloo. Three of these cities – Cambridge, Kitchener and Waterloo – fall under the same regional government: the Regional Municipality of Waterloo (RMOW). The City of Guelph, for its part, belongs to the Wellington County Regional Government. The population of the CTT stands at approximately 450,000.

Among Canada's metropolitan areas, this city-region contains the second highest proportion of manufacturing jobs, with just under one-third of its workforce employed in the secondary sector. One of the important elements of strength is the city-region's industrial diversity. The structure of the manufacturing sector in the region is such that no particular industrial category dominates, with only four sectors (electronic and electrical equipment; fabricated metal products; food products; and industrial and commercial machinery) scoring more than 10 per cent of total manufacturing employment (Filion and Rutherford 1996).

In 1987 Cambridge, Guelph, Kitchener and Waterloo agreed to set up a loose federative arrangement in order to collaborate on economic development initiatives. The event marked the formal birth of the CTT as a 'regional actor' and political construct. From the outset, the CTT initiative sought to capitalize on three pillars of economic strength that became apparent by the late 1980s: (1) the presence of a diversified manufacturing sector with particular strengths in automotive parts and vehicle production, fabricated metal, communication and telecommunication, electrical machinery, food manufacturing and garment textiles; (2) the very strong presence of a financial services sector in the region and, in particular, the presence of several major life insurance companies in the City of Waterloo, as well as a host of smaller general insurance companies; and (3) the recognition that the region's universities and community college provide a unique advantage. This has led public officials to express their desire to create a stronger network between academic institutions, firms, training initiatives and local government. In particular, the University of Waterloo's traditional strength in the technology-oriented disciplines and its permissive approach towards the commercialization of fledgling technology, Wilfrid Laurier University's strength in business management studies and the University of Guelph's emphasis on agrisciences and biotechnology were increasingly seen as providing a unique comparative advantage to the region. It was thus hoped that the CTT could foster the creation of an associative institutional environment which is said to lie at the heart of successful regions.

As an institutional artefact, the invention of the CTT has projected the possibility of constructing an inter-urban economic development network comprising the governing institutions of four cities which had been known for their rather active individual local economic development efforts and, importantly, for their 'healthy' competitive attitude towards each other. In short, the ambition was to be able to collaborate internally in order to compete more effectively externally. It was also hoped that private sector interests would become more closely involved in a region-wide partnership for economic governance, thus combining the strengths and resources of the public and the private sectors in securing the position of CTT (as a technology-oriented 'district') within an emerging spatial division of labour. As one economic development officer mentioned in an interview, the possibility of engaging the private sector in a closer partnership raised considerable excitement at the time because of the potential of business leaders to 'market the region and its assets in a most effective way'. The fact that some formal institutional structure for collaboration had been cultivated was thought of as innovative and it enjoyed a measure of success, not least because of its ability to raise the profile of the region as a whole in a context of intensifying inter-regional competition for investment:

> For ten years the CTT has been fairly successful in raising awareness of this area as a growth area, and a number of spin-off initiatives and organizations have been created out of the CTT, such as the Plastic and Wire Consortium which has

> been very successful in creating technology; investment delegations were hosted by the CTT; and we began to get a very strong reputation outside of the area as a very dynamic growth area (Economic Development Officer).

The process of institution-building in the CTT coincided with increasing excitement shown by Canadian policy makers and commentators as to the potential benefits associated with local action and proactive city-regions. Such an excitement – resembling what Lovering (1999) has termed the 'New Regionalism' – has often translated itself into somewhat uncritical appraisal of the 'regional state' as an active agent of economic change and prosperity.

The celebratory undertones which accompanied the invention of the CTT tended to downplay the historical legacies of municipal localism and 'anti-regionalism' that had characterized much of the recent political history of the cities of Cambridge, Guelph, Kitchener and Waterloo. Those legacies included local conservatism borne out of traditional concerns for social stability held by civic leaders in the region; the important role that municipal governments have played in maintaining the particular identities of the different communities comprising the triangle; and the continuation of territorial differentiation and fragmentation within the region, based on historical, cultural and socio-economic characteristics and encompassing different institutions of both the state and civil society (McLaughlin 1990). More specifically, the modern political history of the region exemplifies the tension between 'local' and 'regional' issues which has become one of the key stumbling blocks in the process of institution building. On closer inspection, therefore, considerable institutional barriers to meaningful collaborative forms of governance have remained.

## Overcoming parochialism? Local versus regional politics

By 1998 there was a considerable degree of optimism in the CTT region about the ability of local authorities and local actors to exhibit associational principles in economic governance by forming, at long last, a formal Regional Economic Development institution, which would provide for frequent dialogue between various stakeholders and help build a regional identity. Despite the celebratory undertone of media commentaries, the evidence suggests that progress towards the *institutionalization* of the key components of social capital and associative governance – trust, engagement and reciprocity – might be hard to achieve. One of the key elements of institutional development in the CTT has been the establishment of capacity for economic governance at the *regional* level in the form of a network of public institutions and private sector interests. As a result, the notion of *re-scaling,* or *jumping scales,* has taken a prominent position within the politics of institutional change in the region. In particular, in building new institutions for economic governance, the prominent problem has been one of *integration.* As the discussion below reveals, creating structures for meaningful network relations – of frequent dialogue, trust and collaboration – between different jurisdictions, power bases and private interests has proved challenging and tentative.

## THE CONTINUED RESILIENCE OF SMALL-CITY POLITICS

The first issue regarding the operation of 'networking institutions' is the extent to which inter-municipal collaboration has indeed materialized within the CTT, given the increasingly strong voices calling for closer networking relations between the cities. Indeed, in the Canadian context the CTT region might be an interesting example of inter-municipal relations within a city-region context because networking and collaboration have been largely attempted on a voluntary basis, rather than being a statutory requirement imposed by the provincial government (Leibovitz 2004).

The significant legacy of local identity and concern for local autonomy are important historical layers which have had important implications for contemporary patterns of inter-jurisdictional relations in the CTT. In particular, the central position of the local state in identifying community needs and community identities, as opposed to cultivating *regional* identity, has lent itself to competitive, rather than collaborative, attitudes towards regional economic development and governance. It remains questionable whether such attitudes have changed in response to the seeming imperative to jump scales in order to create a stronger geoeconomic and geopolitical entity, given the increasing tendency of regions to engage in economic competition. As one important player within the CTT eloquently summarized the regional dilemma:

> There is a certain degree of competitiveness [here] that will always continue. That, in essence, is the kind of major leap that we've got to be able to get to: can we, in a seemingly competitive environment, get to a complementary and supportive environment? And frankly, that's the big question, and I don't know the answer.

Thus, barriers to collaboration have been reinforced by strong sentiments of local identity and resentment towards regionalization. In this context, Economic Development as a municipal function has become (perhaps inadvertently) a symbol of local autonomy, as expressed by the fact that the Regional Municipality of Waterloo has never assumed an economic development function of its own. In 1997, the parameters of regional politics in Waterloo changed somewhat with the first ever direct election of the Regional Chair, thus strengthening the political legitimacy of RMOW. Seeking to capitalize on the stronger position of RMOW as a result of this process, Regional Chairperson, Ken Seiling, has sought to establish the Regional Government as a new 'champion' of regional economic governance. This effort has met with considerable resistance by the local municipalities, especially Guelph and Cambridge. As a senior official at the City of Guelph argued, 'it is the wrong champion for economic development in a regional context because it carries with it a heavy political bag … Guelph does not want to be seen as subservient to the Regional Government'.

The barriers to intra-regional institutional development have also been evident in Cambridge's position within the Region of Waterloo, a position that has always been problematic and uneasy. For example, no direct public transit connection exists between Cambridge and the cities of Kitchener and Waterloo. In addition, the different ethnic background of the communities, the German heritage in Kitchener-Waterloo and the British origins of Cambridge, continue to serve as a cognitive barrier to collaboration in the mind of public officials and business representatives, who refer to Highway 401, which separates the communities, as the 'Sauerkraut Line'. As the Mayor of Cambridge commented in an interview: 'We've always had a strong Economic Development Department ... and I think there's a feeling here that you begin to lose your autonomy if [Economic Development] isn't here'.

The argument here thus concerns the political and symbolic aspects of local autonomy. Given the rather restricted legal and political autonomy that municipalities enjoy in Ontario, any concession by local authorities of whatever powers still remain in their jurisdiction to higher levels of government seems like a genuine loss of local power. Given the region's political history, the resentment towards the process of 'creeping regionalization' has been particularly pronounced, despite the apparent economic imperative to create a 'regional voice'.

In addition, practical considerations of economic competition between the four cities have tended to overcome a presumed and vaguely articulated regional imperative. The incentives for individual municipalities to maximize economic activities within their own jurisdiction, and thereby expand the local tax base, have remained, apparently, too strong to resist. When a 'soft' argument in favour of the often intangible benefit of regionalism is pitted against hard realities of municipal finances, the project of region-building becomes exponentially more thorny. As a senior staff member at the City of Cambridge conceded:

> The CTT is hampered by political boundaries ... it's a major stumbling block when you have political boundaries ... let's face it, if I'm working in Cambridge and I'm trying to attract investment into the community, I'm going to try and encourage people to locate their operations here for a couple of reasons: we're going to sell land and use the revenues to finance other activities, and we're going to expand our tax base which will help the city to grow and finance its services. It just doesn't make sense for us in the City of Cambridge to say 'I will gladly show you a piece of land, and by the way it's up here in Huron Business Park [in Kitchener]'. Why would I do that?

Voluntary, trust-entrenched, inter-city coalition building at the regional level has therefore remained a fragile and largely unstable undertaking in the CTT. The internal tension between the need for regional cooperation in face of 'globalization', on the one hand, and the inherent tendency to compete due to financial/tax base concerns, as well as the protection of local autonomy, on the other, have not been adequately resolved.

## THE REGIONAL POLITICS OF PUBLIC–PRIVATE RELATIONS

Beyond the re-shaping of inter-municipal relations in the CTT, the process of recasting economic governance in the region has also involved attempts at creating a greater degree of association between private sector interests and local government actors. In 'Canada's Technology Triangle' attempts to create collaborative economic development governance framework on a regional scale have entailed a re-visioning of public-private relations which would be institutionalized, according to some actors in the community, so as to provide a strategic direction to the city-region. The re-configuration of public-private relations was seen by most actors involved in the process of institutional development in the region as an important step in providing the CTT region with policy and strategic capacity bent on anticipating change and challenges. Such regional partnership has been perceived as being able to secure a working and flexible framework for (admittedly modest) collective action.

The changing regulatory environment in which Ontario municipalities have found themselves since the mid-1990s, with the reforms introduced by the Conservative government at the time, provides an important context for the dynamics of regional politics. The agenda of fiscal conservatism has induced local and regional governments to seek private sector partners in order to streamline financial expenses and seek greater efficiency in service delivery. Key in this process has been the growing involvement of the business community in local and regional politics, initially via the lobbying efforts of the Kitchener-Waterloo Chamber of Commerce (KWCC) for a new regional economic governance structure and, later on, through the appearance of high-technology associations such as Communitech on the regional political map. Some local actors have, therefore, been attempting to recreate the institutional geography of the region in order to advance the (perceived) economic governance needs of a relatively small region within Ontario which, some would argue, lives in the shadow of the much more economically and politically powerful Greater Toronto Area.

There have been at least two problematic issues at the centre of institutional development in this respect. Firstly, mobilizing private interests to a higher degree of civic engagement proved to be problematic. For all the discourse surrounding the 'partnership imperative' it has been a challenging task to create *stable* institutional mechanisms through which state and elements of civil society could engage in dialogue, exchange of information and mutual learning, and develop the sort of trust which, according to authors such as Fukuyama (1995), Putnam (1993, 1995) and Coleman (1988), is essential to economic renewal.

The relatively weak local presence of private sector interests in governance issues in the CTT region has been a key factor in the tentative engagement between state actors and private sector interests. For a considerable period of time the CTT concept and economic development initiatives have failed to capture the excitement of the private sector, perhaps because they have failed to see their importance for their own growth. The view of an executive of a local manufacturing firm may be instructive in that regard:

> There has been traditionally lack of organization, and as a result weak involvement, on behalf of the business community in this region. The private sector doesn't participate enough. They leave it in the hands of organizations like the Chamber [of Commerce] to be the spokesperson. None of the major firms here, like ATS in Cambridge, have been part of the CTT organizational structure. It hasn't been the CTT helping them to grow.

The relatively weak private sector involvement in local institutions was also confirmed by public officials:

> We don't really have what you may call strong 'policy groups', certainly not strong business groups that drive issues and initiatives. I would suggest more that it's politicians and departments' staff that spark it, and then business would pick up on it rather than drive it (Senior official, City of Kitchener).

While this view may reflect the tendency of public sector officials to exaggerate the weight of their own role in policy formation and downplay the importance of other actors, it was nonetheless a typical view, not only among other public officials, but – importantly – also among private sector representatives.

This, then, leads to a second problematic issue in designing institutional capacity in the CTT region, namely the extent to which civic engagement in evolving institutions has presented itself in a way that might suggest collective learning and collective action among various partners. Indeed, the relationship between the process of 'governance' and institutional change entails an examination of the role of values and norms (the 'soft' side of institutions) in the coordination of a multi-actor environment. As Pierre (1999: 390) argues, 'Institutional theory is a critical component in any understanding of urban governance, not least because it highlights systems of values and norms that give meaning, direction, and legitimacy to such governance'.

In addition, the associative approach to regional economic governance, as developed by Cooke (1995), for example, takes this idea further by arguing that networking implies 'flow processes' between actors and 'a collective setting of direction'. Furthermore, a key element in associative economic governance, according to Cooke, is 'a recognition that best practice is transferable through *learning*' (1995: 14). Importantly, such learning could (and should) take place between actors involved in institutions of economic governance. The potential synergy between actors thus provides a strong theoretical motivation for new institutions and, in our context, for the creation of regional institutions 'above' existing local political agencies.

However, the evidence from capital-state relations in the process of institutional development in the CTT suggests the limits of mutual learning. In reality, the formation of economic governance in the region has been characterized by mutual *distrust* between local state and private sector actors. Furthermore, institutional change has been suggested (most pronouncedly by the Kitchener-Waterloo Chamber of Commerce and the Conservative Business Association) as a result of a fierce critique of local government in the region and this has not been accompanied by genuine commitment for collective and synergetic learning. Consequently, despite the rhetoric of 'partnership' which has been built into the KWCC's vision of restructured regional economic governance, it became quite clear that the desire was to replace one form of governance (supposedly bureaucratic, cumbersome and public-sector led) with another (seemingly flexible, innovative and privately led).

At the very basic level, the critique of local government by the business community reflects antagonistic relationships between the private and the public sector:

> The business community is frustrated with local government leaders from the point of view that these leaders are very parochial. They are very much into protecting their turf, even to the point of forfeiting well-documented savings that potentially exist when rationalization of services can happen (KWCC official).

However, with regard to economic governance and state-capital relations as part of it, the idea of the business community has been, in essence, to 'privatize' the decision-making procedures within newly created institutions:

> Our concept says that municipalities have to be at the table … within this new economic development organization. But from a decision-making perspective *they should be a minority* … Our fundamental belief is that business can do a better job of creating business than governments can (Executive, local financial firm in Kitchener).

Business belief in the incompetence of government thus militates against the possibility of policy synergy between different actors. Thus, the notion of partnership is not born out of mutual trust, but is induced rather by an atmosphere of tension between the public and the private sectors. Such an atmosphere does not bode well for a 'networked' form of economic governance.

In addition, another private sector firm representative commented on the realization by the business community that the 'rules of the game' have changed and that the scales of governance are shifting in important ways:

> The Chamber is taking a leadership position on a number of issues such as economic development, education, health care and skilled labour shortages. I think that by and large we too often ran to government as a solution provider, when in fact the ability of government to come up with solutions is extremely limited, if not obsolete … I think that an important lesson of this whole thing is that we're being forced to look beyond the conventional structure of government because they are obsolete. And in a larger, global context, we know that two levels of government are important now: the supra-national and the city-state. Everything in between becomes mush.

Thus, partnership formation in the CTT points to relatively little willingness on behalf of the business community to engage in a process of exchange and learning with other actors, primarily local government. Rather, the struggle to re-shape the institutional foundations of economic governance in the CTT has taken an adversarial form and has exposed the ambition of several leaders in the regional business community to mould a process of governance which would fit 'business ethics', loosely defined.

Consequently, it was not surprising that KWCC's proposal to create a new regional 'partnership' for economic governance met with substantial resistance. Interestingly, this resistance came from both the public sector (the individual city councils) and from other segments of the business community (see the next subsection), which dismissed the proposed model as politically unrealistic. Public sector reaction to the KWCC proposal of creating a new institution of regional economic governance was sceptical from the beginning. Given the long-entrenched tradition of local autonomy such reaction is not surprising, as the delegation of authority to a new, regional body would signify a loss of autonomy. Beyond that rationale, however, the involvement of KWCC in articulating a refined model of economic governance and a new public-private partnership, has left many within the public sector sceptical about the 'workability' of the proposed model, indeed its very desirability, as well as the genuine motivation behind the Chamber's newly found local leadership.

In particular, the intertwining of governance and economic development issues, as was articulated by the Kitchener-Waterloo Chamber of Commerce, left many within the region with the impression that the 'real' agenda behind the business community's recent sense of local activism was the initiation of a municipal amalgamation process. Indeed, such concerns are not unfounded given that some segments of the private sector in Kitchener and the City of Waterloo have long viewed the existence of two separate municipalities as inefficient, costly and unwarranted. The prevailing line of thinking in that regard has been that municipal amalgamation, at least between the cities of Waterloo and Kitchener, should result in significantly lower business taxes due to savings, thus encouraging greater competitiveness and a better business climate. This view was almost unanimously rejected by local officials and, consequently, the particular institutional form relating to regional economic governance has remained a subject for ongoing debate and disagreement. As one senior official in Kitchener has reflected:

> The Chamber's model of governance … I think it's a very simplistic model they're looking at. They haven't thought out enough the difference between what a marketing body should do and what a governing body should do, and they were sort of mixing the two thoughts without a clear direction of who was still going to deliver what.

The arrival of Communitech – the high-technology business association – to the regional political scene in the late 1990s may have changed some of these dynamics, at least in a small way, because of the greater clout and prestige associated with a high-technology-led institution. Here, emerging local actors may have contributed in a limited sense to the changing regional development trajectory by forming a visible and prestigious alliance which other actors, from both the private and the public sectors, could not afford to ignore. As a Communitech member put it bluntly:

> By virtue of our size and by virtue of the impact we have on our economy, we get their attention. And every politician wants to be part of a success story. If you talked to the politicians who have been part of it, they'll give you different stories why they're in it, but the bottom line is they want to be associated with this type of industry.

Again, however, the evidence points to uneasy public-private relations, despite the growing clout of Communitech and the high-technology sector:

> Communitech and CTTAN and some of the other organizations sat with us when we were getting the organizational structure of the CTT, but then they go their own way. There's not a cooperative kind of thing happening. One of the things that's happening is that there's an expectation that the municipalities would continue to fund these organizations. Well, if we have no say in what they do … and it's a little difficult when you see them not really cooperating with the CTT and us as we try to restructure the institutions here (Senior official, City of Kitchener).

So, the process of altering the structure and scale of economic governance in the CTT region has exposed the limitations to 'collective synergy' which seems to exist in the region. The involvement of KWCC as a political actor and, later on, the appearance of Communitech, may represent a growing desire of certain territorial interests to re-shape the contours and scale of governance in the CTT. The re-scaling project in the region has been perceived to be of importance in order to synchronize the CTT with emerging governing rationalities according to which 'city/region-states' are becoming increasingly important agents in a globalized economy. However, the evidence suggests that the coordination of different actors and their institutionalization into coherent associative forms have proved problematic because of the vastly different perspectives and interests held by different power bases within the region.

## CONCLUSION

'The New Regionalism' has its supposed attractions. For economic geographers and economic sociologists, regions have been the subject of a rejuvenated belief in the resilience of territoriality in economic life (Ward and Jonas 2004). The advantages of size, diversity, economic linkages, proximity and agglomeration factors have been complemented by research which has pointed towards the significance of social ties, engagement, identity and collaboration in cementing trust-based relations. Normatively, the New Regionalism, at least according to some versions, has also offered a way of enlivening democracy and participation. For Amin,

Putnam, Hirst and others, the democratic aspects of regional renewal have been no less important than economic prosperity and in many instances inexorably linked to issues of well-being and welfare.

This chapter demonstrates, however, that the task of region-building in liberal political economies, where long-term and collective trust-based relations are relatively confined to limited spheres, involves more than just accepting the theoretical underpinnings of the New Regionalism. The tension between small-city politics and traditions and the regional imperative, at least in the case of Canada's Technology Triangle, has raised important issues related to the realpolitik of institutional change and the prospects for collaborative forms of governance between small and mid-sized localities. Such localities have their own histories and political cultures to contend with and, when thrown against some intangible ideas – as attractive as they may be – about the regional advantage in a globalizing world, it is those histories, traditions and localized interests that receive a sharper edge. The tension between historical communities and the regional imperative has remained one of the relatively unexplored areas within the context of the supposedly new, hyper-competitive/neo-liberal regionalist order (Rossi 2004; Ward and Jonas 2004).

In Canada's Technology Triangle, the discourse of regionalism has found its practical expression in the limited power, resources and authority that regional functions of economic governance eventually gained. Associative regionalism, at the end of a long and arduous process, has been reduced to a rather idiosyncratic place-marketing body. It had very little in common with the vision of a strategic, reflexive, engaging and synergetic regional 'social brain' (Hirst 1994) envisioned by some members of the community. In the end, of course, the compromise between small/mid-sized places and regional voices might be the one that is more optimally suitable for the four cities' needs and their economic and political context, rather than the grander expectations brought about by more theoretically inclined work. The gap between the political dynamics and regional imperatives of small cities thus raises questions about the limits to the new regionalism that are beginning to be addressed by research (Christopherson 2003), and there is little doubt that further evidence is required in order to address the multifaceted nature of regional versus local politics.

## NOTE

1 The research involved 54 semi-structured interviews with key members of local, regional and provincial government agencies, firms in different industrial sectors, and a wide range of industrial associations, trade union representatives and community organizations. The research also involved collection and analysis of documentary material from a range of sources, including policy documents, official statements, minutes of meetings, annual reports and newspaper articles.

# 5 Tourism in a reluctantly small city-island-nation

## Insights from Singapore

T.C. Chang

Singapore's economic survival, its quest to be a global city and its ambition as Asia's pre-eminent business, transportation and cultural hub underscore the country's relentless push beyond its constraints of size and resource endowment. The pursuit of globalization was guided by a government plan in the mid-1990s called 'Singapore UnLimited', the goal was to pursue economic expansion under the slogan of 'Bringing the World to Singapore, Bringing Singapore to the World'. Developing Singapore as an 'international business hub' and 'cosmopolitan city' were the agendas of Singapore UnLimited (Economic Development Board 1995a).

This chapter highlights one particular dimension in Singapore's global political-economy: *tourism*. By its very nature, tourism is an industry which thrives on resource diversity and spatial expansiveness. Yet as a city-island-state with virtually no natural scenic attractions and only a small land area, Singapore's long-term prospect as a travel destination is questionable. Although Singapore boasts fine weather, good shopping and excellent business facilities, how attractive and sustainable are these conditions as new destinations like China and Vietnam open their doors? In this chapter, I argue that Singapore's vision to be a 'tourism capital' is predicated upon overcoming its size and resource limitations in a number of ways: through development policies and strategic tourism imaging. In terms of policy, Singapore has outlined clear goals to be a regional tourism hub for businesses, people and events ('bringing the world to Singapore'), an exporter of tourism skills and services ('bringing Singapore to the world') and a tourism partner to neighbouring countries ('bringing Singapore and the world together'). Its triple goals to be a tourism hub, exporter and partner supplement its traditional role as a tourist destination.

In terms of imaging, tourism marketing strategically projects a country much larger than itself. Promotional campaigns of 'Instant Asia' in the 1960s/70s, 'A Magic Place of Many Worlds' in the 1980s and 'New Asia-Singapore' in the 1990s conflate the city-state with the Asian continent, thereby projecting an image of size and diversity. While possessing neither the scenic attractions nor the market size of other Asian countries, Singapore is imagined to possess both the cultural and economic resources that enable access to the region (Chang 2001; Ooi 2002). In this discussion, I have deliberately chosen to emphasize both tourism policy and marketing. While policy insights reveal Singapore's concrete plans at overcoming its resource scarcity, marketing efforts represent symbolic strategies that project visions of expansiveness and inclusion. Collectively, both the material and discursive dimensions offer useful insights into how Singapore has grappled (and continues to grapple) with its small size in its quest to be a global tourism centre.

This chapter is divided into four sections. The first introduces some background information on 'Singapore UnLimited' and 'Big Singapore'. This sets the political-economic context to understand Singapore's concerns over size and its globalization imperative. The next two sections deal with Singapore's negotiation strategies in terms of tourism policies and marketing efforts. We shall see how spatial borders are deliberately 'loosened' and how tourism contributes to spatial re-territorialization. In the conclusion, the importance of geographic size is re-assessed. Whether 'small and lean' or 'big is better', size is re-defined to mean not just physical proportions and land area but also, more critically, the ability to command economic functions and cultural images much larger than geographic dimensions may suggest.

## FROM 'SINGAPORE UNLIMITED' TO 'BIG SINGAPORE': IS SIZE DESTINY?

Pronouncements of geographic limitations and strategies of enlargement are regular features in Singapore's government speeches, policy implementations and development plans. As an island-city-nation of 4.1 million people and with 685 square kilometres of land area, its preoccupation with size (or lack thereof) is understandable. Limitations of land, however, do not necessarily constitute economic destinies (or so the Singapore state and planning authorities would have us believe). Geographic borders are constantly stretched and reconfigured to overcome tyrannies of size, scale and distance. In 1993, the then Prime Minister Goh Chok Tong outlined an ambitious economic plan to bring the country into the twenty-first century. Called 'Singapore UnLimited', the goal was to create a global city fuelled by the twin engines of 'Bringing Singapore to the World' and 'Bringing the World to Singapore' (Economic Development Board 1995a).

In bringing Singapore to the world, locally owned companies and 'made-in-Singapore' skills are exported to regional countries. Services and technical skills such as infrastructure planning, sea/airport management, human resource development and tourism consultancy are packaged and sold to regional markets in China, India, Vietnam, etc. (Kanai 1993; Regnier 1993; Hiebert 1996; Yeung 1998, 2000a). Local businesses and entrepreneurs are also encouraged to undertake overseas ventures and investments, aided by government funding and advice. Bringing the world to Singapore, on the other hand, refers to attracting foreign companies and investors to the country. International companies in finance, communications, education, medical care, software development, leisure and tourism are specifically targeted, not only as sources of employment and investment but also as possible collaborators with local companies. In its vision statement, *International Business Hub 2000*, the Economic Development Board (EDB) explained that

SINGAPORE'S SEVEN-HOUR 'HINTERLAND'

*At least 67 cities lie within seven hours of flight from Changi Airport, each one a potential market for Singapore. These are some of the cities which Prime Minister Goh Chok Tong had urged Singapore in his National Day Rally speech to regard as part of its hinterland.*

◆ **Malaysia**
**Capital:** Kuala Lumpur
**Population:** About 21.8 million
**Economy:** Dynamic export sector, main goods being electronic equipment, petroleum and liquefied natural gas, chemicals, palm oil, wood products, rubber and textiles. Imports machinery and equipment, chemicals, food, fuel and lubricants.
**Singapore link:** Was Singapore's top trading partner ($82.6 billion) and export market (18 per cent) last year.

◆ **Vietnam**
**Capital:** Hanoi
**Population:** 78.8 million
**Economy:** Exports crude oil, marine products, rice, coffee, rubber, tea, garments and shoes.
**Singapore link:** Singapore is one of its main trading partners, with trade reaching $5 billion last year.

◆ **China**
**Capital:** Beijing. Other major cities include Guangzhou, Shanghai, Xiamen.
**Population:** 1.26 billion
**Economy:** It is expected to grow by 7 per cent over the next five years.
**Singapore link:** Bilateral trade reached $21.6 billion last year. Singapore is also helping to develop container ports and airports in Guangzhou and Xiamen.

◆ **India**
**Capital:** New Delhi. Other cities like Bangalore have earned a reputation as a global technology hub.
**Population:** 1 billion
**Economy:** Ranging from traditional village farming to modern industries and support services. Booming trade in export of software services.
**Singapore link:** Second-fastest growing Asian economy this year after China, according to the Economist Intelligence Unit. Trade with Singapore hit $6.7 billion last year, with Singapore firms investing in its telecommunication, IT, logistics and light manufacturing sectors.

◆ **Seychelles**
**Location:** Eastern Africa, group of 115 islands in the Indian Ocean, north-east of Madagascar
**Capital:** Victoria
**Population:** 79,300
**Economy:** Growth has been led by tourism and tuna fishing
**Singapore link:** Singapore is one of its main import partners, besides South Africa, Britain, China, France and Italy. Local companies like Banyan Tree and PWD Consultants have resort and airport projects there.

◆ **Hongkong**
**Population:** About 7 million
**Economy:** A free-market economy highly dependent on international trade. Main industries include textiles, clothing, tourism, electronics, plastics, toys, watches and clocks.
**Singapore link:** One of Singapore's top five export markets last year, accounting for 7.9 per cent of total exports.

◆ **Maldives**
**Location:** Group of atolls in the Indian Ocean
**Capital:** Male
**Population:** About 300,000
**Economy:** Tourism is the largest industry, followed by fishing. Most staple foods must be imported.
**Singapore link:** Singapore is one of its main import partners, and is also helping to develop the Male International Airport.

◆ **Japan**
**Capital:** Tokyo
**Population:** 126.5 million
**Economy:** Second-most technologically powerful economy in the world after the US and third largest economy in the world after the US and China.
**Singapore link:** One of Singapore's top export markets (7.5 per cent) and trading partner ($57.9 billion) last year.

*Figure 5.1* Thinking beyond Singapore's borders. (© Singapore Press Holdings.)

Singapore's quest to be 'Asia-Pacific's business hub' is underpinned by international companies investing in the country and using it as a 'value adding gateway to the region' (Economic Development Board 1995a: 1, 1995b).

The concept of 'regionalization' was introduced by Singapore UnLimited. As a small country with a limited market, the surrounding Asian region is re-interpreted as Singapore's resource and market hinterland. Regionalization may be defined as an integration process in which Singapore and regional countries are economically integrated to facilitate capital and labour transfers, create common markets and undertake collaborative projects. While the term was originally applied to cross-border relations with its immediate neighbours Indonesia and Malaysia, it has been expanded to embrace the entire Asian region. Urging entrepreneurs and investors to foray into the region, Prime Minister Goh explains that Singapore's future depends on transborder connections and economic ties. He suggested a 'seven hour flight radius' as a natural catchment for Singapore businesses, with an estimated market of 2.8 billion people (*The Straits Times*, 21 Aug. 2001; see Figure 5.1). Not only serving as a market for Singapore products and services, the hinterland also abounds with investment prospects for Singaporean entrepreneurs, as well as for potential enterprises interested in investing in Singapore.

How successful regionalization is depends on the willingness of entrepreneurs and policy planners to 'go regional'. Regional operations bring manifold opportunities but also challenges. For example, problems have arisen in Singapore-owned industrial projects in China and Indonesia because of a different work culture and political commitment between Singaporeans and host communities (Grundy-Warr *et al.* 1999; Yeung 2000b). According to the then Singapore Minister of Trade and Industry, George Yeo, a 'Big Singapore' mentality is essential if regionalization is to succeed. In his words:

> Managed well, we can be much larger than what we are geographically … A Small Singapore mentality finds the region with all its problems uncomfortable and our diversity a constant source of friction and irritation. A Big Singapore mentality engages the region, celebrates our diversity and uses it to access economic and cultural spaces all over the world (*The Straits Times*, 1 May 2001).

The elasticity of 'big' and 'small' Singapore is worth emphasizing. Size is in the eye of the beholder and scale is socially constructed. Centrally located in South-East Asia, Singapore is imagined to possess both the cultural and economic resources that enable access to the region. This flexibility of size is epitomized by EDB's concept of *shakkei*. *Shakkei* is a Japanese landscaping strategy in which one's garden is enhanced by incorporating surrounding sceneries such that the combined landscape is more attractive than each garden is on its own (Economic Development Board 1995b: 6). In 'borrowing' the attractiveness of surrounding economies, Singapore's size limitations may be overcome. First, by attracting foreign companies to operate in the country, Singapore's investment pool and employment base are enlarged. Second, by establishing strategic partnerships with selected countries, political ties are strengthened, which will aid in future economic relations. Finally, by creating cross-border economic zones with neighbouring locales, individual limitations are overcome by leveraging on one another's strengths through mutually beneficial projects.

An example of a cross-country economic zone is the Indonesia–Malaysia–Singapore Growth Triangle (IMS-GT). Originally mooted in the late 1980s, the IMS-GT was formally constituted in a 'memorandum of understanding' in 1994. A cooperative framework was forged by the three countries which enabled them to benefit from each other's comparative strengths while overcoming individual weaknesses. In the IMS-GT, Singapore benefits through access to abundant land, labour and water in the surrounding countries, while Indonesia and Malaysia benefit by leveraging on Singapore's global communication networks, advanced infrastructure and management expertise. Originally conceived by Singapore as a way to manage its expanding economy, the success of the Growth Triangle ultimately will depend on whether all member countries benefit reciprocally. *Shakkei* must empower all three countries to 'harness external economic space' to create 'products and services beyond traditional concepts of space and resource constraints' (Economic Development Board 1995b: 6).

While I have presented the 'Big Singapore' vision in rather positive terms, the challenges and problems that emerge from regionalization should not be discounted. In as much as borders may be crossed at will by capital, problems are at times very much locked in space and place. Sparke *et al.* (2004), for example, have looked at the hyper-mobility of capital in the IMS-GT vis-à-vis the relative fixity of certain classes of labour. Unskilled Indonesian labour, for example, is regulated in its movements across national borders, while tourists enjoy full mobility. What results are rural-urban migration problems confined to the Indonesian islands of Batam and Bintan, exacerbated by unwieldy development brought about by capital and tourists moving in freely from Singapore and Malaysia. In contrast, the problems of overdevelopment are less obvious in Malaysia and Singapore. In many ways, therefore, the triangle is not borderless at all, but 'transacted by all kinds of divides and disjunctures that represent a veritable efflorescence of boundary drawing' (Sparke *et al.* 2004: 496). Although I will not be focusing too much on the negative repercussions in this discussion, it must be remembered that the imaginative geographies of regionalization are not uncontested and unproblematic (Grundy-Warr *et al.* 1999; Yeung 2000b; Colombijn 2003).

To come back to the original question of whether size is destiny, a case study of Singapore's tourism may offer some answers. Having briefly sketched the processes and objectives of regionalization, the remainder of this chapter will explore the role of tourism in Singapore UnLimited. First I will look at policies of enlargement aimed at developing Singapore as a tourism capital. Here, Singapore's position as a tourism hub, exporter of leisure services and a collaborative partner are analysed. Second, at a discursive level, tourism images are investigated. As we shall see, tourism marketing reflects a 'Big Singapore' ideology in order to project images of diversity and spaciousness. In both accounts, the question of 'size as destiny' is raised and the role of tourism in spatial re-territorialization is discussed.

## POLICIES OF ENLARGEMENT: CONFIGURING NEW SPACES FOR TOURISM

Since the mid-1990s, policies of enlargement have been aggressively pursued as a means of transcending Singapore's limited tourism geography. In 1996, the Singapore Tourism Board (STB) outlined a vision to develop Singapore as a tourism capital (*Tourism 21* masterplan, Singapore Tourism Board 1996). A tourism capital can be defined as an attractive tourist destination, a base for the headquarters of tourism enterprises and a gateway to a region. Just as Paris is a fashion capital and New York is a cultural capital, a tourism capital is recognized for its innovation in tourism development concepts and global influence. The slogan for Tourism 21 is 'Tourism UnLimited: Bringing Singapore to the World, Bringing the World to Singapore'. The semantic coincidence with 'Singapore UnLimited' is deliberate, revealing a strategic parallel between tourism policy and national agenda. Being a tourism capital is thus considered to be part of Singapore's ambition as an international business centre and global city.

For Tourism UnLimited to be realized, new configurations of space are required. Novel ways of recalibrating tourism borders demand looking at Singapore as more than just 'an island of 680 square kilometres'; instead Singapore is re-imagined to be part of a much larger, economically dynamic region. One of Tourism 21's strategic thrusts is appropriately called 'configuring new tourism space'. The STB explains that for Singapore to be a tourism capital, it needs to break free from 'traditional thinking which limits our tourism activities to the resources we possess ... adopting a transborder approach in going beyond physical boundaries, to participate in the growth of the Asia Pacific region' (Singapore

Tourism Board 1996: 16). Singapore's greatest constraint is size and the concomitant problems of resource, land and labour scarcity. Tourism 21 perceives the surmounting of space to be a solution. Consider what STB says about *shakkei* and the sharing of tourism resources: *Shakkei* offers a 'new way for Singapore to look at itself, as well as the world ... In particular, it calls for greater partnership with our neighbouring countries, working in a borderless manner, and creating new economic space for everyone through leveraging resources regionally and globally to overcome each individual country's natural limitations' (Singapore Tourism Board 1996: 16). Towards this end, the STB identified four areas for regionalization: South-East Asia, North Asia, South Asia and Oceania (Tham 2001).

Three specific strategies in Singapore's spatial re-configuration may be identified:

1 Developing Singapore as a regional *tourism hub* for international businesses, events and visitors ('bringing the world to Singapore');
2 Exporting *tourism skills/services* to Asia-Pacific countries ('bringing Singapore to the world'); and
3 Becoming a *tourism partner* with proximate countries through collaborative projects ('bringing Singapore and the world together').

As a tourism hub, Singapore aspires to be more than just a travel destination, it will also be a centre where international companies can locate their Asian headquarters and a regional gateway. As a business hub, Singapore's tourism revenue streams will be broadened beyond its traditional reliance on tourist arrivals. Tourism enterprises bring in investments, create jobs and enhance Singapore's reputation as a global business city. Singapore is currently the Asia-Pacific (or South-East Asian) headquarters for such renowned leisure/lifestyle companies as Club Med Hotels and Resorts, Hilton International, Sotheby's and Aman Hotels and Resorts.

As a tourism hub, the traditional constraints of market size, labour scarcity and resource endowment are mitigated. Instead, qualitative dimensions such as workers' skills, standards of living, infrastructure and spatial connectivities are emphasized. As a key node in a global economy, a tourism business centre is less concerned with 'lower-rung' attributes often associated with manufacturing/production sites. Instead, higher-level, value-added attributes are emphasized. Surveys conducted with tourism enterprises reveal that while operational costs in Singapore may be higher than cities like Jakarta, Kuala Lumpur and Manila, it has retained the headquarter functions of many international companies because of its skilled labour, political stability and transport/communications connectivities (Low and Toh 1997; Chang and Raguraman 2001). The focus on qualitative rather than quantitative dimensions coincides with shifts in the regional division of labour in Singapore in the 1990s, when low-skill, export-processing assembly operations were increasingly 'off-shored' in preference to more remunerative managerial, finance, research and service-oriented work (Ministry of Trade and Industry 1986; 1998; Sparke *et al.* 2004).

The spatial fix offered by being a regional centre negates the qualities of a conventional tourist destination – large size, ample resources, limitless land and labour – focusing instead on Singapore's nodal connections. According to a survey of tourism companies in Singapore, its key strengths are its geographic centrality, political stability, communication networks, language environment and business/financial amenities (Chang *et al.* 1998). While local attributes such as 'politics', 'language' and 'finance infrastructure' are important, geographic connectivity was identified as the most important. Relentless improvements to Singapore's air and sea access, development of new ports, expansion of cruise infrastructure and the availability of hi-tech communication technology in most homes, hotels and business environments enhance Singapore's nodal position in the global economy. The importance of Singapore as a tourism node is further emphasized in the STB's new tourism plan for 2005. Singapore is to serve as a 'services centre of Asia' and a leading 'convention and exhibition city in Asia', leveraging on its workforce, economic connections and infrastructure. High-skill, relatively labour-unintensive services such as specialized health and tertiary education sectors are to be developed to attract high-end medical tourists and international students.

In addition to being a tourism service hub, a second enlargement strategy pertains to the export of tourism skills and services. Under Tourism 21, tourism was redefined not only to mean 'tourist market' but 'knowledge' as well. Traditionally, tourist destinations have always depended on visitor numbers and revenues. With the new focus on tourism businesses mentioned above, a related strategy is to encourage both the public and private sector to 'go regional'. This could be either in the form of overseas direct investments (for example, developing hotels and theme attractions) or by exporting tourism services, expertise and skills (for example, skills in tourism training, consultancy and hotel management).

While the STB had depended on foreign consultants in the 1980s (often from the West) to advise on tourism development, since the mid-1990s it had begun serving as a consultant to emerging markets in Asia. In 1995, the STB established its Regional Tourism Division, devoted to facilitating overseas expansion of Singapore companies and advising on regional investment possibilities. The STB also set up its in-house consultancy division devoted to selling its expertise in feasibility studies, tourism planning and institution building (Lee 2004). Regional consultancy projects have been undertaken in Cambodia, China, India, Laos, Mauritius and Vietnam. Singapore's main English broadsheet, *The Straits Times*, has graphically captured Singapore's regionalization attempts using a vegetative metaphor (see Figure 5.2). In keeping with the book's theme of 'small cities', we might extend this metaphor by suggesting that Singapore has grown out of room to expand economically. Further growth can only take place if new territories and foreign soils are explored. The region abounds with new possibilities and, with appropriate skills, talents and capital, new room for growth may be created for Singapore.

Following the lead of the STB, private sector operators in Singapore have also established their own consultancy services. A good example is Wildlife Reserves Singapore (the parent company of the Singapore Zoo, Night Safari and Jurong Bird Park) which set up its consultancy company in 1995, specializing in zoo design, management and landscape construction. Since its launch, Wildlife Reserves has

# 'Think beyond boundaries'

ST 21/8/2001 p. H4

**Numerous potentially lucrative markets lie within easy reach of S'pore, notes EDB advisory council**

SPREAD your wings and see the wider Asian region as your oyster, rather than simply focus on the immediate neighbourhood.

That was the advice a panel of global business leaders gave Singapore. It proposed the "seven-hour flight radius" concept as Singapore's hinterland, which would encompass 2.8 billion people in at least 67 cities and countries, a huge potential market.

The leaders — chairmen and chief executive officers of the world's leading multinational companies — are members of the Economic Development Board's (EDB) International Advisory Council.

The EDB appointed the council in 1994 to advise Singapore on how it can maintain its international competitiveness. The panel meets once a year.

At a meeting in February, the business leaders urged Singapore to look beyond the South-east Asian region, at places that lie within a seven-hour flight radius of the Republic.

These places include China, India, Japan, South Korea, Taiwan, parts of Australia, the Maldives and the Seychelles.

Within this greater East- and South-Asian region, there are 500 to 800 million people with an average annual income of US$4,000 (S$7,000) each.

Singapore, the business gurus said, should target its products and services at these potentially lucrative markets.

But not everyone agrees with the "seven-hour flight radius" recommendation.

Earlier this month, Harvard Professor Michael Porter advised Singapore to focus on integrating with its more immediate neighbours — like Malaysia and Indonesia — instead, saying that the seven-hour flight radius concept was not very efficient.

But in his National Day rally speech on Sunday, Prime Minister Goh Chok Tong urged companies to exploit business opportunities in cities and countries within the seven-hour flight radius.

"We should regard all the countries and cities which are within seven hours of flying time from Singapore as our hinterland," he said.



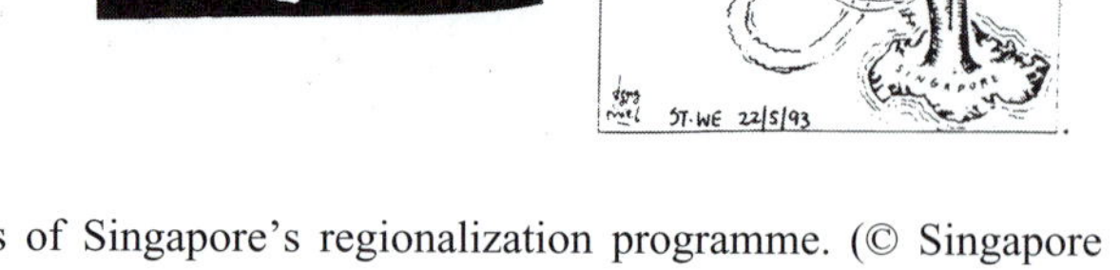

*Figure 5.2* Media depictions of Singapore's regionalization programme. (© Singapore Press Holdings.)

conceived plans for a new zoo in Surabaya, Indonesia (1996), conceptualized masterplans for Bintan's Wildlife Sanctuary (1996), Cambodia's Phnom Tamao Zoo (1996) and Chennai's Dizzee Animal Kingdom in India (1998), as well as provided landscaping advice for wildlife reserves in China and Seychelles. Without regionalization, the company's job scope is limited to managing and operating its three local attractions. In an interview, Bernard Harrison (then Chief Executive Officer of the company; October 1999) mentioned another economic benefit. Well-designed foreign zoos are a good advertisement for Singapore. Many international tourists come to know about the country through visiting its 'extended family' of zoos throughout Asia. Singapore's reputation as a leader in tourism concept planning is thus enhanced.

A third approach in regional tourism involves strategic partnerships. Here, 'win–win' relationships are forged as collaborating countries leverage on one another's strengths while overcoming individual limitations. Partnerships help to enlarge tourism geographies by conjoining proximate destinations as a single destination area. Singapore is involved in collaborative tourism in two ways. At the formal level it is a member of ASEAN (Association of South-East Asian

Countries) and is engaged in various regional initiatives such as Visit ASEAN Campaign, the ASEAN Tourism Agreement Plan (2002) and ASEAN Tourism Vision (2004–2010) (Ghimrie 2001; Timothy 2003). The emphasis here is very much on creating consistent standards and practices across member countries and on encouraging intra-regional travel.

Apart from ASEAN, strategic partnerships such as the IMS-GT are also forged at an intra-regional level. Unlike ASEAN, which involves ten member countries, a 'smaller-scale/lower-level' region allows for the sharing of resources in a more immediate way. With its efficient airport and harbour, Singapore serves as the transportation gateway to the IMS-GT; with their ample land and labour, Malaysia and Indonesia provide the resources for development of beach resorts, marinas and golf courses. Through collaboration, new niche areas in tourism are also created. For example, access to Indonesian and Malaysian coastal resorts has allowed Singapore to market itself as a cruise centre. Cruise tourism was non-existent in Singapore in the 1980s; by 1992, cruise passengers numbered just over 190,000 with 350 ships calling at port and, by 1998, there were over 1.05 million passengers and 1691 ships on call.

Singapore-owned companies have also spearheaded tourism development in the Indonesian islands of Batam and Bintan. The mega-scale Bintan International Beach Resort (BBIR), for example, is operated by Bintan Resort Corporation, a consortium of Indonesian and Singaporean companies. Tourism advertisements regularly conflate regional geographies by depicting Bintan as an extension of Singapore (Simmonds 1997), usually followed by a small footnote declaring Bintan to be '45 minutes by hydrofoil from Singapore'.

Collaborative tourism projects are intended to circumvent harmful competition between countries and costly duplication of infrastructure (Lee 1993). However, collaborations are never uncontested and regionalization brings potential conflicts. In the IMS-GT, Singapore's economic dominance and harmful effects on locals (Indonesian small businesses and landowners) have been questioned (Chang 2001). With tourists spilling into Batam and Bintan (over 70 per cent are Singaporeans), rising crime rates, congestion, poor housing and environmental deterioration have emerged (Lim 1999; Colombijn 2003). Bunnell *et al.* (forthcoming) also note problems of unethical land claims when local communities were cleared from beach fronts for the development of BBIR. These claims have culminated in public demonstrations and violent demands for increased compensation. Such episodes proffer cautionary warnings on the effects of 'Big Singapore' and should be carefully monitored in future research.

## METAPHORS OF LARGENESS: RE-IMAGINING GEOGRAPHIES OF SINGAPORE

Similes and metaphors articulated by the Singapore state provide a way to chart the country's progress as it moves from colonialism to independence, from 'insecurity to a sense of its place in a globalising world' (Ban 1992: 9). The use of similes, Ban (1992: 10) explains, provides an 'organizing rhetorical move' to galvanize the local community in a certain direction. In a similar vein, Kwok (2001) contends that Singapore is a nation born of imagination. Its transformation from 'WW2 to WWW' (from a war-ravaged country to an 'intelligent' island) is the outcome of urban planning imaginations of what the country can and should be. Economic imaginations of globalization, technological imaginations of a wired island and cosmopolitan imaginations of a multicultural society provide ways forward to chart Singapore's future development.

Similarly, metaphors can help us to understand the evolution of tourism development. The STB has consistently marketed Singapore as an important part of Asia. In the 1960s/70s, Singapore was 'Instant Asia'. In the 1980s, it was portrayed as a 'Magic Place of Many Worlds' and since 1997, it has been re-imagined as 'New Asia–Singapore' (Chang 1997; Chang and Lim 2004). The deliberate re-drawing of Singapore's geographic contours around those of the continent accentuates its Asian identity; at the same time an impression of spatial expansiveness and diversity is conveyed – a form of *shakkei* thus takes place. By borrowing imageries of Asia, the world's largest continent, Singapore (South-East Asia's smallest country) can stake claim to the cultural and historical allure of the larger arena.

Singapore was marketed as 'Instant Asia' in the 1960s/70s. As Instant Asia, it promises a holiday destination combining the sights, tastes and cultures of Asia's dominant groups: the Chinese, Malay and Indian communities. For the time-strapped and cost-conscious traveller, Singapore offers the convenience of an all-in-one destination. The STPB (Singapore Tourism Promotion Board, later renamed Singapore Tourism Board) explains that a visit to Singapore offers 'an insight into the land mass and people of Asia … It provides an ideal holiday for the discriminating tourist from the West or distant countries who has neither the time nor money for extended travel' (Singapore Tourism Promotion Board 1966: 24).

The conflation of scales here – local/Singapore with continental/Asia – may be seen as a celebration rather than negation of size. Inasmuch as Asia offers a variety of cultural experiences, it is impossible for the average tourist to sample everything. By presenting Singapore as a 'snapshot' of the continent, it can appeal to mass travellers hoping to sample a slice of Asia. Here, Singapore's size is seen to be an asset rather than a liability. Because of its compact urban form, tourists can experience Chinatown, rural Malay *kampongs* (villages), Little India and the Colonial and Civic District in a single holiday. The STB promotes Singapore as the 'crossroads of Asia' and 'home of the three main races of the area – Malay, Chinese, Indian – and the home of their culture and traditions' (Singapore Tourism Promotion Board 1964, cited in Chang 1997: 550). The emphasis on 'home' and 'Instant Asia' conveys the positive attributes of smallness – cosy, compact and convenient.

We should, however, be wary of marketing rhetoric. As some have pointed out, Singapore is visibly a Chinese rather than multicultural city (Vasil 1992). The numerical and economic strength of the Chinese, and the stable but uneasy alliance among Singapore's ethnic groups in the 1960s, belie the rosy depictions of Instant Asia (Chang 1997). McKie (1972: 5) adds: 'the brochures will tell you that Singapore is multiracial and cosmopolitan, and within the limitations of these words, both are true. But its cosmopolitanism is superficial and multiracialism is so diluted that three of every four people you pass are Chinese'. Singapore's compactness also means that it is never possible to escape into an imaginary 'India' at Serangoon Road or a romanticized Malaysia in 'Arab Street' (despite what marketing brochures suggest) without encountering either the dominant ethnic group or the material presence of a global city.

In 1984, a new marketing image was introduced – 'Surprising Singapore: A Magic Place of Many Worlds'. The themes of expansiveness and diversity, first alluded to under Instant Asia, were reinforced yet again. With rapid development since the 1970s, Singapore has been transformed by highways, modern shopping malls and high-rise buildings. Rural *kampongs* and old buildings in historic Chinatown have made way for new retail, commercial and residential development (Huang 2001). When tourism arrivals dipped in 1983 followed by a period of stagnant growth, Singapore's loss of cultural appeal was regarded as the chief cause. In a policy U-turn, urban planners began to focus on Singapore's architectural heritage and urban demolition gave way to conservation and restoration. Both the STPB and Urban Redevelopment Authority (URA) embarked on new development plans. 'Surprising Singapore' represents a way to re-brand the destination as an intriguing mix of modernity and historicity. Consider the following STPB excerpt (Singapore Tourism Board 1983, cited in Chang 1997: 550), which curiously blends 'Asia with America', 'modernity with century-old traditions', 'orderly metropolis and Asian city':

> Behind the façade of a well-groomed and orderly metropolis, Singapore remains an Asian city to its very core. It's Instant Asia. And more. As if by some grand design, much of what's rich in Asia thrives here – the customs, the traditions, even the buildings. At first glance, Singapore may look like some bustling American city transplanted along the equator. But beneath the towering skyscrapers the visitor will find much of Singapore as it has been the last 100 years or more.

While the 1960s/70s were marked by a predominance of Western tourists, the 1980s saw an increase in Asian visitors to Singapore. The two markets have very different interests (Teo 1982) – while Western visitors delight in Singapore's multicultural heritage, Asians favour modern amenities and attractions. 'A magic place of many worlds' has something to offer everybody. In contrast to Instant Asia, which appealed to an oriental conception of exotic Asia, 'Surprising Singapore' emphasized a multi-dimensional product. State-sponsored guidebooks now highlight Singapore's international hotels and world-class attractions (such as the zoo and Sentosa Island) as well as historic Singapore River (the birthplace of colonial Singapore) and heritage sites like Chinatown and Arab Street.

How does size figure in 'Surprising Singapore'? I would argue that 'a magic place of many worlds' repudiates Singapore's minuscule size by underscoring its multi-dimensional appeal instead. Because Singapore is small, it presents itself as extra-special and appealing. The emphasis on 'small but special' is clearly evident in 1993 when a new tagline was introduced – 'a multi-faceted jewel'. Small but precious, Singapore is described as possessing the 'qualities of a multi-faceted jewel, able to appeal to visitors from East and West with equal ease' (Singapore Tourism Promotion Board 1993a: 7). A new tourism plan was devised, *Strategic Plan for Growth 1993–1995* (Singapore Tourism Promotion Board 1993b), outlining different avenues for product enhancement ('polishing the jewel'). To be a multi-faceted destination, Singapore has to be 'multi-interpreted' as different products to different users (Ashworth 1994: 23). The *Strategic Plan* identified various niche groups: honeymooners, youth travellers, medical visitors, cruise tourists and the MICE market (meetings, incentives, conventions and exhibitions). Specific policies were also outlined to appeal to each group as part of a market diversification strategy.

The metaphor of a gemstone is poignant. Singapore's smallness is acknowledged and the importance of tourism is highlighted. The latter point is worth noting because many Singaporeans often consider tourist attractions as benefiting only foreigners (Teo and Chang 2001). Like a small but valuable jewel, tourism is an economic asset to Singapore. In 1996, the tourism industry generated $S11.1 billion revenue (about $US6.2 billion) which constituted about 17.3 per cent of Singapore's services export (the figure has since risen to about $US8.7 billion in 2004 with 8 million visitors, double Singapore's resident population). Tourism is also exhorted as benefiting Singaporeans through development of heritage sites and preservation of cultural activities (Teo and Chang 2001).

Since 1997, a new tourism campaign has been launched to showcase 'New Asia–Singapore: So Easy to Enjoy'. New Asia portrays a destination which is modern and progressive and, at the same time, uniquely Asian. Unlike Instant Asia, New Asia eschews stereotypes of Asian exotica, highlighting Singapore as a city of the new millennium instead. Much has been heralded of the new Asian renaissance and the Pacific Century spearheaded by the People's Republic of China (PRC) and other Asian economies. As a dynamic, fast-changing and progressive city, Singapore is seen to embody the socio-economic transformations in Asia. The STB explains, 'New Asia is catching up with the rest of the world. We wanted to tell the world that Singapore is part of that change and, indeed, Singapore is at the forefront of that change' (Singapore Tourism Board undated, unpaginated).

While Instant Asia embodies Asia's composite charms in a self-sufficient Singapore, New Asia takes a slightly different approach by representing Singapore as integrally connected to the continent. While not possessing either the scenic resources or the market size of other countries, Singapore is imagined to have both the cultural and economic resources that enable access to the region. With efficient transportation links and business infrastructure, Singapore hopes to be the gateway to Asia. Born of its continental connections and cultural affinities, 'New Asia-Singapore' projects an ideal starting point for tourists visiting Asia, an investment hub for leisure companies and the region's entertainment capital. Consider the following caption from a 'New Asia-Singapore' print advertisement (see Figure 5.3):

*Figure 5.3* 'New Asia-Singapore': an Asian city blending tradition and modernity. (© Singapore Tourist Board.)

> Can a city of efficiency still find time to celebrate? Must a year have only one New Year's Day? Are the works of Leonardo da Vinci found only in Italy? Why can't Jacky Cheung, Elton John and Pavarotti perform on the same stage? Can different races and religions celebrate one festival?

'Singapore = New Asia' constitutes an imaginative strategy to tap the tremendous potential of the region. As Beijing prepares to host the Olympics in 2008 and as China, India and Indo-China are poised for dynamic economic growth in the next decade, Singapore similarly portrays an image of an exciting and eventful city. As New Asia globalizes and turns increasingly to the 'finer' aspects of life such as arts, culture and entertainment (Ibrahim 1996), Singapore similarly represents itself as a cosmopolitan city embracing global talent, culture and people.

Singapore's emphasis on Asia in its marketing campaigns reveals its obsession with size and the economics of survival. As the perpetual 'other', Asia provides an image by which the country projects itself. Instant Asia conveys an image of self-sufficiency vis-à-vis the largeness of Asia. As a compact destination, Singapore offers a substitute for those hoping to enjoy Asia without the time or money to traverse the continent. Rather than a liability, smallness is marketed as a strength. A different strategy is employed for New Asia, in which Singapore is integrally connected to the continent. Here, the continent and the city are co-marketed as dynamic and forward-looking. In as much as Asia needs a tourism capital and gateway, Singapore also needs a regional hinterland to expand its tourism industry. By laying claims to the economic, cultural and scenic resources of Asia, the narrow vision of Singapore's self-identity is broadened. From 'Instant Asia' to 'New Asia', discursive rhetoric of a 'Big Singapore' thus closely parallels policy turns towards regionalism and cross-border development.

## CONCLUSION

Small cities and micro-island states are often assumed to be disadvantaged in terms of tourism. With comparatively scarce resources, smaller land area and fewer opportunities for investments, such locales are limited in the scale and scope of development they can undertake. What is evident from the discussion above, however, is that while geographic size is an ever-present challenge, it need not be an immutable hindrance. Regionalization offers a way for small countries like Singapore to overcome their limitations and diversify their development options.

Since the late 1990s, Singapore has developed itself as a business hub in which tourism enterprises can operate. With skilled labour and effective transportation/communications networks, it hopes to serve the region as a gateway and tourism investment centre. Singapore has also begun exporting tourism skills and services to developing countries in Asia-Pacific (Lee 2004). More recently, Singapore has formed partnerships with neighbouring countries (Indonesia and Malaysia, as well as those of ASEAN) to undertake joint marketing and collaborative projects (Timothy 2003). Above and beyond its traditional role as a tourist destination, therefore, new functions have been created at the regional scale for Singapore as a hub, service exporter and partner. Discursively, Singapore's marketing identity has also evolved from a self-sufficient 'Instant Asia' to an awareness of its place within a larger 'New Asia' continent. In Singapore, size is an ever-present consideration in tourism planning and marketing. Through regionalization, a small city-state like Singapore can be transformed into a network-state, enmeshing it into a web of investments and information flows across the globe.

According to Poon (1989, 1996), a 'new tourism' model has emerged in the 1990s; characterized by new principles and philosophies, new tourism is a reaction to old tourism. While old tourism espouses economies of scale and quantity (mass tourists, international chain hotels, etc.), new tourism focuses on economies of scope and the quality of touristic experience. Under this new mindset, mass production of standardized artefacts and services gives way to batch production of specialized goods catering to niche groups. Instead of 'going it alone', new tourism firms and enterprises are also entering into alliances with each other in order to benefit from joint marketing and reduce harmful competition (Poon 1989). In Singapore, the turn towards regionalism may be similarly regarded as a transition to new tourism. Regional collaborations supplement single-country development; niche groups rather than mass markets are courted; and geographic dimensions are regarded as flexible and transgressible.

Under new tourism, size is also being redefined. Size does not refer to physical proportions alone but, more importantly, to the ability of a tourist destination to harness economic functions and project social images much larger than itself. From 'Singapore is too small' to a belief that 'there are no real limits or constraints in this borderless world' (Singapore Tourism Promotion Board 1993a: 16), physical size may be substituted by regional connections. To be a tourism capital, therefore, Singapore's influence must be disproportionately greater than its geographic dimensions. With spatial reterritorialization, scale need not be the ultimate nemesis in small city-island-states. To return to my original question of whether size is destiny, tentatively it might be concluded that it is not. Regional tourism policy and discursive marketing practices in Singapore suggest that inasmuch as scale is socially constructed, so too are borders, regions and place identities. Regional connections provide new room and positive hope for Singapore's tourism expansion. Future research on small islands, cities and countries should therefore explore the potential as well as the challenges that accompany regional tourism and transborder developments.

# Part 2

# The urban hierarchy and competitive advantage

# 6 The festival phenomenon

## Festivals, events and the promotion of small urban areas

Andrew Bradley and Tim Hall

Selling or marketing a particular geographical locality has emerged as a central part of the contemporary process of inter-urban competition for global capital (Chang 1997; Young and Lever 1997). In this competition, place attributes and local cultural identities are often used in the form of 'cultural capital' to project an alluring image to potential residents, investors and visitors (Ashworth and Voogd 1990; Kearns and Philo 1993; Kenny 1995). The marketing of these images has been seen as a response by policy makers to adjust to the changing nature of the economic structure of Western Europe and North America (Holcomb 1994) as the economies attempt to adjust from heavy industry and manufacturing to a reliance on services or high-tech industries. This in turn has an impact upon a place's appeal to potential investors and residents. The older virtues of a central location, cheap labour and low taxes are not abandoned, but much more emphasis is placed upon quality of life issues (Bovaird 1995).

Places are now marketed as a great place to live, as well as a great place to do business. As firms increasingly rely on recruiting and retaining highly-paid managers and with technical innovation making business less dependent on the supply of unskilled or semi-skilled workers, so must the city be seen as habitable or consumable by upscale executives and middle-class professionals. This creates an important distinction within the literature on place marketing, as it is being increasingly recognized that 'urban promotion involves the selling of a location not only for business but also as a place to live ... these images of lifestyle tend to be predominantly anchored around two things, culture and environment. The use of leisure time is considered an increasingly important aspect of the decision-making process for both long-term relocation decisions and short-term (for example, convention location decisions) business or tourist decisions' (Hall 1998: 127) and, as Page (1995: 217) notes, 'the development of festivals and special events may make an important contribution to the image of a destination' and therefore may have a significant impact on the economic development of a location.

Despite all of the rhetoric concerning the supposed saliency and centrality of place promotion to contemporary urban change, there is, to date, little, if any, empirical evidence that this is the case. Put simply, despite the great attention paid to place promotion by academics, we know little of the actual images of towns and cities held by key economic actors whom cities target with promotional materials.

This gap has implications both for the academic understanding of place promotion and for decision makers in, for example, local authority planning and economic development departments. This chapter addresses this by examining the media coverage that festivals receive and how this can contribute to images of place.[1]

## SMALL URBAN AREAS

The emphasis that has been placed on place marketing as a response to a declining industrial base and a shift towards a service industry-based economy within much of the academic literature may not, however, represent the experience of all locations involved in promoting positive images of place with a view to encouraging economic development. There are a great many locations that are using the promotion of place image that have little or no industrial base that has declined. Perhaps the best examples of this are the 'new' towns within the United Kingdom (for example, Milton Keynes and, to a slightly lesser extent, Telford because it was the 'birthplace of the industrial revolution'). This emphasis on industrial and large metropolitan areas is also present within the academic literature on place marketing as, 'putting it bluntly, we really do not know what has happened outside of the big cities' (Millington 2002: 40).

This dependence on using large metropolitan areas and areas that were once dominated by manufacturing industry has the effect of distorting the picture of place promotion activities. This is not to say that the decimation of the manufacturing base in these locations has not played a crucial role in the growth of place marketing activities, but it needs to be recognized that place marketing is not exclusive to these areas. This, ironically, may be due to the changes in manufacturing that have occurred and the rise in importance of service industries; with the location of offices and production facilities being freed from the past constraints of access to raw materials and so on, businesses are now able to pick and choose their location based on a different set of location criteria. Therefore, locations that previously may not have had access to the resources necessary to attract industry are now pitched into the place promotion battle to attract industries that are increasingly footloose.

The dominance of large cities in the academic literature has the effect of overgeneralizing the way in which locations have pursued place marketing initiatives as it takes little or no account of the experience of small and medium-sized cities. It could be argued that small and medium-sized cities have also suffered the impacts of de-industrialization and the associated closure of component manufacturers or material extraction facilities. Also, the experience of small and medium-sized towns may not be similar as regards the marketing initiatives available to them as they do not possess the same cultural capital of images that can be accessed by larger locations. The decentralization of economic activity, however, makes small and medium-sized towns important prospective locations for receiving economic activity. Therefore the place marketing policies they adopt need to be recognized as an important facet of locational decision making.

## THE FESTIVAL PHENOMENON

The contemporary promotion of festivals can be defined as one strand of local economic policy that is used to cushion the negative effects of the painful transition from an industrial to a post-industrial economy (Booth and Boyle 1993). In this model, festivals are incorporated within the language of economics, with the attendant measurements applied to policy analysis: investment, leverage, employment, direct and indirect income effects, social and spatial targeting and so forth. Festivals become a strand of place marketing, with cities vying against other cities to flaunt their ownership of top quality events, museums and galleries, fine architecture, symphony orchestras or rock musicians. Depending on the audience, a city's festivals are packaged and re-packaged to become an incentive for the potential investor, property developer, potential tourists or residents. A recent study showed that 79 per cent of local authorities had recently established arts, cultural or sporting events. Further analysis of the data showed that the geographical scale of attraction of these festivals was evenly spread from international through national to local (Millington 2002).

Despite the obvious ubiquity of festivals, serious academic studies of them have been surprisingly neglected (Waterman 1998). A conventional approach to the study of festivals in human geography has tended to take the view that the arts festival was little more than a transient cultural event with a measurable impact on the landscape, event and economy which has simply been mapped or modelled (see Leyshon *et al.* 1995; Nash and Carney 1996). However, festivals are not simply bought or consumed (in an economic sense) but are also accorded meaning throughout their active incorporation into people's lives. They epitomize the representation of contemporary accumulation through spectacle and consumption in an era of flexibility (Harvey 1987; Zukin 1990).

## FESTIVALS AND QUALITY OF LIFE

Recent analyses of relevant location factors (see for example, Smeenk 1992) reveal a growing importance of the perceived quality of the working environment (Bovaird 1995) in the choice process. Other factors affecting location and investment decisions relate to the availability of highly skilled personnel, high-tech knowledge and an attractive social climate. The city's cultural supply, even in the narrow sense, has a direct impact on these aspects (for instance, historic environment, prestigious working location), but may also indirectly affect them through, its influence on the local working population characteristics (quality of personnel, purchasing power). Although culturally orientated aspects do not represent the prime motives in the choice of living accommodation (Brabander and Gijsbrechts 1994), recent research (Blommaert *et al.* 1992) points to an increased importance of culture in location decision making. It is also interesting to note that the use of culture as an attractant to a location is not equally important to all sub-segments of the population. People belonging to the upper socio-economic groups in society seem to value cultural supply more than other segments of the population (Voye 1985; De Lannoy 1987; Ebels and Ostendorf 1991).

However, the use of festivals as a method of attracting new businesses should not be overestimated. Whilst quality of life factors, such as a location's cultural provision, have undoubtedly become more important as the locational restrictions placed upon businesses have become relaxed due to advances in technology, it goes without saying that the potential to attract companies remains largely conditional on other factors, such as the availability of reasonably priced location sites, transport infrastructure and local markets. However, there is also some evidence to suggest that the availability of artistic and cultural activities can, in certain cases, be a contributing, although rarely decisive, factor in plant and location decisions (Cwi 1992; Johnson and Rasker 1993, 1995). There is also a suggestion that perhaps the more crucial role played by the festivals in economic development lies in the category of business retention as, 'while culture is a positive influence on employment, it is better at holding people than pulling people to an area … while culture helps people to stay, it does not persuade them to want to move' (Rodgers 1989: 65).

## CHELTENHAM

Cheltenham lies mid-way between Bristol and Birmingham on the edge of the Cotswold Hills in England. It is the second largest population centre in the county of Gloucestershire, after Gloucester. Cheltenham's relatively good accessibility also means that it serves as an extensive catchment and travel-to-work area for central and eastern Gloucestershire and the South Midlands. The industrial composition of Cheltenham reflects the changes that have taken place in the local economy in the past 30 years. There has been a significant decline in the traditional manufacturing base which has been centred on the aerospace and defence industries since the end of the cold war. As a result, the local authority instigated a number of strategies during the 1980s to attract jobs in the financial, business services and public administration sectors of the economy. These policies have been relatively successful as several major national and international companies have set up headquarters buildings in the town including; Chelsea Building Society, Dowty Aerospace, Smiths Aerospace, Eagle Star, the government's General Communications Headquarters (GCHQ), and the Universities and Colleges Admissions Service (UCAS).

### Place marketing campaigns in Cheltenham

Currently, Cheltenham does not engage in any premeditated conventional place marketing campaigns aimed at the business market. There are a few brochures, however, that are aimed solely at tourists which are available through branches of the tourist information office. The local authority does, however, have some information such as land availability and price details that are available should they be contacted. Therefore, the principal way in which Cheltenham seeks to promote itself, in view of the paradox of increased development ruining the very thing that has spurred growth within the town, is through the quality of life that it can offer. A crucial part of this is the many festivals that take place within Cheltenham that are designed to offer a varied menu of sporting and cultural entertainment aimed to appeal to visitors of all ages and interests. These festivals have been instrumental in leading Cheltenham to proclaim itself to be the 'festival town of Britain' (Cheltenham Borough Council 2001) and the year-round series of regional, national and international festivals has sustained Cheltenham's appeal to both visitors, businessmen and residents, and it compensates for the town's lack of a major historical or geographical attraction such as a river, cathedral or castle. These festivals allow Cheltenham to 'culturally punch above its weight ... [and are] a significant factor in retaining and attracting businesses to the area' (Cheltenham Borough Council 2002: 4). Moreover, one of Cheltenham Borough Council's stated aims is to develop Cheltenham's reputation as a festival and event town (Cheltenham Borough Council 2001). Central to this aim is maintaining and protecting 'the image that the town has in offering a high quality of life ... [as] ... employers relocate to, and stay in, Cheltenham because the profile is good for business and is attractive to employees ... [and Cheltenham] ... would lose this image at its peril'. (Cheltenham Borough Council 2002: 9).

However, whilst there are a large number of festivals that take place in Cheltenham that are of international standing, a sample of three festivals was used as the basis of this study. The three selected festivals were:

- National Hunt Festival
- International Festival of Music
- Cheltenham Festival of Literature

These festivals were chosen as they offer a balance between cultural festivals and sporting events and thus may not preclude certain individuals who have no interest in one of these categories from having been exposed to the media coverage that they receive. They were also chosen as they represented festivals that draw competitors/participants on both a national and international scale and therefore, it could be argued, have a profile which would attract media attention from the national press. This was verified by an unsystematic analysis of online newspaper archives for previous years.

## Measuring the impact of Cheltenham's festivals

Whilst a great deal of prominence has been placed on the use of festivals within Cheltenham as a means of projecting and maintaining a positive and enticing image for the town, little is known about how effective this strategy is. None of the bodies that are responsible for the management of festivals or for local economic development have any mechanism to measure the efficacy of these policies. Whilst this is often a criticism of place marketing techniques generally, the decision to place so much faith and money in a strategy without any way of measuring its outcome needs to be called into question. The only economic assessment report that has been published in relation to one of Cheltenham's festivals was one

*Table 6.1* Brief histories of three selected festivals

*The Cheltenham International Festival of Music*

The Cheltenham International Festival of Music (hereafter known as the Festival of Music) was the first of Cheltenham's post-war cultural festivals, beginning a mere five weeks after the end of the Second World War, and, as such, it is the longest running festival of its kind in Britain. The initial mandate for the festival was for it to be a showcase for British contemporary music but it has expanded to include more music from the classical and romantic repertoire, building around the presentation of world premieres and commissions for which the festival has become famous. The festival, which takes place annually in July, opens with a free concert entitled 'Picnic in the Park' with live music and fireworks and runs for two weeks.

*The Cheltenham Festival of Literature*

The Cheltenham Festival of Literature (hereafter known as the Literature Festival) began in 1949 when Gloucestershire writer John Moore organized a gathering of writers to celebrate the written word in Cheltenham. The festival, which takes place annually in October, is widely acknowledged as the very first purely literary festival to be set up in the United Kingdom, and has grown considerably from the first festival in 1949 which contained just nine events, to its current status as a huge and varied festival of international repute.

*The National Hunt Festival*

The National Hunt Festival takes place every March at Prestbury Park, although the event can trace its origins back to 1819 when a course was marked out on the nearby Cleeve Hill and a three-day event took place under the patronage of the Duke of Gloucester. The festival has been held annually since then apart from a break of 16 years from 1829 to 1845, when racing was suspended after a large group, led by Cheltenham's parish priest, the Reverend Francis Close, hurled rocks and empty bottles at the horses and riders, claiming that gambling was immoral. Twenty years later, in 1865, the meeting was moved to its present location. In the 1930s, the meeting acquired its current status as the climax of the National Hunt season when the feature races, the Gold Cup and the Champion Hurdle, were added. For many years, the festival has also been linked with the Queen Mother and the Wednesday of the meeting is now known affectionately as 'Queen Mum's day' when she used to be on hand to present prizes for the Queen Mother Champion Chase.

produced for the Network Q Rally of Great Britain (Lilley and DeFranco 2000). Whilst there has only been one published economic impact assessment in relation to Cheltenham's festivals, the postponement and subsequent cancellation of the 2001 National Hunt Festival due to the foot-and-mouth crisis led to several media reports which estimated the amount of revenue that would be lost to local businesses. For instance, 'The outbreak led to the postponement of the Cheltenham Festival, the climax of the National Hunt Festival season and an event crucial to the economic wellbeing of the town. Around 150,000 people visit annually, pouring around £10m into local coffers' (Vasagar *et al.* 2001: 32). The estimate of a negative economic impact of ten million pounds is also referred to in several other articles (Savill 2001; Weaver 2001; Brown and Laville 2001). However, the source of this estimate or an identification of the affected businesses is not given within any of the articles.

## Newspaper coverage of Cheltenham's festivals

For the purposes of this study, newspaper reports have been used as a surrogate for the media in general. Newspapers have also been used within previous studies in place marketing as, 'the newspaper is one historic vehicle for this marketing, as documented by studies of nineteenth-century boosterism' (Myers-Jones and Brooker-Gross 1994: 196; see also Belcher 1947; Abbot 1981).

Six national newspapers were chosen:

- *Financial Times*
- *Daily Telegraph* (*Sunday Telegraph*)
- *The Times* (*The Sunday Times*)
- *Independent* (*Independent on Sunday*)
- *Guardian* (*Observer)*
- *Daily Mail* (*Mail on Sunday*)

The media coverage was dominated by the National Hunt Festival, which accounted for 73 per cent of all articles analysed and 83 per cent of the total words published.

## Qualitative analysis of media reports

The section analyses the key characteristics, constructs and stereotypes that are contained within the newspaper coverage of Cheltenham and its festivals. Much of the coverage sought to portray Cheltenham's festivals as being imbued with quality. This was achieved in a number of ways such as references to the quality of performances within the festivals, the quality of individual performers and the quality of the events themselves.

## Quality of performers and performances at Cheltenham's festivals

One of the principal mechanisms by which the quality of Cheltenham's festivals was portrayed was through the many references to the fact that Cheltenham's festivals were hosting a number of premieres at varying geographical scales. For instance, Anonymous 1 (1999: 82) notes that, 'the programme also included the European premiere of Mark-Anthony Turnage's Silent Cities'. There were also premieres at a world level at the Music Festival as 'one of Britain's finest ensembles, the Vellinger Quartet, visits the festival, with vintage Mozart and a world premiere by the young composer Huw Watkins' (Anonymous 2 1999: 39). National premieres are also mentioned, such as the British premiere of Roland Caltabiano's *Marrying the Hangman* (Fairman 1999). Not only is the quality of the individual performances singled out, but it is also acknowledged that Cheltenham's festivals are of a significant stature to attract an infrequent performance from a highly accredited artist, with a performance of the BBC Philharmonic Orchestra being referred to as making a 'rare excursion from their Manchester home' (Clements 1999: 21).

### The quality of Cheltenham's festivals

It is not only the individual and groups of performers that are singled out as personifying the quality of festival that Cheltenham has to offer. The festivals themselves are lavished with praise and are referred to as:

- Pinnacle of its field
- Unique
- Anticipated
- Oldest
- Attracts famous names

Many reporters refer to Cheltenham's festivals as the pinnacle of their particular field of endeavour. It is claimed that 'a marketing guru would describe it as the "market leader"' (Anonymous 3 1999: 19) when referring to the Literature Festival. It is also reported that the National Hunt Festival is not only the pinnacle of its field but the event 'dominates national hunt racing more than any individual event dominates any sport in these islands' (Anonymous 1 1999: 35). Various adjectives are used to further reinforce the position of the National Hunt Festival as the highlight of the jump racing calendar, with commentators claiming that it has 'an incomparable parade of jump racing', 'the hallowed Cheltenham festival' (Scudamore 2000: 69), 'national hunt's defining moment' (Edmonson 2000: 26), 'as good as national hunt racing gets' (Oaksey 2000: 40). Reed (2000: 6) draws an analogy between the national hunt season and the flat racing season in that 'flat racing's extended and increasingly international season has numerous peaks from 2,000 guineas day in May to the Breeder's Cup and the Japan and Melbourne Cups in November. But jump racing's lines converge all winter in the glittering terminus of the festival'.

### Cheltenham's old/traditional image

It would appear that some reporters hold the view that there is a consensus amongst their readers as to the image of Cheltenham: 'the Gloucestershire town, as readers will know, is rather refined and genteel' (Anonymous 4 1999: 11). This view is reinforced by a number of adjectives that are used in conjunction with references to Cheltenham, such as 'the stylish town' (Anonymous 2 1999: 40), 'the elegant regency promenade' (Reed 2000: 6), 'quietly civilized' (Maddocks 1999: 8) and 'the old spa town' (Anonymous 3 1999: 10).

### A new image for Cheltenham: innovation and hedonism

Set against this 'traditional' image of Cheltenham as a stylish, civilized, conservative spa town are a number of images which are radically different, the first of which being references that claim that events at Cheltenham's festivals are innovative and different and go against what could be termed 'conservative'. For instance, *The Course Inspector* (2000: 37) claims that 'the key to Cheltenham's phenomenal success is that the place never stands still. Year after year innovations appear to make the flustered customer feel more comfortable'. One of these innovations, perhaps prompted by the conspicuous consumption of alcohol at the National Hunt Festival, is 'the first chemist to open on a British racecourse' (Armytage 2000: 42). The reviews of some of the performances at the Music Festival also indicate that the conservative ethos of its organizers may have been overstated. Various performances are referred to as 'striding out from twentieth century tradition' (Maddocks 1999: 9), 'wildly daring' (Dove 1999: 22), 'courageous' (Larner 1999: 46) and 'an adventurous evening of music theatre' (Fairman 1999: 34).

## THE SIGNIFICANCE OF THE NEWSPAPER COVERAGE OF CHELTENHAM AND ITS FESTIVALS

Each of the key constructs that emerge from the depiction of Cheltenham and its festivals is positive in nature and tends to create an image of the town that could, potentially, be attractive to new investors, residents and tourists. The main feature that is present in both the reports on festivals and on Cheltenham generally seeks to portray an image of 'quality' for the town. This quality is implied with reference to not only the festivals themselves, but also the festivals' participants. This quality image gives the reader the impression that Cheltenham's festivals are of a significantly high stature to be a draw to the biggest names in their particular fields to partake in the event and also to draw festival audiences from celebrities and other prominent figures. This association with the 'best' in their particular field and the endorsement of major celebrities, be they participants or members of the audience, only serves to highlight and reinforce Cheltenham's quality image. The analysis of the images contained within the coverage of Cheltenham generally and its festivals and events indicates that the constructs used are different from those found in more conventional promotional campaigns. It has been argued that promotional packages rarely tend to be original and in fact draw upon a very limited range of themes and motifs (Hall 1998).

Whilst the images of Cheltenham that are portrayed within the media coverage of the town and its festivals also draw upon a similarly limited number of key components, these components are very different from those contained within the glossy brochures of traditional place promotion campaigns. The images of Cheltenham contained within its media coverage may also be more effective as they are not part of an overt advertising campaign and are grounded within the socio-cultural characteristics of the town, not as part of a 'managed' image produced by an advertising agency or marketing department.

Cheltenham's festivals appear to be an excellent mechanism for the creation and maintenance of positive images of place. Their coverage in the media positions Cheltenham in the spotlight to a far greater extent than it would receive without coverage of its festivals. The analysis of the coverage of Cheltenham suggests a complex and contrasting image that is constructed through the media. This coverage, however, is generally complimentary to the town and it creates an image of quality that permeates the boundaries of individual festivals and is also contained within the reports of news events within the town. Whilst there does seem to be a tension between the stereotyped traditional image of Cheltenham as a home for retired colonels, conservatism and refined young ladies, and that of a 'modern' town and the revelry associated with the National Hunt Festival, these images only serve to highlight Cheltenham's cosmopolitan nature.

However, whilst the coverage of Cheltenham's festivals helps to generate a great deal of media attention that could help to create a strong positive image for the town, the analysis of the coverage on its own gives no indication of the extent to which these images and associations are transferred to the way in which the town is perceived by key economic actors. Also, the differentiation between the images of Cheltenham and the images of Cheltenham's festivals may, however, be purely theoretical, as the recipients of these images may not create this distinction. Moreover, as was previously identified, much of the research into place marketing, particularly amongst practitioners adopting a semiotic approach, has failed to assess the extent to which the recipients of place marketing interact with the images of place that they receive and the subsequent images that they create. The neglect of this issue has serious ramifications for the understanding of place marketing and its effectiveness in producing positive economic benefits.

## THE IMPORTANCE OF THE MEDIA IN CONSTRUCTING AN IMAGE FOR CHELTENHAM

There was a unanimous belief within Cheltenham's business community that the media was one of the principal mechanisms in the construction of images of Cheltenham by external audiences. For example, 'the power of the media to create an image is very important and the media take this image to a much wider audience' (Informant: Chelsea Building Society). Images of Cheltenham are believed to be transmitted through the media via the use of a limited number of stereotypes which serve to reinforce the traditional image of the town.

The construction of an image of Cheltenham by the business community, delineated along an internal–external axis seems to indicate that there is a belief that people who do not have regular or direct experience of the town construct an image that largely corresponds to its traditional image. The traditional image of Cheltenham that was identified within the media coverage, that portrays the town as the home of retired colonels, affluent residents and the venue for various festivals, is the image that the business community believes is held by its suppliers, customers and potential employees. These images are largely associated with affluence, a key construct of Cheltenham's traditional image, in that they draw heavily upon the retirement of the upper echelons of military personnel, the countryside and high-profile cultural and sporting festivals.

The business community seems to believe that it is only the young and those with direct exposure to Cheltenham who have a new image of Cheltenham as a vibrant, cosmopolitan town which is generally comparable to the new, innovative image contained within the media coverage. However, a key component in the construction of this 'new' image would seem to be directly related to Cheltenham's night-time economy. This was something that was not identified within the media coverage, as it seems to be viewed as the antithesis of Cheltenham's traditional image of being conservative and affluent. The elements of the new or innovative nature of Cheltenham's image, which are largely associated with Cheltenham's festivals, are only introduced when informants seek to counter some of the inaccuracies that they see in the town's traditional image. This delineation of image may indicate that Cheltenham's festivals are just an ephemeral gloss whilst the old traditional image endures. This, in turn, raises questions about the effectiveness of image-modifying initiatives, such as place marketing, which seek to change people's perception of place. Whilst, Cheltenham's traditional image does not have the negative connotations associated with heavy industry as have some of the locations that have used place marketing as a mechanism to assist in changing people's images (Kearns and Philo 1993; Ward 1998; Bradley *et al.* 2002), its traditional image seems to be resistant to change.

Cheltenham's image, however, would appear to be something which is of direct concern to the business community as it has instigated its own research into this area to identify the constructs that people associate with the town. The results from these studies are generally comparable with the images of Cheltenham that are contained within the media coverage relating to the town and its festivals and also to the external images of Cheltenham that the business community believes people have. There is some evidence, however, that younger people may be less aware of Cheltenham's traditional image and their image may be more related to the new, innovative image that was found within the media coverage.

An integral part of Cheltenham's image is its association with a number of festivals: 'The festivals brand Cheltenham, the literary festival, the music festival, etc. they help this elitist image that people buy in to' (Informant: Cheltenham Chamber of Commerce). 'I am a strong believer that quality counts, if the town is associated with quality events there will be a spin-off from that' (Informant: Gloucester Development Agency). The principal spin-off from Cheltenham's hosting of festivals is believed to be the symbiotic relationship that they have with the image of Cheltenham.

> Cheltenham needs events/festivals to give Cheltenham a profile … If you think of what other places have done, Birmingham has improved its image through culture and events, Bristol will get its act together sooner or later, it hasn't yet but it will. What would Stratford be if it didn't have what it has? Any town that is of any size is trying desperately to promote itself for direct tourism or profile to get businesses in and we can't just let go as it is a very competitive business and we have to keep investing (Informant: Cheltenham Borough Council).

Festivals are therefore seen as being not only an integral part of Cheltenham's image, but also a key component of the town's economic fortunes.

The number and type of Cheltenham's festivals, however, can also go some way to challenging what is seen as the traditional image of the town:

> [The amount of festivals] refers to my image of Cheltenham being like a mini-London. To a certain kind of person I think that it means a lot, we have a good mixture, we have the book festival, the open air music, but then we have other things like the cricket festival, but you also have the football team that are doing quite well which helps to bring a broad band of entertainments for people. People feel that Cheltenham is a place to be and that there are things there to do (Informant: Smiths Industries).

## THE IMPORTANCE OF THE MEDIA COVERAGE OF CHELTENHAM'S FESTIVALS

Whilst it is generally recognized that Cheltenham's festivals play an important part in terms of creating and maintaining the image of the town and the associated economic development benefits, it is also recognized that the media play a crucial role in the dissemination of images of the festivals. 'Undoubtedly the media coverage of Cheltenham's festivals plays an important role in maintaining the town's image' (Informant: Krone UK Technique).

In part, the images of Cheltenham that are transmitted through the media coverage of its festivals are deemed to be especially important due to:

> the media that cover it. For instance, would you see the Cheltenham festival of literature in the *Daily Star*? This is all stuff that is in *The Times*, the *Telegraph*, the *Observer*, the *Independent*, but it is that sort of image. People who read the broadsheets are considered to be of a certain class or standing and having that amount of coverage in these papers undoubtedly maintains Cheltenham's image. Even for the National Hunt Festival, the broadsheets put it in perspective of the town that it is in, whereas the tabloids just concentrate on the form and the racing gossip (Informant: University of Gloucestershire).

There is some disagreement, however, as to whether the images that are contained within the media coverage of Cheltenham's festivals serve to reinforce the town's traditional image or whether they serve to challenge people's perception about the town.

Cheltenham's festivals and the media coverage thereof are seen as an important part of the way in which images of the town are transmitted to external audiences. There is, however, a concern that the images contained within this coverage tend to reinforce the traditional image of the town rather than expanding or updating the town's image. This, in part, is related to the holding of high-profile cultural festivals of music, literature and jazz which, when coupled with associations of educational excellence and links to the royal family, are seen to fit into the stereotypical image that is transmitted by the media of an affluent, upper-middle-class town populated by retired colonels.

## CONCLUSIONS

Cheltenham's festivals are something that have become synonymous with the town's image which help to brand the town. It is also recognized that Cheltenham's festivals are important in marketing the town to external audiences and they are an important part in the competition between places for economic development. Cheltenham's festivals are seen as an important mechanism for keeping the town in the public's eye and as a result this has positive economic benefits for the town. There is also a suggestion that the media coverage of the Cheltenham festivals can go some way to counter the stereotyped image of the town and bring its image into line with that which the business community see to be more representative of modern Cheltenham. The selective nature of the festivals that receive media attention may go some way to explain this, as it is believed that, should the coverage be extended to include some of the more 'peripheral' festivals, the images of the town used may not draw so heavily on the stereotypes.

The media coverage of festivals produces a complex and contrasting image of place and this has impacts on its economic development which are not easily explained by the existing literature. However, small and medium-sized towns possess far less capacity to advertise and promote themselves within the arenas of inter-urban competition. Therefore, festivals are important media through which small and medium-sized towns are able promote themselves to external audiences. Festivals generate a far greater amount of media attention than would normally be received by such urban areas, something which has been largely ignored in media studies, studies of place marketing and local economic development.

With the significant growth of festivals, it is essential that small and medium-sized towns adopt an identifiable niche within the festival marketplace. One way in which this can be achieved is through holding a small number of 'quality' festivals that are the leaders in their field which could, potentially, lead to the creation of an identifiable brand image for a location. Without this, the impacts of festivals in small and medium-sized towns may become diluted or lost amongst the plethora of newly organized festivals and events.

## NOTE

1 This was achieved by undertaking a quantitative and qualitative analysis of the text of the media coverage and 17 semi-structured interviews with key business personnel to identify the perception of the importance of the media coverage of the festivals to local economic development.

# 7 Gentrifying down the urban hierarchy

## 'The cascade effect' in Portland, Maine

Loretta Lees

For some time now, 'a focus on medium and small-scale communities has been neglected in favour of large metropolitan areas in North America ... this bias has obscured the possibility of observing important connections between local-scale physical and social changes and non-locally based regional-scale political and economic developments' (Knopp and Kujawa 1993:124; for an exception, see Markusen *et al.* 1999). This bias towards research on large metropolitan cities has long been evident in the gentrification literature. As Dutton (2003: 2558) states, 'much of the empirical and theoretical research in the 1980s and early 1990s, either explicitly or implicitly, considered gentrification in the context of cities occupying strategic positions in the international urban hierarchy'. But this is changing as a growing body of gentrification research in smaller cities provides the beginnings of a much-needed empirical mapping of the development of gentrification outside large metropolitan cities (Bridge 2003; Dutton 2005). This research into gentrification further down the urban hierarchy makes a welcome contribution to my earlier calls for a 'geography of gentrification' (Lees 2000). In the UK in particular, there is now debate about the relationships between gentrification in smaller, provincial cities and in larger, metropolitan cities.

In seeking to explain the historical and geographical patterns of gentrification, several scholars have invoked the hydrological metaphors of saturation diffusion and cascade. For example, Atkinson and Bridge (2005: 16) argue '[a]s gentrification has become generalized so it has become intensified in its originating neighbourhoods ... This has led to a cascade effect down an international and regional set of urban hierarchies in which the saturation of investment motives in gentrified cities like New York and London have pushed towards neighbourhood changes in new regional modes'. Dutton (2005: 213) complains that this image of provincial gentrification caused by the downward trickle from saturated global cities implies a single mechanism for gentrification everywhere and ignores the role of local context in 'mediating gentrification processes'. In its place he proposes a dual model of gentrification to better distinguish its manifestations in core cities from that in peripheral cities. However, despite its recognition of local contingency in mediating the flow of gentrification down the urban hierarchy, Dutton's more nuanced model retains the same diffusionist logic as the cascade model, along with such associated hierarchical metaphors as 'pulse' (p. 212) and the purely negative role for local contingency as a buffer against the 'external factors leading to gentrification in provincial cities' (p. 224). At their core, then, both these models of gentrification suggest a one-way flow of influence percolating down the urban hierarchy. This diffusionist ideal in turn presumes the temporal and processual superiority of primate city gentrification: what happens first in New York or London then cascades outwards to affect other cities nationwide. But what processes might be involved in such a downward diffusion of influence?

Analytically, it is possible to distinguish at least three possible mechanisms for such a cascade effect. The first is economic: the idea, as Atkinson and Bridge suggest above, that as the good investment opportunities (rent gap) are exhausted in primate cities such as New York and London, so capital seeks out new frontiers lower down the urban hierarchy. This presumes tightly integrated real estate and labour markets and the easy diffusion, nationally, of information about new opportunities. Dutton (2003, 2005), for example, demonstrates how the diffusion of gentrification from London to Leeds was driven both by individuals cashing in on the appreciation of property in the south-east and buying larger properties in Leeds and by the development there of back-office financial service facilities which took advantage of the lower labour and other costs in provincial cities. The informational and other barriers to the downward cascade of intercity investment into new gentrifying areas are easier for corporate investors to overcome. In Dutton's study of Leeds, for instance, corporate investment has been a driving force in inner-city regeneration (2003: 2559). It is likely that any such cascade of intercity investment flows will tend to come later in the process, after an initial phase of pioneer gentrification has established a secure enough beachhead to attract attention (and money) from investors elsewhere. This is typically led locally, increasingly with government support. In any event, such a diffusion is more likely to occur in the UK, France and the Netherlands than the USA or Canada, where migration fields and property markets, like the urban system more generally, are less centralized around a single capital city.

A second possible cascade mechanism is cultural: the diffusion of a gentrification lifestyle or identity from centre to periphery and from the local to the global (or vice versa). Podmore (1998), for example, discusses the role of the mass media in reproducing the values and meanings of gentrification from one metropolitan context to another; in her case the habitus of loft-living in New York to the 'secondary city' of Montreal. She examines how the urban episode of loft-living is transferred beyond the boundaries in which it was produced (Manhattan) to be reconfigured and reconstructed in a different context (downtown Montreal) through a series of negotiated and recursive socio-spatial practices undertaken by agents. Rofe (2003) discusses how gentrifiers are projecting their identity from the scale of the local onto that of the global. He argues that 'gentrifying areas provide the conduit through which the template of a gentrification-derived transnational identity is focused and projected onto the scale of the global' (p. 2521) and that 'as the pioneers of new lifestyle practices in their quest for distinction, the gentrifiers act as the arbiters of taste thereby transmitting them to the wider community' (p. 2522). As Rofe and others before him (see for example, Mills 1993) point out, real estate agents and developers mobilize the 'gentrifier lifestyle' in their marketing ploys. The inner city and the gentrifier lifestyle thus become a consumable product, what Mills (1993) calls the 'gentrification commodity'. The irony is that the diffusion of the 'gentrification commodity' reduces its symbolic value as a distinctive lifestyle or habitus.

Finally, a third mechanism is that of serial reproduction: the reproduction of regeneration policies, plans and ideas across cities and, in particular, from bigger cities to smaller cities. Think of the way that waterfront redevelopment, re-packaged by those people who first did Faneuil Hall in Boston then South Street Seaport in New York City, sold the idea of putting the old commercial city back in touch with its waterfront (see Edensor, this volume). These ideas have been taken on board by practically every post-industrial city that has a waterfront or even a canal to redevelop. In my discussion of the British government's *Urban Task Force Report* and *Urban White Paper*, I comment on the 'gentrification blueprints' based on London and Barcelona being taken to cities such as Liverpool and Manchester (Lees 2003a). McCann (2004) argues that this 'serial reproduction' of policies (Harvey 1989) or 'policy transfer' (see Dolowitz and Marsh 2000) tends to foster weak competition and crowding in the marketplace 'that works to the detriment of most cities by fostering a "treadmill" effect in which every city feels an external pressure to upgrade continually its policies, facilities, amenities and so on to stave off competition and maintain its position in the competitive urban hierarchy' (McCann 2004: 1910).

Though useful as an analytical heuristic, the actual geographies of gentrification are likely to be more complex than these cascade models imply. First, these mechanisms of cascade are by no means exclusive; more often than not they will operate in tandem. Second, the uneven establishment of gentrification across the urban system may, in turn, generate its own eddies and secondary sources of cross-cutting and cascading influence. This suggests the importance of investigating the operation of the so-called 'cascade effect' in more detail.

To that end, this chapter considers the particular case of Portland, Maine, USA, and its place in the gentrification cascade. Located 115 miles north of Boston (see Figure 7.1), Portland has fared better than many small and mid-sized cities across the North American 'rust belt'. Over the past 40 years, its historic 'Old Port' has gone from a dilapidated warehouse district serving dying maritime industries to one of the major engines of economic growth in the metropolitan region. Several waves of gentrification have transformed not only the built environment of the Old Port and surrounding areas of the downtown, but also their place in the space-economy. In addition to providing back-office services for Boston-based banks and insurance companies, Portland's downtown now also boasts a lively arts and entertainment scene, which is not only a significant economic draw in its own right, but also features centrally in the place marketing efforts of Portland officials seeking to position their city in the emerging knowledge economy. In turn, this kind of arts- and entertainment-led regeneration is now being promoted as something of a model for other mid-sized cities to follow across the north-east. Thus the case of Portland suggests a more complicated

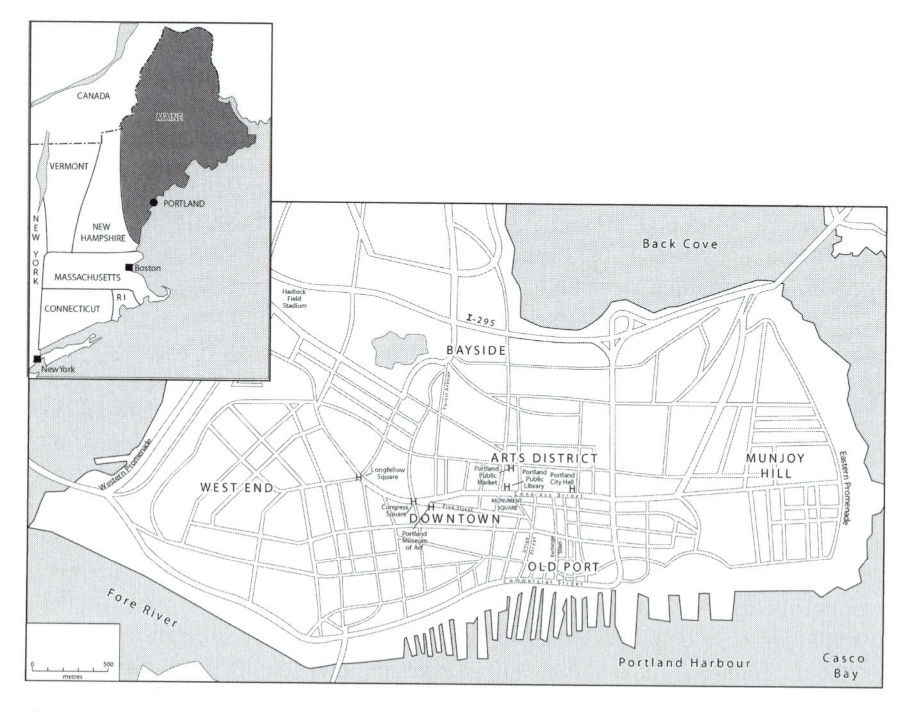


*Figure 7.1* Map of Portland, Maine. (Source: courtesy of Loretta Lees.)

geography of gentrification than implied by often throw-away references to a cascade effect.

The chapter is divided into three parts. The first explores the historical geography of Portland's urban regeneration. Far from lagging behind New York and Boston in the initial gentrification of historic districts, Portland was right there with them from the start. In this respect, as in others, the operation of the three cascade mechanisms I identify above was locally mediated and context dependent. The second part discusses how the city's plans for arts- and entertainment-led regeneration emerged and successfully took root in the wake of the last recession. Finally, I conclude by highlighting how Portland's success depended on a number of locally specific features of the city. Thus ongoing efforts to replicate its arts-led regeneration strategy in other mid-sized cities across the region will depend very much on local circumstances.

## FROM DECLINING MARITIME CITY TO THE HOTTEST REAL ESTATE MARKET IN THE UNITED STATES

Long the regional service centre for northern New England, the city of Portland was hard hit by post-war suburbanization and economic restructuring. While the population of the metropolitan region increased by 51 per cent between 1960 and 1980 to just over 180,000, the population of downtown Portland fell by 51 per cent and that of the city as a whole by 16 per cent to 61,572, as middle-class families fled the city for surrounding towns and suburbs. Such demographic shifts were tied up with significant economic restructuring as Portland's manufacturing sector, and in particular its shipbuilding and other maritime related industries, fell into relative decline. City officials responded to these challenges with a bold programme of urban renewal. In 1946, the newly appointed Portland City Planning Board issued a Le Corbusier-inspired redevelopment plan for downtown Portland that would drive city policy for the next 30 years. Slum housing in and around the downtown was demolished to make way for highways to speed traffic around the downtown and also for construction of modern public housing (Violette 2005).

The 1961 destruction of Union Station was a major catalyst for the emergence of the historic preservation movement in Portland. In 1964, local citizens organized the Portland (later Greater Portland) Landmarks Association. In this respect, Portland was far from lagging behind New York and Boston, as advocates of the cascade effect might expect. Drawing on a long, largely patrician pride in the city's 'Yankee' heritage, the Portland Landmarks Association was actually founded a year before a similarly white and middle-class coalition succeeded in getting the City of New York to establish the New York Landmarks Commission. The Boston Landmarks Commission was not established until 1975, though it was not until 1990 that the City of Portland would create a comparable local government body with formal statutory powers. As well as lobbying city government, the Portland Landmarks Association was quick to exploit the creation, in 1969, of the National Historic Register on which a large number of historic structures in downtown Portland were formally listed as significant to the nation's heritage.

The first stirrings of gentrification in Portland during the 1960s and 1970s were largely contemporaneous with what Hackworth and Smith (2001) identify as the 'first wave' of gentrification in New York City. As in New York, the capital to renovate Portland's historic properties came largely from individual private investors, supported by low interest rates and a favourable federal tax regime. In Portland, as elsewhere, demand for historic preservation and inner-city living was driven largely by the so-called new middle classes employed in the professional services industries, such as media and the law, numbers of whom doubled in Portland between 1970 and 2000 (SOCDS 2005). The first wave of residential gentrification in Portland, then, was part of wider national trends, rather than the effect of any cascade from leading metropolitan areas.

In other ways, though, Portland's gentrification was dependent on an influx of professionals from out of state. As one informant, interviewed in the mid-1970s by Sanders and Helfgo (1977: 25), put it:

> Portland was an unchanging city for a long time. In the last ten years there has been an influx of people looking for a better community in which to bring up their children. These newcomers are not nomads but high quality professional people. The University has gone from a zero to a major center.

Likewise, Knopp and Kujawa (1993) argue that the city's historic and other amenities, along with its reputation for liveability, were key to attracting – and keeping – these new middle-class workers to Portland.

To that end, the transformation of Portland's Old Port district became one of the city's distinctive selling points. As the first gentrifiers were investing sweat equity into renovating historic residences, the warehouse district of Portland's historic Old Port was also being transformed into something of a counter-cultural refuge. Attracted by its historic aesthetics and cheap rents, artists and craftspeople rented storefronts and transformed disused warehouse space into lofts, galleries and workshops. This commercial and retail gentrification soon spread and by the late 1970s the area had developed into a lively retail and entertainment centre, peppered with bars, restaurants, and small, independently owned speciality shops, which remains its strength today. A number of scholars have emphasized the important links between gentrification and this culture of urban lifestyle and aesthetic sensibility (Zukin 1982; Ley 1996).

This early gentrification of Portland was a grassroots phenomenon, led by small entrepreneurs and residents. City government was relatively uninvolved in its early stages. Instead, the city, like the big business-dominated Chamber of Commerce, was focused on keeping local taxes down and on the federally supported modernist urban renewal agenda with its associated concern for traffic management, slum clearance and flagship developments. But the threat posed to downtown retail by the opening of the Maine Mall in suburban South Portland eventually spurred the city government and its business leaders into action. Building on the early success of the small business-led revival of the Old Port, the city made some modest investments in improved street furniture, sidewalks and lighting in and around the Old Port. However, this was a far cry from the levels of local (and federal) government involvement in the contemporaneous South Street Seaport and Faneuil Hall waterfront redevelopments in New York and Boston or from how much more involved the city would become, during the 1990s, in actively promoting and providing formal support for arts- and entertainment-led urban redevelopment.

The early commercial success of this small-scale redevelopment in the Old Port provided the base for much more extensive redevelopment across downtown Portland during the 1980s. As in New York, this second wave of investment had a more corporate component and there is also some evidence in metropolitan-area-level data that the appreciation in Portland property values lagged behind that in New York, as would be predicted by an investment cascade outwards and downwards from larger metropolitan markets (see Figure 7.2). This gentrification of historic downtowns was significantly aided by the Economic Recovery Tax Act of 1981. In addition to its reduction in marginal income-tax rates for high earners, a feature of this Reagan-sponsored bill, inserted by north-eastern lawmakers to advantage their region in competition with the Sunbelt, that was less appreciated provided large tax credits – up to 25 per cent of total capital costs for the restoration of historic structures within districts like the Old Port formally listed on the National Historic Register – to encourage developers to renovate rather than demolish historic structures. With the majority of downtown Portland's building stock consisting of low-

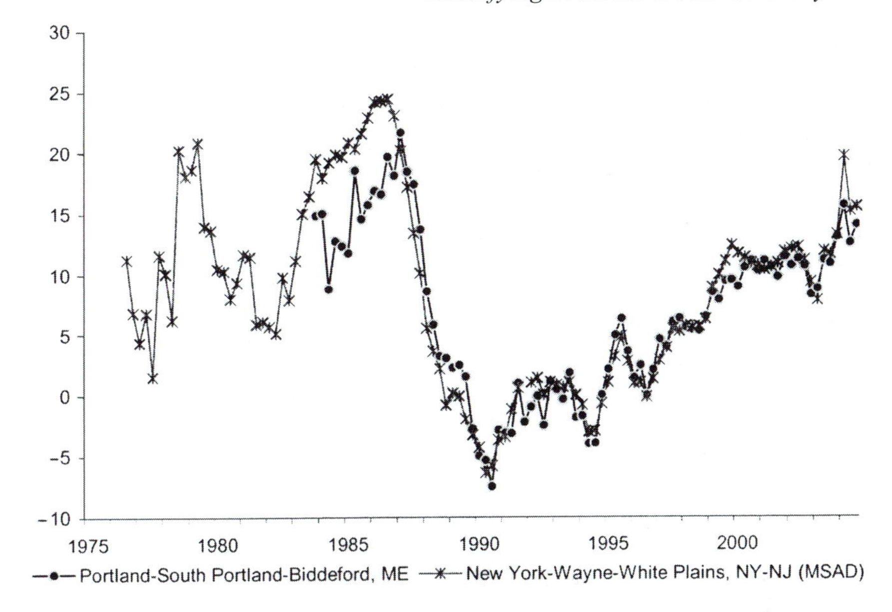


*Figure 7.2* Annual change in quarterly OFHEO house price index. (Source: courtesy of Loretta Lees.)

rise warehouses, office blocks, and storefronts dating from the period immediately after the 1866 Great Fire, developers there were well placed to exploit this tax loophole. According to one developer, it was the single most important factor in the revival of the Old Port, providing 'the financial incentive to carry the development of the Old Port beyond the sweat-equity stage' (quoted in Harr 1985: 39) insofar as larger corporate capital and institutional investment increasingly came to overtake the individualist sweat-equity of pioneer gentrification. Hackworth and Smith (2001: 467) call this stage the 'anchoring of gentrification'. Between 1985 and 1988, over 1,000,000 square feet of downtown office space was constructed in Portland, increasing the total supply by almost one-third (Schlosberg 1988: 45).

Hand-in-hand with this commercial redevelopment of the Old Port was a residential one driven by in-migration, particularly from large urban centres in the north-east. In 1990, up to 30 per cent of the residents of the most heavily gentrified downtown wards were recorded in the census as living out of state five years previously, compared with just 12.6 per cent in the Portland Metropolitan area and 12.5 per cent in the state as a whole. Data from new driving licence applications indicate that migrants to Maine during the 1980s came disproportionately from professional and managerial occupations (Ploch 1984, 1985). This is further evidence that the second wave of gentrification in Portland was the result of a migratory cascade from larger urban centres, itself no doubt driven by the success of the city in attracting back-office insurance and financial service work from Boston.

Residential gentrification in and around the Old Port had a number of effects. It halted a half-century of population decline that left the older neighbourhoods of peninsular Portland with disproportionate numbers of elderly, low-income, minority and otherwise socially marginal residents (City of Portland 1991). Although per capita incomes are now close to the city-wide average, this figure conceals significant income inequality within the gentrified downtown (for details, see Lees 2003b). As a result, affordable housing for low-income residents has become increasingly scarce. Housing costs in Portland increased dramatically during the 1980s. Adjusted for inflation between 1979 and 1989, the median contract rent increased by 47 per cent and the rise in median home value was even more dramatic at 79 per cent (City of Portland 2000). In 1986, Portland was labelled one of the hottest real estate markets in the United States (*Real Estate Investment Journal* 1986, cited in Knopp and Kujawa 1993).

Against this backdrop of feverish redevelopment, the city began preparing a new plan for downtown. But by the time that its *Downtown Vision* plan was finally published in 1991, downtown redevelopment had temporarily ground to a halt in the face of recession, which hit the city and region particularly hard. Measured by payroll employment, Maine's employment did not return to its pre-recession, 1989 levels until 1996 (City of Portland 2000). Thus, what began during the late 1980s boom as an effort by the city to manage growth and protect its historic character ended up, in the midst of deep recession, as an effort by the city 'to position itself for future growth' (City of Portland 1991:1).

## AN ARTS- AND ENTERTAINMENT-LED URBAN RENAISSANCE

Together, the late 1980s recession and 1991 plan for downtown mark something of a turning point in Portland's redevelopment. Henceforth, the city government would be much more actively involved in steering redevelopment and in promoting the creative and cultural industries as the engine of the city's future growth. Especially in the first respect, Portland's trajectory tallies nicely with Hackworth and Smith's (2001) arguments about the increasing involvement of the state in the recent, 'third wave' gentrification in New York City.

In 1991, the city created the Downtown Portland Corporation (DPC) to direct and organize downtown redevelopment. Modelled on downtown improvement districts in other cities, the city council-appointed board of directors of this public–private partnership was given the authority to spend monies raised by a tax on property within the boundaries of the district. As well as lobbying and place marketing on behalf of downtown business, the DPC manages a revolving programme of low- and no-interest loans to retain existing businesses and attract new ones. While there are important tensions within the downtown business community between, for instance, the desires of restaurant and bar owners for more liberal liquor licensing laws and of other business owners for a less raucous and more family-friendly atmosphere downtown, for the most part business leaders have been united and here enjoyed considerable support from city government in trying to restrict and control the 'undesirable' uses of downtown public space by 'deinstitutionalized mental patients, loitering groups of youth and abusers of drugs or alcohol' whose 'highly visible presence ... was having a chilling effect on pedestrian activity and business investment' (Herbert Sprouse Consulting. 1995: III, 5–6; see Lees 2003b).

Securing the public space of the city was central to the city's redevelopment strategy. Like its increasingly feverish place-marketing activities, the city's 1991 plan for downtown sought to capitalize on Portland's potential to offer urban amenities 'more common to larger urban settings while still offering people the scale, friendliness and sense of participation of smaller communities' (City of Portland 1991: 6). Though it makes no reference to developments elsewhere (apart from a passing reference to the destructive urban renewal projects 'many cities have pursued ... which dissipated their sense of community and of place'), the *Downtown Vision* plan was tapping into a wider celebration of the vitality of the city and street life (see Lees 2004). In particular, the plan called for the creation of a 'pedestrian activity district' encompassing Free Street, the Old Port and the Congress Street corridor, and the use of zoning regulations to reserve street-level property therein for retail or arts and entertainment establishments that generate foot traffic. These efforts to promote pedestrian activity were central to the city's strategy to 'set Downtown apart ... [as] an exciting living, working, and recreational environment' (City of Portland 1991: 3).

Both as an end in itself and as a means of revitalizing the downtown economy, the city has championed its cultural institutions and artistic community. In 1996, with support from the DPC, the city formally created the Portland Arts District – an area loosely stretched out along Congress Street. The idea of creating such a district to tie its cultural institutions more firmly into the regeneration agenda had been floating around City Hall for a number of years. Relative to other cities of similar size, Portland boasts a wealth of well-established cultural institutions including several colleges and universities, a symphony orchestra and a number of large and financially well-endowed museums, which city planners hoped to capitalize on as a 'magnet for bringing in tens of thousands of visitors into the Downtown who would likely not otherwise come here' (City of Portland 1991: 36). To increase employment in what it bullishly named the arts-related 'industries' (City of Portland 1991: 36), the 1991 plan had called for the city to 'promote the Downtown as the local statewide, and northern New England centre for arts and culture' (City of Portland 1991: 39). A subsequent consultants' report for the city estimated that arts institutions in Portland were responsible for attracting cultural spending with an estimated direct impact of $US33 million. The report promised that a formal Arts District to organize and promote these institutions would not only improve their economic performance and impact but also 'create a more favourable business climate in the District and in downtown Portland' (Herbert Sprouse Consulting 1995: 3). In this way, arts policy was also quite self-consciously economic policy (see Waitt, this volume). As the 1991 city plan rhetorically asked, without the arts 'would this be a city that attracts and retains residents, visitors, and businesses?' (City of Portland 1991: 37).

By explicitly linking public promotion of the arts to its economic benefits, the city pre-empted criticism of fiscal irresponsibility with its plans to back local arts groups with public money. Through direct grants as well as property tax rebates, the city supported numerous artistic and cultural institutions located in and around the Arts District. For example, it provided funding to enable the Children's Museum of Maine to relocate downtown, while in 1997 it sold a condemned building on State Street to the Portland Performing Arts rather than to a commercial tenant who would have paid a higher price as well as property taxes. In 2000, it also created the Portland Public Programme and allocated 0.5 per cent of its annual capital budget for the restoration or acquisition of permanent public art to beautify the downtown as well as other parts of the city. Following the model of Dallas and Tucson, Arizona, the city also designated the Portland Arts and Cultural Alliance, a non-profit umbrella group of local arts agencies founded in 1985, to oversee the District and spend funds allocated to it by the city and raised from private benefactors for physical improvements (Sutherland 1995).

Repeating a now-familiar cycle, cheap rents, as well as government support, attracted a number of small art galleries and related facilities to the Arts District. A 1995 cultural census of Portland counted more than 340 arts employees in the Arts District (Barringer 2004). Though rent increases accompanying the general economic upturn of the late 1990s drove several arts organizations out of the area (Blom 2000), the DPC directory of downtown organizations currently lists 37 arts-related organizations in the Arts District, while 144 of the 200 'nonprofit cultural organizations, creative industry businesses, and individual artists' in Portland recorded by the much more comprehensive New England Cultural Database (assembled by the New England Council for the Arts to document the economic impact of the cultural industry) have zip codes that place them around the Arts District in peninsular Portland. Indicating its growing importance between 1997 and 2002, employment in the arts, entertainment and recreation industries in Portland increased by 89 per cent within the Portland MSA, compared with just 16 per cent nationally over the same period.

Thus the arts and creative industries have provided one of the mainstays for growth in downtown Portland since the last recession. Under the auspices of the Arts District, investment in downtown Portland began to spread up the slope from the Old Port and along Congress Street during the mid-1990s. Formerly the hub of the city's retail and shopping activities, the Congress Street area of downtown had been particularly hard hit by the recession in the late 1980s. In 1992, retail vacancy rates reached an all time high of 40 per cent (DPC 1997: 4) and between 1990 and 1994 its tax-assessed valuation fell by more than 50 per cent (City of Portland, 2000).

Though large-scale commercial investment has since followed, it was initially private philanthropy and publicly supported arts institutions that filled this rent gap. With a loan from the DPC and a major donation from Portland resident and Intel microchip heiress, Elizabeth Noyce, the Maine College of Art was able to expand by purchasing the boarded-up Porteous building, previously home to the city's largest department store before its 1991 move to the suburban Maine Mall. Noyce also made large donations to support the construction of expanded facilities by the Maine Historical Society and the Portland Museum of Art at either end of

Congress Street, as well as providing the seed money for the $US5 million Portland Public Market. Modelled on Pike Place Market in Seattle, the Portland Public Market opened in 1998 and won the 'Downtown Achievement Merit Award' from the International Downtown Association for 'its demonstration of how philanthropy, small business, and government can create a forward-looking commercial enterprise' and contribute to 'downtown revitalization' (Portland Market 2001).

Noyce backed her philanthropic donations, made through the Libra Foundation she established, with substantial commercial investment in downtown real estate. In the early 1990s, she bought a number of large office buildings along Congress Street. She invested heavily in renovating these buildings and attracting office tenants whose workers provide a key customer base to sustain a revival of retail along Congress Street. Through her personal connections, Noyce was also instrumental in persuading the renowned Maine retailer, L.L. Bean, to locate an outlet on Congress Street. For many in the city, the arrival of L.L. Bean was a symbol that the downtown had turned a corner (Lees 2003b). For instance, the Portland City Manager declared, 'Downtown has certainly been brought back from the dead … to the land of the living' (quoted in Vegh 1996).

Whether these policies were as instrumental in its post-recession recovery as the City of Portland (2002) and others sometimes claim (Clark 2000), there is no denying that Portland did relatively well throughout the long boom of the late 1990s, though compared with the Portland-South Portland Metropolitan Statistical Area as a whole, its economic performance was much less impressive, as the city's latest comprehensive plan of the City of Portland (2002) ruefully notes. In this respect, Portland is far from alone. Recent growth in the United States has been concentrated on the suburban fringes of large metropolitan areas (see Garreau 1991). Despite the best hopes of urban planners and policymakers, inner cities still lag behind the rest of the country.

## EXPORTING THE PORTLAND MODEL OF URBAN RENAISSANCE?

As far back as 1985, Portland was being touted as 'an augury of New England's postindustrial future' (Harr 1985: 35). While New England was one of the slowest-growing regions in the United States in the 1990s, Portland stood out as an exception. The city was the only metropolis in the region to make the list of the top 20 receivers of young, single, college-educated adults from 1995 to 2000 (Miller 2004). Such rankings are important in anointing some cities as successful and worth emulating (McCann 2004):

> According to the US census, each of the six New England states lost more young, single, college-educated adults than they gained from 1995–2000. But Portland emerged as a model that could reverse that trend. It has gained recognition for everything from distinctiveness – its working waterfront and converted warehouses – to world class eateries.
>
> (Miller 2004)

Increasingly, other small cities in New England, such as Pawtucket, Rhode Island, and Burlington, Vermont, are following the Portland model on how to attract young professionals.

However, Portland as a model of success may be hard to replicate. Because politicians and consultants want easy off-the-shelf solutions, they tend to ignore the way that the success of the Portland model depended upon the particular historical geography that I have sketched out above. Portland's urban renaissance was influenced by three important contingent factors. First, unlike many other 'rust-belt' cities, Portland's economy was based on merchandizing, services and tourism. As a result, it had (and has) a more diverse labour market and its employment base did not suffer the severe economic decline that other more industrial and manufacturing cities did. Indeed, during Portland's urban and economic decline, banks and insurance companies remained in the city and preserved its place as a regional service centre in the urban hierarchy (Schlosberg 1988). Some even acted as a further catalyst for growth; for example, in 1970, Casco Northern Bank built a new headquarters in Portland, the first major construction project in the city since the last Casco Northern building went up in the 1930s. It was a risky venture, as John Daigle, chief executive of Casco Northern Bank said: 'We thought it might stimulate some economic activity, but we didn't really know what to expect' (cited in Schlosberg 1988:45).

Second, downtown Portland has gained significantly from the philanthropy of millionaire Elizabeth Noyce. Noyce began buying up major downtown buildings and properties in 1995, playing what Vegh (1996) calls a 'fairy godmother role'. Her most notable purchase was the former Maine Savings Bank plaza, the Maine Bank and Trust building and the 'time and temperature' building, all on Congress Street. Suddenly, she controlled one-quarter of all the city's office towers and one-tenth of all the office space in the city. As Vegh (1996), notes '[i]t was one of the city's biggest real estate deals ever and came at a time when vacant storefronts and empty offices marked a downtown hit hard by the recession'. Although Noyce died in 1996, her other projects were completed (for example, a downtown parking garage and the Portland Public Market). Many of the shopkeepers who subsequently invested in Congress Street only did so as a result of Noyce's pioneering reinvestment.

Third, Portland has an unusually strong grassroots entrepreneurial tradition. The renaissance of the Old Port was largely the result of individual initiative and entrepreneurial energy rather than city or state encouragement. It wasn't until 1997 that municipal officials actually sat down with Old Port leaders, at what became known as the Old Port Summit, and committed public money for Old Port improvements such as public safety, cleanliness, marketing and so on. Rather than directing the redevelopment of downtown Portland, the City has been in the position of trying to steer, through the planning process, and capitalize on, through its place-marketing efforts, a commercially led and substantially grassroots process of reinvestment (Lees 2003b). But this has led to criticisms that, in the absence of effective strategy, stasis or even erosion may occur in Portland (Barringer *et al.* 2000). Portland appears to lack some of the leadership, networks and strategies necessary to capitalize on its success. For example, the city does not fully fund a cultural liaison for the arts community and most organizations such as the Convention and Visitors Bureau exist with volunteer membership. This in itself could well be a problem for the development and sustainability of the so-called 'creative economy' in Portland.

More recently, as well as providing a model of successful urban renaissance for other small towns and cities in New England, Portland is also being used as 'a test study' of how to embrace Richard Florida's (2002) 'creative economy' (Miller 2004). Embracing the creative economy makes sense to Portland's politicians and policy makers, for 'the creative class' is the demographic that attracts upscale restaurants and stores and art galleries: in essence, the renaissance features that Portland possesses but could just as easily lose if this particular demographic were to leave the city. In May 2004, almost 700 people from throughout Maine and New England came together at the Bates Mill complex in Lewiston, Maine, to explore the creative economy in Maine. The occasion was convened by Maine's state governor, John E. Baldacci. In his keynote address at the meeting, Richard Florida praised Maine's creative and entrepreneurial spirit, quoting his mentor Jane Jacobs who said 'new ideas require old buildings' (Maine Arts Commission 2004: 8). He gave new credence to, and indeed has extended the life of, the processes of gentrification that have occurred in Portland for the past 40 or so years. The Governor and his advisers have taken on board Florida's persuasive argument that city (and indeed regional) competitiveness is connected to the central location of not just the professional middle classes but also of bohemians and gays, in other words through gentrification. The belief that a creative workforce will lead the way in terms of urban and economic regeneration and development is so strong that, in late 2004, Governor Baldacci accepted recommendations from a state-wide committee to foster 'Maine's creative economy' (see Maine Arts Commission 2004). Governor Baldacci believes: '[t]he Creative Economy is a catalyst for vibrant downtowns, expanding cultural tourism, encouraging entrepreneurial activity and growing our communities in a way that allows us to retain and attract creative workers … an investment in a stable workforce and competitiveness' (Maine Arts Commission 2004: 3). This is the state support for post-recession gentrification that Hackworth and Smith (2001) discuss, even if 'gentrification' is disguised as part of a wider agenda. But the fostering of the creative economy in Portland and, indeed, throughout Maine must be the subject of further study.

To conclude, the story of urban renaissance in a small city like Portland complicates the cascade idea in at least three ways. First, historically, Portland did not lag behind New York and Boston in the urban renaissance game but was right there with them, perhaps even ahead of them. Second, although not a high-order or first-tier city, Portland has become the regeneration model for towns and cities elsewhere in New England. Third, Portland's success, in terms of its urban renaissance, is due to a series of historically and geographically contingent reasons – its place in regional/state city hierarchy; its strong local entrepreneurial base; its success in the regional service economy and in attracting back-office services from Boston; and as the State of Maine's only metropolis – giving a strong economic base on which to grow the arts. If these features aren't in place elsewhere there is no reason to expect that city governments seeking to replicate the Portland model will succeed.

# 8 The re-construction of a small Scottish city

## Re-discovering Dundee

Greg Lloyd, John McCarthy and Deborah Peel

Devolution has provided an opportunity for Scotland to challenge its status as a 'stateless nation' (McCrone 1994). It has been described as involving the establishment of quasi-autonomous government in those regions that have what may be considered to be a national cultural identity, whilst not involving the transfer of full legal sovereignty (McNaughton 1998). In practical terms, devolution has involved the transfer of specific but differential legislative, executive and financial powers from the United Kingdom government in Westminster to elected bodies in Scotland, Wales and Northern Ireland, with these elected bodies remaining subordinate to the legal framework defined by the centre (Bogdanor 1999). Since devolution, the Scottish Parliament and Scottish Executive have defined a distinctive Scottish political agenda which has asserted a number of key issues which are held to be core to modern Scottish identity. Importantly, the underpinning principles of devolution and decentralization are manifest in Scotland in the wider discourse of subsidarity and the search for identity (Bond and Rosie 2002). Moreover, place-marketing strategies and city-cultural identities are being promulgated against a backcloth of globalization.

There is considerable emphasis on the broad policy arena concerned with social justice; indeed, this is the policy context responsible for discharging land-use planning and development priorities. Thus the promotion of social inclusion and social cohesion has highlighted the urban agenda. In recent times, this has focused on addressing the causes and issues associated with area-based regeneration and identifying and highlighting client-based priorities in Scotland's principal cities. A parallel national policy priority is that of promoting economic competitiveness and growth. Together, these main national political objectives have placed the urban agenda centre stage within the broader discourse of sustainable development. Yet, how do individual cities, faced with the challenges of addressing social exclusion and promoting economic growth, rationalize and accommodate these policy priorities?

The experiences of the city of Dundee, which is located on the north-eastern seaboard of Scotland, offers some pertinent insights (see Figure 8.1). With a local population of 143,000 it is Scotland's fourth largest city. Dundee City Council covers a geographical area of 24 square miles and it is the smallest local authority area in Scotland (Dundee City Council 2004a). Many of the city's traditional

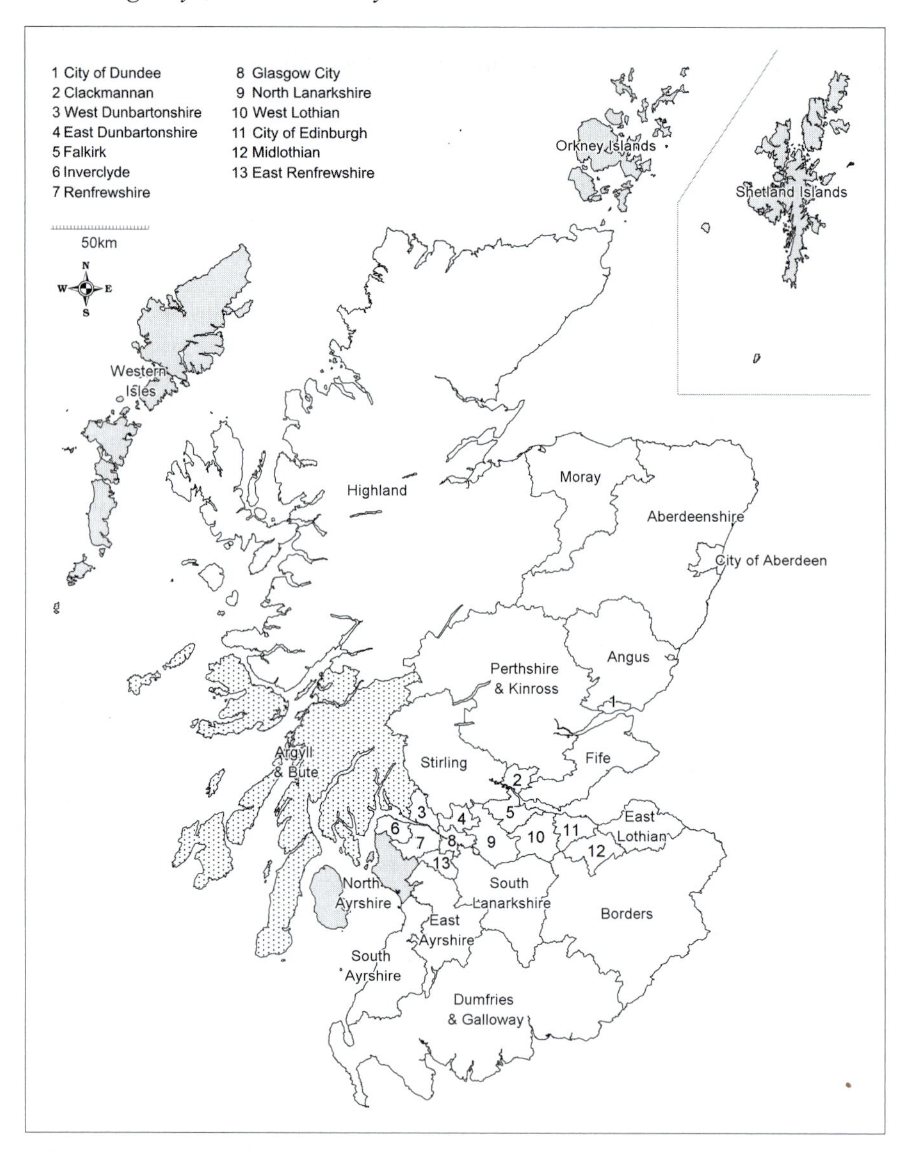


*Figure 8.1* Dundee in its economic context. (Source: courtesy of Greg Lloyd, John McCarthy and Deborah Peel.)

industries have had difficulties in adapting to new technologies and markets. As a consequence, the city continues to experience absolute population decline and relatively high unemployment and it demonstrates evidence of social deprivation, together with significant areas of vacant and derelict land (Scottish Executive 2004a). Significantly, however, the municipal authorities, along with the business and voluntary sectors, have engaged in a major improvement in the city's external image. This has focused on the revitalization of the city centre, with long-term investment in Dundee's retail and cultural facilities. Moreover, clusters of 'new economy' sectors, including biotechnology, medical science and multimedia software development are emerging. What can we learn from the city's attempts to rediscover itself?

In part due to the lack of sustained success of urban regeneration policies and programmes over the past 30 years, the idea of sustainable development has become an important theoretical framework for considering long-term, integrated and strategic regeneration policy. As a consequence, the principles of sustainable development underpin all the Scottish Executive's policies (Scottish Executive 2002b). In this context, Carley and Kirk (1998: 3), for example, have argued that 'poverty, social alienation and urban dereliction are incompatible with sustainable development' and that developing a sense of community and high-quality public goods may help to overcome these problems. Moreover, they also identify democratic participation as integral to successful processes of management and governance, along with the 'sophisticated interaction between public, private and voluntary sectors' (p. 4). In terms of promoting sustainable urban regeneration, they advocate longer-term solutions, steady and strategic processes, positive reinforcement and innovation and the physical, economic and social reintegration into the broader townscape of hitherto disadvantaged estates or neighbourhoods. This highlights the importance of taking account of the broader economic and structural changes taking place within particular cities and reconciling these with the local impacts on communities, whilst simultaneously being aware of the opportunities for reconstruction being put into place by the cities themselves. Further, this introduces a duality of social and economic trajectories which any arrangement for city governance has to work with, for and towards.

This chapter explores the contemporary processes of city reconstruction in Dundee, a segmented city characterized by economic restructuring and industrial contraction, and indices of high social deprivation. This relatively small city therefore offers an example of different planning and development agendas being executed in very specific circumstances. In particular, we consider how the city is shaping its institutional identity and place image in the context of its relationship with the other principal Scottish cities. We address the significance of institutional sponsorship, coalition and partnership working; processes of visioning; and particular interventions, based on cultural regeneration and learning. Reflecting wider debates associated with the reconfiguration of governance arrangements and the re-scaling of policy regimes (MacLeod and Goodwin 1999), the city of Dundee is responding to contemporary influences of change at a number of levels. There are three principal scalar dimensions to these processes of reconstruction which bring together (1) national aspirations for the city, notably through the National Planning Framework; (2) the development of a city-regional perspective; and (3) an emphasis on the promotion of localized city development. This scalar sensitivity to place-making provides an important layering for understanding the nature of the city's rebirth and provides the structure for this chapter.

## THE CONTEMPORARY URBAN AGENDA IN SCOTLAND

Although the Scottish government asserts that '[g]rowing the economy is our top priority' (Scottish Executive 2003a), social justice is also held to lie at the heart of the contemporary political agenda. This appears to place an important emphasis on the fairness of outcomes and the spatial redistribution of opportunities and life chances (Bailey *et al.* 2003). Effectively, this political priority is presented as being about 'closing the gap':

> We are committed to building a better Scotland, where a child's potential, and not their background or postcode, will decide their future. We want to see a Scotland where every neighbourhood is a safe, attractive place to live, work and play, where public services meet the needs of people, not the demands of the organizations which deliver them, and where social justice is a right, not a privilege (Scottish Executive 2002c).

The implications of this agenda, however, are that the synthesis of different policy sectors (for instance, housing, education, employment or transport) will differ depending on time, timing, context and circumstance. These differential interdependencies will then impact on how individual cities work towards making the practical connections and high-quality places demanded. Community planning is an example of such integrated practice which was put in place under the Local Government in Scotland Act (2003b). Importantly, community planning provides the statutory basis for the management of community regeneration and the delivery of improved public services. The main aims of community planning are described as 'making sure people and communities are genuinely engaged in the decisions made on public services which affect them; allied to a commitment from organizations to work together, not apart, in providing better public services' (Scottish Executive 2003b). This highlights two important strands of contemporary policy thinking, namely, participation and partnership.

The refinement and implementation of national economic development policy was informed by the deliberations and recommendations of *The Cities Review* (Scottish Executive 2002c). This argued the case for a more explicit reliance on Scotland's cities as economic engines of national growth and development in a European context, particularly with respect to the asserted policy intentions of securing greater competitiveness, regional balance, social and community justice and environmental sustainability. The primacy of cities within Scotland has been reflected in subsequent key policy documents, including *A Framework for Economic Development* (Scottish Executive 2004b) and the *National Planning Framework* (Scottish Executive 2004a).

### Planning at a National Scale

The production of the *National Planning Framework* is highly significant for individual cities since it provides the strategic spatial framework for public policy in

Scotland. In the foreword to the *Framework*, the (then) Minister for Communities asserted that its publication:

> marks an important first step in recognising the challenges we face in Scotland's long-term territorial development. But it is a perspective, not a prescriptive master-plan or blueprint. It is a planning document that analyses the underlying trends in Scotland's territorial development, the key drivers of change and the challenges we face. Describing an issue in spatial or territorial terms does not remove the dilemmas we face in policy or spending decisions. The framework is, however, one of the factors we will take into account in coming to difficult decisions on policy and spending priorities as well as providing a context for development plans and development decisions.

Significantly, the *National Planning Framework* drew specific attention to the issues and implications for Scotland arising from the enlargement of the European Union (EU), the European Spatial Development Perspective and the requirement for a spatial planning framework to inform future EU funding regimes (Lloyd and Peel 2003). Significantly, the key elements of its development strategy rested on cities operating as the main economic drivers in national economic space, whilst taking full account of Scotland's peripheral location in an expanding European spatial economy. The repetition of the economic mantra, allied to the sustainable development rhetoric, sets particular parameters to city development in Scotland.

The *National Planning Framework* set out the policy ambitions for Dundee as they impact on the national strategic agenda. This suggested that the

> challenge for Dundee is to reverse population loss. Great strides have been taken in the past decade in improving the quality of the city centre, enhancing cultural facilities and establishing new knowledge-economy clusters. Many young people come to the city for further education. The challenge is to encourage a higher proportion of them to stay. The strategy for the Dundee city region will be to promote regeneration, neighbourhood renewal and further improvements to the quality of urban living within the city boundary. Priorities include the redevelopment of the Waterfront and the further development of knowledge-economy clusters such as the Digital Media Campus, the Tech Park, the Medipark and the Scottish Crop Research Unit and the improvement of public transport services to these growth areas. Reducing the rail journey time to Edinburgh to under an hour would help to attract high value jobs to the city (Scottish Executive 2004a: 76).

This quotation brings together the difficulties of city development, reconciling inherited issues of industrial contraction and the processes of restructuring and preparing to realize anticipated opportunities and potential growth for the city. Here degeneration and regeneration agendas come together forcefully. These embrace a spectrum of scales and sectors which can include consideration of individual neighbourhoods and streets, the city centre through to the regional functional economic zone of the city. Moreover, development in Dundee necessarily involves communities of place, interest and identity, and issues of connectivity within the national spatial economy, particularly in terms of the city's relationship with Edinburgh.

## Planning at a City-Region Scale

The *National Planning Framework* set out a case for the management of Scotland's spatial economy through the definition of city-regions. The concept of city-regions is currently much in vogue (Tewdwr-Jones and McNeill 2000). It reflects contemporary thinking that links the spatial extent of closely linked economic activity, rather than a more instrumental focus on the city and its jurisdictional boundaries (Simmonds and Hack 2000). This will be highly contested according to circumstance. Thus, the *National Planning Framework* asserts a:

> broad recognition of the need for a more nuanced understanding of the geographies and scales at which policies are played out on the ground in our neighbourhoods, cities and nation state. In developing strategies, in allocating resources and in delivering services, the [Scottish] Executive needs to have both a clear sense of how geography influences the functioning of our economy, society and environment, and knowledge of where [Scottish] Executive policies will impact (Scottish Executive 2004a: 4).

The contemporary prescriptions associated with the contested concept of city-regions present a context for Dundee in which the city is linked to its functional housing and labour markets in its primitively demarcated geographical hinterland (See Figure 8.1). There are difficulties here, however, in articulating this into practical physical expression as it is challenged by the parameters set by local government boundaries, arrangements for local government finance and inherited perceptions of identity and reputation. Here, the old construction of knowledge and fixed geography collides directly with the more contemporary fluid conceptions of spatiality, connectivity, quality and knowledge flows. Nonetheless, it also offers an opportunity to present (and construct) Dundee as a 'regional' centre of activity and this changes the perceived role of the city in a wider strategic and more dynamic setting.

Interestingly, in the Dundee context, a city-regional perspective has been long established in policy thinking and practice. It was first promulgated in the later 1940s and earlier 1950s through the Tay Valley Plan. Its aim was specific: to 'form a background for urban and rural reconstruction physically, socially and economically, which must ensue over the next 30 to 50 years in order to roll back the tide of rural decadence and urban concentration' (Lyle and Payne 1950: foreword). The main objectives of the Plan were to ascertain the physical, social and economic advantages and disadvantages of the region; to demonstrate how the advantages of living in the region may be maintained; and, to secure the cooperation of the statutory planning authorities in the region to coordinate their various schemes. Here, the context was the (then) new 1947 Town and Country Planning Act. The contemporary agenda of spatiality, in the form of the *National Planning Framework*, and the overarching coordination of public services through community planning, provide a very different context.

Today, Dundee is again in the process of reconstructing itself and is seeking to manage the parallel challenges associated with its economic and social duality. In practical terms, this is being effected through a skein of institutional, socio-economic and cultural initiatives. These comprise both hard (physical) and soft (marketing and imaging) interventions planned over the short, medium and longer terms. This requires a more detailed analysis of planning at the local scale and particularly the arrangements for partnership and governance in the processes of reconstruction.

## DUNDEE: CITY AND GOVERNANCE

Dundee has a long-established heritage as a whaling centre, manufacturing city and a trading locale. Its coastal and estuarine location proved to be highly significant in its earlier economic fortunes and was the bedrock of its industrial expansion in the nineteenth and early twentieth century. This was based on the manufacturing of goods based on its trading links associated with flax and jute and the processing of local agricultural produce and soft fruits, and industrial activities associated with whaling, trading, shipbuilding, engineering and ancillary port activities (Whatley 1991). In the earlier part of the twentieth century, however, changing market conditions together with relatively high-cost modes of production and distribution contributed to a process of endemic decline (Whatley 1991: 2). The severity of the economic restructuring of the city was reflected in its social and demographic conditions, particularly with respect to mortality, housing and health (Rodger 1996). This led to an ongoing, even escalating, process of city-wide decline which persisted into the 1990s. Today, those processes of change sit alongside different pulses of regeneration and development.

Since 1945, a number of parallel interventions has been put into place to address the issues associated with the processes of industrial restructuring and economic contraction. These included early attempts at managing Dundee through city-regional planning, which were driven by the introduction of statutory land-use planning legislation (Lyle and Payne 1950); a specific focus on city rebuilding through the preparation of a masterplan for Dundee (Dobson Chapman 1952); a series of physical renewal and environmental improvements to different neighbourhoods in the city (McCarthy and Pollock 1997); the development of specific industrial improvement areas promoted through national legislative opportunity; and a catalogue of activities precipitated by active partnership working in the city (Bazley 1992). As documented in detail by McCarthy (2006), for example, these interventions subsequently consolidated around a more 'official' entity in the form of the Dundee Partnership which was created in 1991.

The Dundee Partnership has remained an informal institutional arrangement amongst the key players in the regeneration of the city's economy and it has provided a link to other organizations such as the universities, the private sector, trade unions and government agencies. Thus, although it is primarily public sector-led and dominated, there are strong links with the business sector. Although individual partners have changed over time (Fernie and McCarthy 2001), the partnership seeks, principally through moral persuasion and active networking, to establish a common agenda for action and spending by the individual participants and partners within the Dundee community (Lloyd and McCarthy 2003).

It is clear that the nature of the Dundee Partnership's action agenda was influenced by two sets of ideas. On the one hand, there was the proven experience of collaboration and joint working between the different stakeholders engaged in very serious and difficult development activities in the city. This pragmatic and experiential learning of 'what works' in addressing urban degeneration led to a realization that partnership working could be effective within tightly-defined regeneration agendas. On the other hand, the idea of the Dundee Partnership conformed to emerging supply-side economic development practice where the 'thickness' of institutional arrangements was held to be an important metric in the regeneration equation (Amin 1999). Thus, the Dundee Partnership mobilized different agencies, partners and organizations onto a relatively more corporative footing to address the common problems of the city. Subsequently, it has played a key and demonstrable role in the regeneration and redevelopment of the city (Lloyd and McCarthy 2003).

In practice, the Dundee Partnership has become the principal mechanism for information sharing, policy prioritization and the establishment of a common agenda for action for the recovery of Dundee. Effectively, it has provided a means by which the key players could be encouraged to sign up to a corporate vision and has formed an important arena for developing a common understanding of the different interventions taking place. It effectively acts as a 'prism' in that all individual actions are taken in the full knowledge of other institutional activities. This suggests that the added-value of the Dundee Partnership is based on coordinating a knowledge base, establishing a more consistent and focused emphasis on the problems of the Dundee economy by creating a momentum of action to address the problems of Dundee as a whole and by establishing a consensual vision for the future. As a city development mediator, the Dundee Partnership has attempted to avoid potential inter-organizational conflicts and to work on the basis of a mutually beneficial strategic agenda (Lawless 1994). A synergistic approach thus forms an important strand of Dundee's resurgence. Thus, individual partners are able to pursue particular agendas, guided in their action by a master vision of the city's redevelopment. Today, the Dundee Partnership continues its activity in the context of community planning. Two principal strands are evident.

On the one hand, the contemporary urban authorities and business leaders in Dundee are addressing the inherited effects of economic restructuring through initiatives associated with the promotion of social justice and a locally sensitive interpretation of community planning that reflects the priorities of its distinct neighbourhoods. This corresponds to a more localized scale of political intervention. These policy measures are attempts at recovery, based on a perceived need to

*Figure 8.2* Desperate Dan, part of the urban design enhancements in the city centre. (Source: courtesy of Greg Lloyd, John McCarthy and Deborah Peel.)

regenerate a sense of community, improve the health and well-being of residents and develop an appropriately skilled workforce. On the other hand, the urban authorities are actively engaging in attempts to revitalize and diversify the economy at the city-regional scale, and in the context of global competitiveness. This is involving processes of re-imaging and re-branding of the city of Dundee, around, for example, its status as a regional retail area, a tourist destination and a centre of academic excellence.

In terms of the implementation of the Dundee Partnership's broad vision for the city's regeneration, this is essentially project-based. There have been particular physical city-centre projects, concerned, for example, with the complete rebuilding of the retail fabric, urban design initiatives and the erection of public art. These reflect the city's industrial heritage in terms of jute and journalism (see Figure 8.2). The promotion of social justice has been addressed through area-based and client-focused projects. This has included, for example, the physical renewal of housing stock in addition to projects to tackle particular social issues such as drug abuse. A principal feature of the Dundee Partnership is that it asserts a clear understanding that the 'corporate' outcome of its work is the overall recovery of Dundee for the benefit of all its residents. Thus, the particular history, demographics and culture of the city have moulded and formed the operational characteristics of the Dundee Partnership in terms of the city's social and economic revitalization. It is attempting to throw off its reputation (and implicit stigma and culture) of decline. In effect, Dundee is in the process of reconstructing itself in a number of physical and psychological ways.

## Planning at the City Scale

The Urban Task Force in England asserted a powerful case for visioning and the role of visions in facilitating an urban renaissance (Department of the Environment, Transport and the Regions 1999). Nonetheless, academic debates have advocated caution with respect to the deployment of visioning and visions in contemporary policy agendas, as the concepts are complex and contested and can be deployed in different ways (Shipley *et al.* 2004). The principal drawbacks centre on the contested interpretations of visions, negative effects of raised expectations or threat-making (apocalyptic visioning) and the potential displacement effects of ongoing policy innovations in specific areas (Peel and Lloyd 2005). Notwithstanding this caveat, however, visioning as a technique has assumed a prominent role with respect to current policy formulation in Scotland.

Significantly, the *Review of Scotland's Cities* (Scottish Executive 2002a) affirmed and informed the growth-oriented view and policy priorities of the Scottish Executive. These included arguments promoting the role to be played by the principal cities within the national economic strategic agenda. Moreover, this reasoning was reflected in the Scottish Executive's Partnership Agreement (2003a) which proposed that the individual cities would prepare 'growth strategies' that would maximize their particular unique characteristics and opportunities. Importantly, this coalition government political statement asserted:

> city policies are not simply about redistributing resources from successful to less successful places, rather, city, or place, policy is also essential in dealing with market and policy failures that limit productivity growth. City policies have to be creative as well as redistributive and they have national as well as local benefits. They support local change, creativity and adjustment which are all essential to wider national progress (Scottish Executive 2002a: 273).

Each city was invited to design its own destiny. Moreover, this process involved the preparation of ten-year 'city visions' which attracted national financial support through a Growth Fund (Peel and Lloyd 2005). Importantly, these particular visions were generated through partnerships between councils, community planning partnerships, partners and other stakeholders taking forward existing collaborative processes (Scottish Executive 2002a).

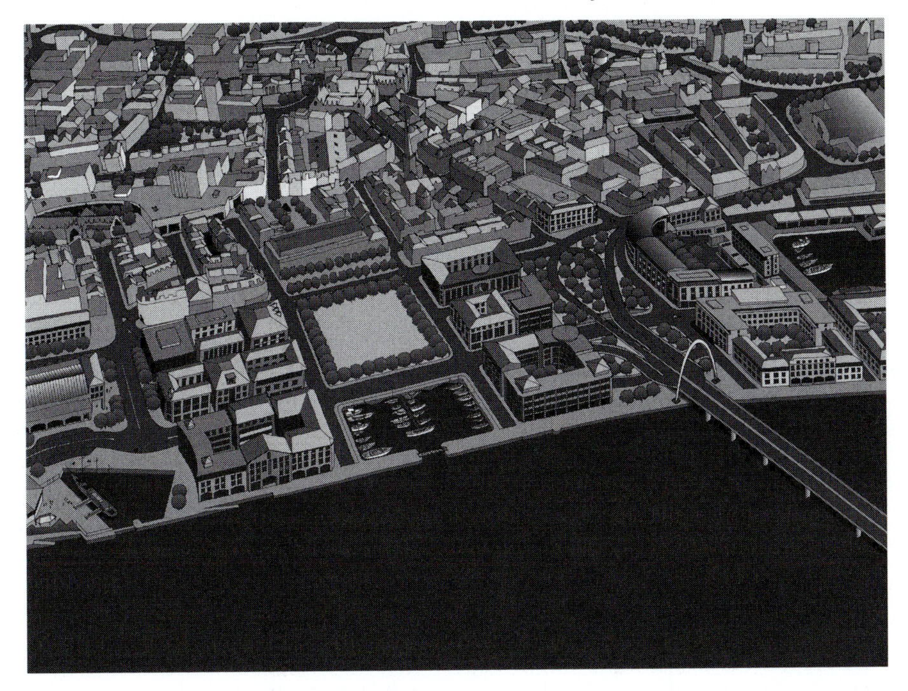

*Figure 8.3* Marketing the waterfront vision in Dundee. (Source: courtesy of Greg Lloyd, John McCarthy and Deborah Peel.)

In launching the city vision for Dundee, the Leader of the Administration stated:

> Our city vision will help accelerate the successful transformation that Dundee has undergone in recent years. We will be working closely with our public and private sector partners to ensure that the whole community is involved in driving the city forward.

The Dundee city vision document explicitly alerted the Scottish Executive to the challenge of building a holistic city-region vision when it is currently isolated from its hinterland by the existing administrative boundaries. Here, Dundee is articulating the city-region zeitgeist and seeking to connect with the realities of its labour and housing markets and journey-to-work patterns. Further, part of its vision is also presented in terms of a vision for a 'whole city' and one that is not hampered by existing governance and funding constraints (Dundee City Council 2003a). In practical terms, the city vision echoes the enthusiasm and established experience of partnership working, the emphasis on delivery and the associated processes of community planning. Dundee explicitly acknowledges the importance of community planning in the preparation (and execution) of its vision, which is presented as a 'framework and the impetus for a successful city'. In particular, it articulates a focused programme for change, which is based on the redevelopment and regeneration of the city's waterfront (see Figure 8.3).

In terms of the reconstruction of the city, the city vision presents a positive agenda for physical and economic change. This celebrates Dundee's past – particularly the trading links associated with the River Tay – and sets out an ambitious refurbishment programme for a key development asset, the waterfront. This reflects global interest in waterfront regeneration (McCarthy 2006). Moreover, the vision challenges any negative perceptions of a city that has only experienced industrial and economic contractions in its recent past, and promotes an agenda for the renewal of the city profile in its strategic regional setting. Importantly, it lays claim to contemporary post-industrial economic ideas, such as a city based on tourism, health and education. Celebrating its marine heritage, for example, the city hosts the RSS *Discovery* and the *Unicorn*. In addition, it is becoming an important destination for cruise liners and marine tourism. Dundee is also the site of the first Frank Gehry building in the UK – the Maggie Centre. The city has the highest number of further education students per capita and the redevelopment of the universities' real estate is indicative of Dundee's commitment to become a world-class city of learning (Peel 2006). Each of these initiatives is important in its own right and forms part of a complex mosaic of projects which is being implemented at a localized scale, by appropriate stakeholders, and within a strategic context.

## Planning at a Quarter Scale

The construction of a post-industrial economic base for the city of Dundee depends, in part, on the creation of a new image and reputation. More importantly, it has required the provision of new infrastructure and buildings and coordination by the city's Planning and Transportation Department. The zoning and marketing of a 'cultural quarter' and a 'digital media park' are clear examples of how the reuse and redevelopment of former industrial sites are being promoted as ways to secure sustainable regeneration. These individual but coordinated initiatives illustrate how Dundee is working across traditional public-sector boundaries to integrate public relations, campaigning and marketing with environmental enhancements, economic development and city planning. The development of the Cultural Quarter epitomizes this inter-professional and partnership approach.

The original impetus for the Cultural Quarter was provided by the City Centre Initiative for Dundee, which emphasized the need to enhance cultural activity in the city centre in the late 1980s, with a key objective being the development of a new arts centre. The development of the Dundee Contemporary Arts Centre (a cinema and arts complex with a restaurant and theatre) in turn prompted the City Council to commission a feasibility study for a formally designated cultural quarter. This was identified as necessary in order to establish the credibility of the city within the wider cultural arena. Moreover, the city's Arts Action Plan (Dundee City Council 1997) underlined the need for a strategic approach to cultural provision. The subsequent feasibility report (EDAW and Urban Cultures 2000) emphasized the significance of ensuring a critical mass of cultural production- and consumption-related activity. Public consultation on the notion of a cultural quarter – where uses were concentrated rather than dispersed – produced some opposition from cultural uses outside the proposed quarter. It was felt that designation would detract from their own viability. Some local residents expressed concerns that large 'theme pubs' might be attracted to the area as a result. Nonetheless, businesses within the proposed quarter were supportive of designation in view of the perceived potential increase in footfall within the area.

The Cultural Quarter was subsequently designated in the Local Plan (Dundee City Council 2003b) which indicates that 'cultural and related leisure and business activities' will be encouraged in the area, in addition to speciality retailing or other small independent uses (Dundee City Council 2003b: 15). Significantly, there are also strategic links to wider policy. For instance, the City Council's Corporate Plan (Dundee City Council 2002) aims to promote key sector initiatives including the creative and knowledge-based industries. Further, the Economic Development Plan (Dundee City Council 2001) prioritizes such industries in view of the potential for expansion. This is linked in particular to the strength of the city in new technology areas such as digital media, with a thriving computer games industry, including companies such as Rage Games and Vis-Interactive, and expertise in computer games technology at the University of Abertay. Importantly, development of the Cultural Quarter is identified as a key strategic goal of the Plan. Following the publication of the *National Cultural Strategy for Scotland* (Scottish Executive 2000a), Dundee City Council was the first local authority in Scotland to adopt a local cultural strategy which seeks to relate its cultural planning activities and processes to its broader corporate objectives. Significantly, this identifies the potential contribution that cultural activities can make to broader economic and social well-being in the city.

The vibrancy and cachet generated by its designation and proactive marketing as a Cultural Quarter would appear to have contributed to developments outside the area. In particular, a Digital Media Park is planned on ex-railway land adjacent to the Cultural Quarter. This strengthens existing links to the teaching and research capacities at the Universities of Abertay and Dundee. Moreover, the city continues to demonstrate a *'hot spot'* reputation in the digital media sector and there are indications that the choice of Dundee as a location by digital media firms is increasingly linked, in part, to progress made in changing the city's image. A masterplan has been drawn up and outline planning permission was granted in 2004. It is anticipated that the new Park will provide up to 300,000 square feet of new-build office and specialist-serviced accommodation over a ten-year period, including flexible provision of a range of 'incubator' uses. Such interweaving of cultural and knowledge uses is clearly stimulating associated entrepreneurial activity together with a shift in economic activity that is consistent with the knowledge economy which is explicitly written into the national economic strategy (Hepworth and Spencer 2004). Thus, past and current interventions have emerged and intertwined to create different scalar initiatives which would appear to offer the potential to create opportunities to contribute to the city's reconstruction.

## CONCLUSIONS

Dundee is undergoing a complex period of change, including continuing industrial contraction, economic restructuring and moves to promote a sustainable future in terms of its economic base, its communities and environment and its particular reputation and culture. Interestingly, devolution and the associated search for identity in Scotland has precipitated new thinking and debate around the appropriateness of different scales of intervention, the importance of connectivity and quality, the nature of spatial relations and, importantly, the image of locality. Often referred to as 'Scotland's fourth city' and sometimes overlooked when compared with Edinburgh, Glasgow and Aberdeen, Dundee faces a particular set of spatial, structural, institutional and cultural issues. What can we learn from the Dundee experience?

First, there is an emphasis on the preparation and implementation of a Dundee city vision to articulate an agreed agenda for change. This represents an action plan based on a positive partnership between the city council, the broader Dundee community planning partnership and other partners and stakeholders associated with city-wide development. This reflects the Scottish policy agenda for the city and rests on an accepted process of coalition building in Dundee's anticipated urban development and management. Community consultation forms an important strand of this municipal visioning process. Moreover, it is clear that the city leaders are making attempts to situate Dundee with respect to global processes of change. Experience suggests that there is an emphasis on action and delivery to sustain the momentum of change in the city.

Second, Dundee is reconstructing itself in the context of its broader city region. This is a significant project and, again, reflects national policy aspirations and intentions, and the wider European agenda of structural change. In the Dundee case, it also involves the immediate challenge of constructing a holistic city-region vision when it is currently isolated from its hinterland by institutional and geographical boundaries. This raises important questions about administrative boundaries, financial arrangements and policy relations in processes of revitalization, and the importance of strategic cooperation in the city regeneration process. Here, considerations of scale are important. Dundee is a relatively small city in terms of its geographical size, its demography and its economic base. The interest in the city-region idea, the identification of the city within the national spatial framework and the opportunity to devise its city vision are ways in which the issue of scale can be countered. The city-region in particular is a way in which Dundee can assert its associated web of functional relations in terms of housing and labour markets. It allows the city to clarify its links with adjacent towns and should allow it to present an image to the world of a much larger-sized urban centre. Its physical location on the north-eastern seaboard of Scotland, effectively at a pivotal position between Edinburgh, Glasgow and Aberdeen, would suggest a potential for greater synergies within Scotland's national spatial economy.

Third, within Dundee there is an emphasis on a post-industrial portfolio of interventions associated with policies for community, cultural, educational and environmental regeneration. Identity is all-important in this line of reasoning. Clearly, the new urban agenda in Scotland has asserted a key role for the four principal cities, and this affords Dundee an opportunity to challenge its image as a dour, post-industrial place. The assertion of the national spatial planning framework clearly promotes Dundee into a position of importance in the national economic agenda. In this context, the city is responding through its attempts to improve its relative economic performance, address its social justice agenda and promote its environmental image. Here it is possible to argue that the national strategic position has allowed a rethinking around the role of Dundee and this is reflected in the different scalar initiatives associated with the city itself. The Cultural Quarter is an example of the forging of a new economic base which spans biotechnology, healthcare, digital media and virtual entertainment. These endeavours form part of the re-imaging of Dundee as a 'learning' city and reflect a wider awareness of the significance of the knowledge economy. Moreover, these interventions target particular locales of the city and reinforce the marketing campaign strategy which uses the slogan 'City of Discovery'. Here, considerable efforts are being made to revise and enhance the city's reputation and this is being re-cast with a rethinking of Scotland in terms of its national spatial economy.

# 9 Urban myth

## The symbolic sizing of Weimar, Germany

Frank Eckardt

Weimar is a small city with a big repuation. Situated in the state of Thuringia, Weimar has remained a small city compared even with its neighbouring cities – Erfurt and Jena. Nevertheless, Weimar has become the only East German city which has been able to keep its number of inhabitants stable while other former cities of the Socialist Republic lost 20 per cent or more residents after the German reunification. Weimar is an example of a small city that has used its 'cultural capital' to enable its integration into national and European urban structures. This cultural redevelopment strategy, however, has not been introduced in a way that is comparable to flagship developments and is only to a limited extent due to 'festivalization policies'.

The significance of Weimar in German history has been assured by its assumed *genius loci* throughout the last two centuries. The urban development of Weimar has to be understood with acknowledgment of the role of myth (Theile 2000). In urban theory, the role of myth has often been discussed with regard to urban tourism and the consumption of places (Urry 1995: 129–162). The strength of the myth, however, goes beyond the tourism sector and is part of a wider cultural economy of Weimar. In this chapter, the influence of the myths linked to the city of Weimar will be analysed as the key factor for a small city in eastern Germany facing double transition – into the reunified German Republic and into the globalizing world.

Myth-driven urban development is understood as a predominately modern phenomenon wherein rationality itself becomes a myth. As Zygmut Bauman (2004) shows, modernity started from the assumption that the enlightened spheres of culture and politics are separate from each other, while the creation of myth overshadows the modern split between both aspects of society (Bauman 2004). Mystification, therefore, leads to a specific form of urban development wherein the ambivalent relationships between culture and power are constantly at stake and perceived as being balanced. In Weimar, the process of societal differentiation did not take place and, instead, the myth of enlightened power was produced. The people of Weimar also do not differentiate between the 'centre' and other parts. Weimar has been too small to develop a modern geography of differentiated spaces. Weimar has been dominated by its function as the court of the Duke and has hosted neither a bourgeoisie nor a working class, which are both regarded as having imposed particular social and spatial patterns on the modern form of urban life (see Häußermann and Siebel 1987).

## PRODUCTION OF MYTH

The discrepancy between the number of inhabitants and the symbolic importance of Weimar can be analysed using a myth model of urban development based on a certain atmosphere – where social and personal relations are dependent on easy encounters enabled by limited opportunities to meet. For example, Weimar is ranked low in the urban hierarchy in terms of economic wealth and industrial production. However, if one were to make a ranking of symbolic places in Germany, the city might be placed after Berlin as the second most mentioned cultural city. Different surveys support the idea that Weimar be placed high at the top in a symbolic urban hierarchy and Weimar has a high concentration of many places of significance for German culture.

The face-to-face communication between intellectuals is the root of the Weimar myth. The most important part of all visits to Weimar has always been the search for a personal nearness to buildings and places which have symbolized the intellectual spirit of the place. When Johann Wolfgang von Goethe left his hometown of Frankfurt in 1775, he described his motives in letters to his mother. At that time, Frankfurt was seen by the young poet as a 'dirty nest' where the industrial revolution had started to take root. The city was dominated by a bourgeoisie with a 'hands-on' approach leaving little space for romantic feelings. Goethe was attracted by Weimar as he could then work with the young Duke Carl August whose mother had started to build up an intellectual milieu in the small Thuringian town. Originally from Russia, Anna Amalia later initiated the famous library with a rococo-style reading room and chose enlightened teachers for her son. When Goethe fled the upcoming development of a German bourgeoisie in Frankfurt to be part of a pre-modern form of government, his biographer Karl Otto Conrady (Conrady 1994: 251) concluded 'his willingness to contribute to the freedom of the people has been lesser than his belief in the possibilities of patriarchial power'. The widespread picture of a clichéd intellectual, who spent most of his time writing and thinking, is even less tenable. It is convincingly reported that the German genius appreciated Weimar because of its lack of industry and trade, the opportunity for a closeness to nature, the vineyards and the opportunities to meet the opposite sex (Merseburger 1999: 64–110).

However, even though Goethe might have had the intention to reform the dukedom Saxony-Weimar, he should not be regarded as one of the fathers of German democracy. For example, his attitude to human rights points strikingly to his anti-democratic personality. Nevertheless, even during Goether's lifetime, he and Weimar were already regarded as a small island of liberty in a largely backward Germany (Wilson 2000). All the same, Goethe, being a writer and politically active in Weimar, created an image of the city that attracted the attention of a wider public in Germany. This was the outcome of a promotion generated by the growing distribution of German literature and the installation of a theatre culture. Friedrich Schiller, fleeing to Weimar from political persecution in the south-east of Germany, probably embodied the myth of the liberal Weimar, the German idealism *pars pro toto*, more than Goethe (Safranski 2004). However, in his later years, Goethe rejected romantic anti-modern movements and entered his 'classical period' with a rationalist attitude. In this period, he reflected on many issues that demonstrated his open-mindedness and modern approach to reflections on natural science, Eastern philosophy or the 'nature of light' (Mandelkow 2001). Goethe's productive years until his death in 1832 are more or less the substance of the mystification of Weimar.

While it is true that many intellectuals visited Weimar during the classical period, the city missed major urban developments that had taken place in other German cities such as the shift to industrial production, massive migration from the countryside to the cities and the political awakening on the eve of the 1848 revolution. In contrast to its neighbouring city, Jena, where Goethe had already ordered his soldiers to fire on politically motivated students on strike, Weimar was calm and devoted to the residence of the ruling Duke. With liberalism based on the freedom of the arts, the royal court of Weimar attracted the composer Franz Liszt to be 'director of music for extraordinary services'. Under his influence, Weimar experienced the so-called 'Silver Period' in which the famous pianist reformed the Weimar theatre and performed, for example, the work of Richard Wagner that was neglected or even forbidden in other parts of Germany during the period of political and cultural restoration (Bahr 2001).

After the death of Liszt in 1886, a 'Third Weimar' emerged. In 1900, the sister of the mentally ill philosopher, Friedrich Nietzsche, bought a villa in Weimar and started to begin misusing her brother's fame. Although Nietzsche remained in Weimar only for a short time, she was capable of linking these two places in the cultural perception of the educated classes of Germany. Falsifying many manuscripts, Elisabeth Förster-Nietzsche published the Weimar-based Nietzsche archives with an anti-Semitic and nationalist fervour (Riedel 2000). Weimar was no longer synonymous with liberal artistic movements and had, by then, already become a place of rivalling interpretations of the city's cultural heritage.

While conservative, nationalist and reactionary artists and writers claimed Weimar as a place of the 'real German', the reformist and modern protagonists had a difficult standing against it. When the Belgian architect, Henry van de Velde, and Harry Graf Kessler came to Weimar, the local authorities and the general public demonstratively showed their hostility against the 'New Style' and Walter Gropius faced strong resistance from the conservative forces in Weimar when he was nominated as director of the Bauhaus School in 1919. For a short period, Weimar attracted the European arts avant-garde of modernity. Names like Feininger, Kandinsky, Klee, Muche or Itten assured the international attention of the Bauhaus and its worldwide reputation in later years. In 1925, however, Walter Gropius abolished the Bauhaus in Weimar to avoid a slow death because of the decision of the regional parliament to cancel all payments and funding (Hüter 1982).

## MYTHS AND THE ANTI-MODERN CITY

The number of inhabitants and the size of Weimar remained small in comparison with the neighbouring city of Erfurt. The forced industrialization of Erfurt during the periods of Prussian rule and the German Democratic Republic kept alive the duality of an 'industrialized' Erfurt and a 'cultural' Weimar. Both cities and the whole region of Thuringia experienced severe poverty until the twentieth century. Nonetheless, Weimar's culturally symbolic position in the urban hierarchy of Germany remained. The expulsion of the Bauhaus from Weimar, however, showed that it had become a place where antagonistic political discourses fought to decide the way in which the cultural heritage of the city was to be integrated in the ongoing processes of modernization in German society. In the 1920s, the first democratic republic hosted its parliament in the national theatre of Weimar. In his inaugural speech, the social-democratic president, Friedrich Ebert, invoked the *genius loci* and the democratic tradition inherent in the 'city of Goethe and Schiller'. Only a few years later, the hostile atmosphere of Thuringia and Weimar and the impracticality of the location led to the relocation of the parliament to Berlin. During the National-Socialist times, Weimar came to be one of Hitler's favourite cities, which he visited 19 times (Mauersberger 1999). Abusing the heritage of Nietzsche and Goethe, persecuting political enemies and creating a culture of terror, the most visible wound of the Nazi dictatorship remains the concentration camp of Buchenwald on the hills surrounding Weimar where, from 1937 to 1945, Jews, 'asocial elements' and political prisoners suffered and where 56,000 prisoners were killed.

The people of Weimar, however, had denied knowledge of the character of the camp, when forced to visit Buchenwald by the American soldiers. As evidence shows, the city and the concentration camps were linked intensively, so that it has to be considered a part of the ambivalent history of Weimar (Schley 1999). Weimar was meant to work as a model for the reshaping of German cities. With the building of the oversized 'Gauforum' (a Nazi administrative institution) in the middle of the city, the Nazi architecture was intended to cut into the historically grown urban mosaic. The Nazis wanted to tidy up the chaotic street structure by implementing a dominating orientation towards their buildings in the city (Loos 2000). During the GDR period, the city, its history and especially Buchenwald were interpreted as legitimation for the socialist regime (Overesch 1995).

## SOCIALIST IMPLEMENTATION OF MODERNITY

As socialist rules restricted freedom of movement, the number of inhabitants remained stable during the GDR period. To avoid the possibility of Weimar becoming a sort of counter-capital city, it was Erfurt which gained the status of 'Bezirkshauptstadt' (regional capital city) and hosted regional governmental institutions. Nonetheless, Weimar's experience of the socialist period was similar to that of other East German cities. Besides neglect for the built environment, the presence of the Russian army on the hills on the opposite side of Buchenwald had a particular influence on the life of the Weimar population. However, as part of GDR society, Weimar experienced a significant transformation during the 45 years of the socialist regime. This meant, in the first place, that Weimar was integrated into an industrial regional complex which had been developed previously only in the neighbouring cities of Erfurt and Jena. Nevertheless, Weimar kept up and maintained its position as a city of national cultural importance and had only partial success in catching up with the overall industrialization of East Germany. However, for most of its inhabitants, the city followed a 'principally different framework of a socialist city' (Häußermann 1996: 7) wherein from 1972 private ownership of land was nearly totally abandoned. Architects and urban planners were powerless, obeying the prerogatives of the higher echelons in the GDR planning system. This became especially visible in Weimar where the University of Architecture and Civil Engineering (HAB) rejected the heritage of the Bauhaus which was seen as bourgeois by the socialist regime (Schätzke 1991).

According to the socialist ideal of a city of workers and a homogenized social society, everyday life in Weimar was organized by integration into the process of production (Niethammer 1990). Especially in the Honecker era, housing politics became the major field of commodification of the East German population. A serious lack of housing units had generated substantial criticism of the entire political and societal system and was then, in 1971, addressed as the primary social objective for the following decades. This led to the building of prefabricated high-rise housing estates which were placed only on the outskirts of Weimar (for example, in Weimar-West and Schöndorf). In the GDR period, these new houses were commonly regarded as being modern because of their water, gas and electrical facilities. Following the distribution policies of GDR companies, families with children were given priority during the allotment of these dwellings. Meanwhile, the older city areas between the railway station and the area of 'classical Weimar' together with the north-east went into decay and were left either to 'asocial elements' who were not members of company associations or to the elderly (see also Hinrichs 1992). Thus, like other German cities, Weimar exhibited a low segregation on the basis of income and social status but was clearly divided by age (Hannemann 1996).

The socialist regime in general respected the significance of the inner city of Weimar and used it as an attraction for the modest foreign tourists that already pilgrimed to it in that period. However, the inner city suffered severe underinvestment because the total economy was not able to fulfil the basic consumption needs of the East German population.

Although Weimar was lucky that the socialists also respected the historical Christian monuments in the city, especially the churches, and did not blow them up as in other East German cities, the principle had to be realized that socialist rule must visibly express the victory of the working class.

## RETURN OF THE MYTH

Changes occurring following German reunification led to a wide range of visible expressions in urban life. The transformation from a 'gray into a colourful city' (Bertels 1995) was accompanied with political transition characterized by the introduction of political autonomy for the city and the restriction of urban planning competences which recognized fundamental rights of land ownership

(Häußermann 1996). In 1992, the urban sociologists Christiane Weiske and Uta Schäfer, affiliated to the HAB Weimar, conducted interviews with inhabitants of the city (Weiske and Schäfer 1992). According to their findings, 74 per cent of the persons interviewed found that Weimar had an 'urban character' while only 13 per cent saw the place as a village and seven per cent stated that the city had to be regarded as a 'capital or metropolis'. Interestingly, only two per cent of the interviewees compared the city to a museum. In this document of early studies about the transformation of Weimar, the authors also applied qualitative studies with regard to a wider perspective of urban life in the city. While doing so, the perspectives of the researchers reflected the return of traditional myths about Weimar. In the introduction to the research report, the authors conclude: 'The myth of Weimar. It exists and it still lives. It is the myth of enlightenment'. In the body of their argument, the 'reality' of the myth is said to be in harmony with both the conscience of the population of Weimar and the urban form. The table company/salon of Anna Amalia, the company of friends led by Liszt, the Bauhaus Weimar – all these are examples of a kind of culture, where communication is the aim and the centre of formal and informal relationships between equal citizens. Social hierarchies have been relaxed and social dynamics and mobility have been enabled. Here, the real nodes between everyday culture and the haute culture of the Weimar tradition can be found. The sociologists are but one of the intellectual elites of the post-reunification situation where 'Weimar' has been idealized.

## From myth to events

After the German reunification, Weimar's population began to fall. The regional government of Thuringia reacted with a policy which intended that small communities at the edge of Weimar be integrated into the administrative borders of the city. This intervention prevented suburbanization, which could result in a loss of inhabitants from the core city and which occurred to a large extent as settling outside on the green meadows was previously forbidden. In particular, the integration of Gaberndorf, an old agrarian village which quadrupled its number of inhabitants in five years as a consequence of Weimar's suburbanization, kept the total number of inhabitants steady. With the shifting of administrative boundaries in 1999, the size of Weimar was changed for the first time since the community border reforms of the Weimar Republic, when new classifications of city limitations were introduced. The question of which city should be the capital of Thuringia was discussed once more after the collapse of the GDR system. A clear decision in favour of Erfurt was taken only weeks after the declaration of the regional constitution. Weimar was seen as being enriched already with its cultural heritage. Moreover, the city was smaller in comparison with other Thuringian candidates. Erfurt, Gera, Eisenach, Jena and Gotha had more inhabitants and a larger geographical scale but were observed to be threatened more by the processes of shrinkage, suburbanization, de-industrialization and social decline than Weimar. Since the second largest city of Thuringia, Gera, had neither been given any governemental office nor a university, Weimar was in a difficult position to protest against the decision to make Erfurt the regional capital. In the regional planning documents, Weimar was accorded the same status as the other forementioned larger cities. By that time, the government argued that Weimar should get lower governmental support because of its smallness and that the cultural institutions such as the university and the theatre should merge with those in Erfurt. The symbolic size of Weimar helped to avoid the loss of these institutions at that point as they could claim to be of national significance.

The re-establishment of Weimar's symbolic size had been generated amidst the turbulence of the transformation of all parts of its life. When the American urban planner, Peter Marcuse, first came as visiting professor at the Weimar 'University of Architecture' in 1990, he was not able to get a taxi from the railway station to his apartment (Marcuse 1990). The physical infrastructure and other obstacles to everyday life in those days of transition hindered the responsible and influential actors in their development of a long-term vision or strategy for urban development. As for the political debates regarding the future development of Weimar at that time, the urban sociologist Johannes Böttner titled his book *What Now, Little City?* which perfectly expresses the lack of direction in public debates of the early 1990s (Böttner 1996). Without giving suggestions, Böttner reflected the confusion and varying ideas about possible strategies. In 1993, the initial enthusiasm for German reunification had waned and Weimar enthusiastically embraced the idea of applying for the status of Cultural Capital of Europe for the year 1999. Although the suggestion was made by political actors from the East, the West German Klaus Büttner, mayor of Weimar since August 1990, was the one who embodied the idea the most. In 1993, the European Commission unanimously voted for Weimar to be Cultural Capital of Europe for the year 1999. With this award, the European Union wants to focus on certain cities which contribute particularly to European culture. The title does not carry any financial support from the European Commission, but it has been regarded by political actors as supportive of urban development projects and the general promotion of the city. The application for 'Weimar 99' resulted from a strategy to reach these objectives through a general approach to the concept of 'event organization' (Schulz 2003). The introduction of the 'Kunstfest' (Arts festival) way back in 1990 was the first major attempt to make use of the positive connotations of Weimar for the development of the city. Until the start of the 'event year', Weimar was seriously overloaded with preparatory tasks and related costs. From 1996 onwards, the Kunstfest was planned as an integral part of the framework for Weimar 99.

In 1994, the city had nearly given up further investment in the upcoming mega-event. Because the city's budget was not balanced, the regional government withdrew Weimar's right to prepare its next financial framework. This extraordinary measure was guided by the installation of an external auditor. In the background of this financial crisis, voices from the broader public claimed that the city should refuse the title. As a consequence, the 'imported' West German mayor was replaced during local elections in July 1994 by the former GDR second mayor of Weimar, Volkhardt Germer, who has been in power since then (Fascher 1996). In 1995, further dedication to Weimar 99 was only safeguarded by external intervention from the Thuringian Government when the West German Bernhard Vogel from the Christian Democratic Party (CDU) took control. Vogel was a strong supporter of Weimar 99 and finalized the much debated structure for organizing the event. With the founding of Weimar 1999 Kulturstadt Europa GmbH in 1995, the Vogel government institutionalized its predominance and influence on the realization of the Cultural Capital Year. The federal state and the city were only given a minor stake in this organization. Nevertheless, the deadlock on the precise planning of the upcoming event was not broken and major controversies were still to come. This was especially true with regard to one key person, the cultural director for Weimar 99. In 1996, the hitherto President of the Classical Weimar Foundation, Bernd Kauffmann, was appointed to the seat and his new position was vested with far-reaching competences (Hammerthaler 1998). His programme was ambitious and criticized for being restricted to high culture and the national intellectual elites. The local population felt widely excluded from the upcoming events and, in the polls, the majority of the inhabitants did not support Weimar 99. The conflict between the population of Weimar and Kauffmann's plans peaked when he suggested dedicating a central square (the Rollplatz), which is still used as one of the scarce parking places, to an art exhibition by the French artist, Daniel Buren. A broad protest by the population, supported by the local media and food outlets, put sufficient pressure on the decision makers to restrain him from this plan (Frank 2003).

Coinciding with the two-hundred-and-fiftieth anniversary of Goethe, Weimar 99 had a programme which focused merely on high culture and the classical German arts. Although claiming to be European-oriented, foreign contributors played a minor role. However, the population soon started to like the cultural events and there seemed to be a substantial switch in public opinion. Even the rather unpopular exhibitions on 'Rise and Fall of Modernity' and 'Official and Unofficial – Arts in the GDR' could not upset the improving relationship between the local audience and external attention to this small city. In particular, the question of how to deal with the heritage of Buchenwald was followed with much interest by external observers. The sensitivity of the organizers and particularly of Bernd Kauffmann led to an approach where the concentration camp was represented in an appropriate way (Roth 2003).

Although Weimar 99 was viewed as a convincing success, the realization of the event year fostered the hypothesis of 'occupation or colonialization' of the East by the West. The famous playwright, Ralf Hochhuth, used this issue in his work *Wessis in Weimar* which found a wider audience within the East and with critical intellectuals in the West. The fictive play claims to reflect the reality that East German elites in the symbolic city of Weimar are replaced by incapable Westerners. In reality, the institutions in Weimar have been only temporarily directed by a few Westerners and some, such as Bernd Kauffmann, have left the city because of opposition.

## THE GEOGRAPHY OF MYTH

The mapping of Weimar is dominated by the attractions of the city which are upheld by the external world. This is most remarkable with the cultural heritage of Classical Weimar, which is under the auspices of the Classical Weimar Foundation where the Federal State contributes the majority of the annual budget. In German federalism, this interference with local and regional cultural affairs is an extraordinary exception and only legitimized by the acknowledged special significance of Weimar for national culture. But the support from the German government could only have been fostered by the national consciousness of the myths and their tracing back to the sacred places of the 'real Germany'.

While seven million visitors came to the small city of 60,000 inhabitants in the period of the Weimar 99 events, the number of tourists in Weimar reached an average of around four million in the following years. With this number, however, the level of tourism has been put on a higher level than it was before. In a survey of 252 representatives from the tourism sector, the vast majority stated they had gained from the European entitlement (Eckardt 2005). In conclusion, Weimar has found its place in the (foremost national) imagination of German culture and is a 'must' for visits by school classes. The only remaining question is the view of the actors that tourism has not created separated places for the local population and the visitors. One of the authentic aspects of Weimar is the convulsive and complex structure of its small alleyways where inhabitants live and numerous restaurants and cafés are situated. The inner city is certainly used by both groups, but the lines of communication for tourists in the city are limited and qualitative studies (Eckardt and Karwinska 2006) on the places they frequent indicate that there is a particular geography for tourists in the city: there is a special central parking area for tourist buses in the city from which an axis leads to Goethe Square. Another route passes from the railway station to the inner city alongside which are situated cafés, museums, restaurants and youth hostels. Within the inner ring of Weimar, cafés and restaurants signify, through their design and gastronomical ranger, a certain preference for either tourists or locals. Major tourist cafés at the Frauenplan (near Goethe's former residence) are predominated by tourists and locals use them only selectively. This becomes most obvious when only a few visitors are present in the city in winter. On the other hand, gastronomic places dominated by locals can be clearly defined. At Theatre Square, one bakery frequented by both groups lies opposite to another where only East Germans spend their time. 'The Westerners do not seem to like the rather narrow entrance and our restricted offer of sweets', is how the owner explained the situation (interview with author 25 March 2004).

The mythological geography of Weimar is also produced by another important group of temporary inhabitants in the city: students and academics from the Bauhaus-University Weimar, which is the second largest employer in the city. When the European Summer School was started in 1997, international students were subject to a mass attack by young neo-Nazis hanging around in the inner city. Only the courageous engagement of the university president was able to mobilize the wider public for the security of foreign students. When national newspapers started to report how these students had to be safeguarded by the police to and from the university, a consensus was reached that the city needed a *cordon sanitaire* where the neo-Nazis would not linger. Even during Weimar 99, artists were attacked and pieces of public art were destroyed by these groups. Nevertheless, the neo-Nazi attacks are widely rejected and a broad union of activists from all parties organizes protests against the manifestations of neo-Nazis who often want to demonstrate at Buchenwald (Eckardt 2003: 158–168).

In subsequent years, the influence of students on the city's life has become even more obvious (Morales 2005). Students of architecture and music are attracted to the city as followers of the Bauhaus and Liszt myths. Approximately 5400 students from both the Bauhaus-University and the Liszt School of Music account for nearly ten per cent of the population derived from West Germany and these young people contribute to all parts of urban life. Most of them have not grown up in the city and are in need of economical accommodation. As the students have only limited financial resources, the phenomenon of shared flats (WG) has led to an incumbent upgrading of many houses that had poor facilities. In particular, the unrenovated houses along the 'Trierer Ring' surrounding the inner city of Weimar offer reasonable housing for students. Regarding the labour market, it is obvious to see how the students contribute to the cultural economy with their low wages and flexible working hours. Many tourist cafés and restaurants depend on their client-friendly attitude and prefer students to unemployed locals. Last but not least, students support many cultural events which would not survive if based only on local support.

As regards the principle of a sacred place, Weimar speaks of the separation of the 'holy' from the profane (see Eade and Sellnow 2000): Weimar is a profane place with broken-down houses beyond the inner city. Weimar-West attracts the concerned attention of political decision makers as the GDR high-rise estates in these areas show a higher concentration of those receiving social benefits and immigrants. However, the differences compared with the rest of the city are only relative. The separation from the rest of the city is generated more by the relative geographical distance (Pippi 2004). The situation in Weimar and other parts of the city reflects the mobility patterns of the inhabitants. While Weimar has been socially homogeneous during the GDR period (Rasche 1992), the income differences are now leading to the establishment of a suburban tract of the better-off in the surrounding region. Another myth, the dream of living in the untouched green, has left the poorer population behind in certain areas, and creates disturbances in the small villages receiving the new population. In some cases, the incoming suburbanites have increased the number of inhabitants in these small places from 500 to 4500. Empirical research, however, has shown that conflicts between insiders and outsiders have not arisen and that the suburban population remains intensively linked with Weimar in terms of work, transport, social infrastructure, consumption and culture (Eckardt 2005: 150–233). In accordance with the requirements of the State government, Weimar reacted to this development with their incorporation into its territory. Since then, Weimar has been the only East German city where the population has remained stable.

## CONCLUSION

Weimar is an example of how small cities can regain significance in a national culture and even beyond for the external visitor and for inhabitants. With regard to its development throughout the different political systems, however, only within a society with freedom of movement and where culture exchanges symbolic values into material ones, does Weimar place itself into a situation wherein symbolic and real size (in terms of inhabitants) are related to each other in a direct way. When the Dutch writer, Cees Noteboom, visited Weimar as one of the first individual tourists after the destruction of the Berlin Wall, he asked himself: 'Why am I here in Weimar? To immerse myself in Goethe, of course'. He took a hotel room at the Frauenplan where the German poet lived and studied essays by Ortega y Gasset about Goethe and Kant. Noteboom pursues his interest in Weimar because, to him, Goethe seems still omnipresent in the German culture. During his stay in Weimar, Noteboom noted two interesting observations: first, he saw two young girls reading poems at the graves of Goethe and Schiller with 'such an emphasis that he was afraid the dead would come crawling out of their final rest'. Second, he went to Theatre Square where a performance by the citizens of Weimar took place and came to the conclusion: 'Goethe and Schiller are not dead. While the exodus out of East Germany takes frightening shapes, the people from Weimar meet their poets here, holding a board saying "We are staying here"' (Noteboom 1991: 283–291).

Like Noteboom, many visitors followed the fame associated with the few public spaces and streets between Goethe's house and the Bauhaus building. The small city of Weimar has remained a nucleus in a shrinking urban landscape which has suffered the loss of more than 1.5 million inhabitants. After 1989, other East German cities, such as the neighbouring Erfurt, lost up to one-quarter of its inhabitants because of demographic change, suburbanization and westward migration. While all East German cities are looking to cope with their shrinkage, Weimar benefits from the direct and indirect effects of the myth economy. The roots of the sacralization of Weimar lie at the beginning of the nineteenth century when the intellectual life in Germany developed its classical form. Since then, Weimar has developed a symbolic value for German identity. Untouched by the processes of industrialization and the challenges of the bourgeoisie, Weimar remained a niche city.

The myth transformation of the city became increasingly anti-modern and, in this sense, Weimar reflects, better than any other German city, the overall development of German society with its crucial antagonism between cultural and economic modernity (Merseburger 1999: 405). The forced modernization in the GDR period seems to have had only a limited effect on the principal dualism between the sacralized heroism of the classical period and the profane everyday life. This dichotomy doubled after German reunification, when the transformation of the city centre was driven and pushed by external factors. With the realization of the European Cultural Capital Year in 1999, these conflicts became visible but, to some extent, were made negotiable. While Weimar suffers from a high unemployment rate like every East German city, many other factors of decline, such as the shrinking population and a lack of urban culture have been avoided by opening up the city to tourists, artists, students and academics. Therefore, it seems that for the first time in the history of Weimar, German haute culture and its representatives have been accepted by the local population. Interactions between East and West Germans have found a home in Weimar which is still an exceptional occurrence in reunified Germany.

# Part 3

# The cultural economy of small cities

# 10 Garden Cities and city gardens

Malcolm Miles

My purpose in this chapter is to reconsider the idea of the Garden City from a contemporary viewpoint. This is appropriate in a book on small cities: the Garden City proposed by Ebenezer Howard in 1898 is a city of 30,000 people, the scale of a medium-sized town. The application of Howard's ideas at Letchworth, Hampstead Garden Suburb and Welwyn, in the 1900s to 1920s, produced what remain small-scale urban settlements. The Garden City informed the development of post-war English town *and country* planning and, more recently and obliquely, the development of Poundbury, a model village planned by Leon Krier near Dorchester in the south of England.

I preface the chapter with a reflection on the possibly sentimental appeal of small cities, noting issues to which to return in the final section. Because it allows me to write from a position in the present, I begin the main part of the chapter with an account of Poundbury and then reconsider Letchworth, in part through an arts project which revisits its original aims. Finally, asking how the radical agendas of the Garden City have, or have not, been sustained and whether there are equivalents today, I look briefly at squatter gardens in New York's Lower East Side. Some questions linger: Was the Garden City regressive or an alternative modernism? Is the model of a Garden City sustainable? Does sustainability extend to socio-cultural formations?

## CONVIVIALITY OR SENTIMENTALITY?

The aesthetic appeal of small cities is easy to see. The idea of the small city, based on a reading of heritage culture, has become an ideal trading on widespread, and mythicized, ideas of dystopia in the metropolises and mega-cities in which it is claimed most people now live.[1] Small is beautiful because large is nasty. The prejudice is ingrained in modernist urbanism. After the Chicago sociologists' celebration of the city as a site of intense mobility and competition in the 1920s and 1930s, Lewis Mumford, in *The City in History* (1968), demonizes industrial and metropolitan cities, asserting that the conditions of such cities are 'a badge of shame' and that rapid advances in technology mark a disintegration, visible in city planning on a day-to-day basis, analogous to 'ultimate plans for atomic, bacterial, and chemical genocide' (Mumford 1968: 551, 553 cited in Agnotti 1993: 8–9). This makes Mike

Davis (1998) seem mild when he links the future of Los Angeles to the script of the disaster movie. Thomas Agnotti sees Mumford reflecting 'a deeper trend in political thought that considers the central planning problem to be the control of technology, analogous to Frankenstein's problem of controlling his monster' (Agnotti 1993: 9). Perhaps planning is itself a monster from the viewpoint of market economics and its calls for deregulation.

Hugh Barton (2000: 125), in *Sustainable Communities*, takes a figure of 5000 (the number of people needed to support a school) as defining neighbourhood size. He notes a study in London which found that neighbourhood centres need to be within 800 metres distance of users' homes; and that local identity is a criterion for coherence. He adds 'This range and variety gives the designer pause for thought. While the general principles are clear the justification for specific standards and sizes is often obscure. There is sometimes an element of sleight of hand' (Barton 2000: 126). I worry that sleight of hand might apply rather widely in the promotion of small cities as icons of conviviality. For example, Herbert Girardet offers the following homily:

> It is a great art to make a city convivial, as the best examples we have inherited show us. Cities such as Florence, Salzburg, and Prague seem to have been purpose-built for lively interchanges between people. Narrow human-scale streets contrast with well-appointed public buildings and wide, open gathering spaces. They are products of good planning but also of organic growth; they are functional but remain on a human scale; they are centres of economic activity but also of social and cultural energy.
>
> (Girardet 1990: 118)

Girardet's nice picture is challenged by Richard Sennett's account of the high incidence of violent crime against the person in fifteenth-century Paris, a city probably comparable to Prague or Florence then. Streets were 'the space which remained after buildings had been constructed', bearing an imprint of 'aggressive assertion' fuelled not least by alcoholism as defence against pain. I would add that the well-appointed public buildings, housing the operations of a class which became the bourgeoisie, line squares in which elites displayed their power.

There may also be emptiness or claustrophobia, rather than neighbourliness, in a small city. Stefan Hertmans observes that Friedrich Hölderlin went mad in Tübingen, a small city. He reads the search for comfort (*Gemütlichkeit*) as driving him mad 'because it promised something that did not exist – deep closeness in an otherwise philosophically unfathomable world' (Hertmans 2001: 32). Hertmans' description of Tübingen is worth citing:

> The whole atmosphere of the old-fashioned and easy-going university, the misty closeness of forests on the hills (apparently idyllic when seen from the town walls), the pure woodland air that penetrates into the narrowest streets in the early morning ... this gives the illusion that what Proust called the lost

> period of paradisial experience can be held on to, can be extended into our age … so many invitations to step out of your life and to start something else.
>
> (Hertmans 2001: 31)

He views this as attractive and grotesque.

Developers use the urban village concept, a step outside the city's frenetic ambience, as a marketing strategy for schemes which tend to re-code a neighbourhood in the terms of upwardly mobile consumers.[2] There seems an assumption after the failure of post-war social engineering that cities must be divided into enclaves to make communities. This suggests that some, at least, of what is said of small cities, old cities, nice cities – other cities than those we have – is sentimental. The negative urban image is generalized either as war stories to appeal to an academic version of the appetite for war comics (Miles 2000: 44–52) or as a state of perpetual conflict in which city dwellers are the dupes of urbanization. In contrast, Elizabeth Wilson remarks that 'To destroy the cities and return us all to small communities would be to capitulate to the puritanical and controlling elements that critics of town planning … have identified as a hidden agenda' (Wilson 1991: 154).[3] Wilson adds that there is no reason to suppose community is found in settlements which 'offer none of the opportunities for escape, anonymity, secret pleasures or even public crowds' of cities (ibid.). Small cities do not offer better conditions than big cities per se, and I would argue that to privilege them denotes a regressive anti-urbanism. Nicholas Schoon writes 'Since the industrial revolution began in these islands, the history of most of its [sic] cities has essentially been one of desertion by people with choice and money' (Schoon 2001: 3). Some of them might go to Poundbury.

## POUNDBURY

Poundbury is a village, not a city, but it exemplifies the scenario in which new urban forms are reproductions of those of a selective past. A walk along a winding, cobbled street of cottages with Georgian-style doorways is like a walk through Hardy's Casterbridge (see Figure 10.1). Poundbury was built on farmland belonging to the Duchy of Cornwall, identified by West Dorset District Council for development to meet the housing needs of a growing population in the south of England. A masterplan was drawn up by Leon Krier, and Poundbury also reflects the views of Charles Windsor, a contributor to debates on architecture and tradition since the 1980s.

Poundbury has a projected eventual population of 5000 and is being built in phases, each surrounded by a mini green belt. Its planning adopts progressive principles such as the mixing of private and rented housing in the same street. Its design is an eclectic vernacularism. Buildings are constructed to a high specification in traditional materials such as Portland stone, Purbeck marble and brick. A brochure says 'Chimney stacks – all of which possess a proper function, if only to ventilate drains – are subtly embellished with patterns of brick. Fine detail is everywhere' (Poundbury 2003). A code does not allow television aerials or

*Figure 10.1* Poundbury street scene. (Source: courtesy of Malcolm Miles.)

satellite dishes but provides cable, requires all services to be grouped in a common trench and regulates exterior colours (white or cream, but traditional pink for the toll house). Compliance is a condition of residence – as it is in the Disney Company's venture into real estate at Celebration, Florida (MacCannell 1999). A brochure states that 'agreeing not to fill the front garden with bathtubs of purple gnomes seems eminently reasonable' (ibid.). I have seen no such bathtubs or gnomes during my visits (but I have not noticed them where I live in Exeter either, and wonder if only people in London would have these). Nicholas Schoon remarks that 'Most visitors love Poundbury. Housing and town planning professionals are impressed, even inspired, and the houses fetch premium prices' (Schoon 2001: 232).[4]

An account of Poundbury emphasizes its adherence to a perceived informality in rural settlements:

> Homes gathered closely round an irregular street pattern, with shops and workplaces within walking distance to reduce dependence on the car, blend happily with their surroundings and can be a positive addition to change.
>
> (Walker 1999: 27)

This is in context of a government policy to expand the housing stock, and the account mentions several other recent schemes in which developers have sought to integrate small-scale, low-key schemes in provincial, semi-rural and urban sites.[5] These are collectively described as a 'version of the parochial community that has been lost in so much of modern Britain' (ibid.). Here, small is clearly correlated with community. Of those who dislike Poundbury, the writer says they 'surely miss the point'. But I wonder if there is a mismatch between an appeal to a parochial community and the eclectic vocabulary of building styles and types in Poundbury, which suggests an architectural vocabulary more characteristic of an urban elite. These styles range from a Georgian toll house, a terrace of Regency houses (see Figure 10.2) and Victorian cottages to an Arts and Crafts nursing home, an octagon, a severely neo-classical graveyard with grey obelisks and a village hall which looks as if it might be in Saxony. Around Pummery Square are a village store, a pub called The Poet Laureate and a hairdresser, and nearby is a business park for light industry and a chocolate factory. Windsor says:

> Poundbury is a scheme which in the broadest sense is about integration, not segregation, where the conventional distinctions between rented and owner-occupied housing have gone, where tenants and owners are next door neighbours and where, quite against the grain of recent convention, homes are cheek by jowl with workshops and factories.
>
> (*Westcountry News* 17 July 1999: 27)

He continues that the style is not old-fashioned but revisits timeless principles that create a sense of community. There is an intelligent system of traffic management with few signs and cars do not impede the right to stroll. Everything is within walking distance, and Dorchester is a few minutes' bus ride away. Whether Poundbury is a traditional village, a suburb of Dorchester, or a gentrified enclave in the countryside raises other issues: it may not be viable to create a settlement from an idea rather than economic necessity; a community might fail to be engendered without either history or a material basis for common interests; and cosmopolitanism is simply absent.

There *are* positive aspects to Poundbury. Its homes score well on energy efficiency. As a non-driver, I appreciate its pedestrian-friendly streets. I admire the initiative to mix private and social housing and the mixed-use zoning. The development of the site in phases allows adaptation; phase II is 'more "urban" than "village", with more formality, a predominance of three storey housing, wider streets and more civic space' (Duchy of Cornwall n.d.). But I would not move there. My reasons are complex and I do not say they should be of interest, but they include the ersatz quality of Poundbury's mix of vernacularisms; and the combination of progressive principles with a code which makes all decisions about inhabited spaces aesthetic ones. More than that, I doubt I could stomach the benign patriarchy:

> The voice of Poundbury is heard through its associations. The Prince of Wales hears at first hand on his regular visits here to monitor the health and sanity of his and Leon Krier's creation.
>
> (Poundbury 2003)

The voices of civic associations, or philanthropic societies, may be those of class interests. Similarly, the claim that Poundbury is not an artificial construct but has the depth and texture of an old community 'that has been attending to its own business for a very long time' (ibid.) assumes certain, parochial values. But, as Tracey Harrison wrote in the *Daily Mail*, 'Every detail in this honey-coloured development is aimed at pleasing the eye' (brochure for Poundbury produced by C.G. Fry & Son). Such statements could be juxtaposed to Nancy Fraser's argument in *The Phantom Public Sphere* that historically, in Europe, 'the elaboration of a distinctive culture of civil society … was implicated in the process of bourgeois class formation' producing a discourse of publicity which claims accessibility, rationality and equality but is 'deployed as a strategy of distinction' to create new elites (Fraser 1993: 6). Siegfried Kracauer draws attention to the non-validated exchanges which take place in the hotel lobby where, as in a church, one is a guest, yet 'those dispersed in the lobby submit without question to the incognito of the host. They are simply people lacking a relationship to one another' (in Frisby 1988: 128–129). An aesthetic distance from everyday life, as David Frisby summarizes, leads to a relation to a void enforced in a voluntary code of behaviour. The code then becomes the site of sociation, but inhibits it.

## THE GARDEN CITY

Issues of regulation and informality characterize the Garden City, for me at least as much as those of town and country. Howard's first act of regulation was to limit a Garden City's size to a population of 30,000 (plus 2,000 living in surrounding farmland), living in 5500 houses on a 5000-acre site set within a green belt. As its population grew, satellite cities would be built. Howard knew of Edward Gibbon Wakefield's proposal for the planning of the colony of South Australia and James Silk Buckingham's 1849 plan for a model town of 10,000 inhabitants (Hall and Ward 1998: 12). The context for Howard's plan also includes model industrial villages such as Port Sunlight (1888), which follows a nearby precedent set by Prices Candles in 1853, and Bourneville (1879). Both were philanthropic models providing workers with decent housing as a means of ensuring loyalty and productivity as well as moral fibre, and they combined residential and industrial with green and recreation spaces. The look of Port Sunlight, overshadowed by the bulk (and smell) of the soap factory, is that of a Tudorbethan English village around a green. In some ways, with a historical adjustment and uniform instead of eclectic architectural style, Poundbury is not unlike Port Sunlight. Barton writes that, after Robert Owen's New Lanark, 'it was widely assumed by nineteenth century philanthropists and urban reformers that the village, or something like it, was the ideal settlement … to replace large cities' (Barton 2000: 21).

It seems size was correlated with dis-ease. The first principle of Howard's *Tomorrow: A Peaceful Path to Real Reform* (1898)[6] is that towns are blighted by overcrowding, pollution, absence of nature, expensive drains and gin. Their attraction is in offering better-paid employment than agriculture together with amusements. The countryside is characterized by fresh air, space and sunlight, but offers long working hours or unemployment (after the eighteenth-century agricultural revolution) and lacks society and public spirit. Howard tries to fuse the best of both worlds in a proposal for a spatial reorganization of society no less radical than Charles Fourier's plan to replace the cities of France with equal-sized phalansteries (Beecher and Bienvenu 1983).

The Garden City would consist of six boulevards radiating from the centre, a garden surrounded by public buildings and a Crystal Palace, or circular glass arcade resembling a winter garden: 'This building is in wet weather one of the favourite resorts of the people, whilst the knowledge that its bright shelter is ever close at hand tempts people into Central Park' (Howard 1989, in LeGates and Stout, 2003: 313).[7] The buildings are to be of varied architectural styles; factories, warehouses, dairies, markets and other industrial and commercial spaces are spread along an outer ring. Goods are to be transported by rail to and from source. The owners of factories are exhorted to provide proper sanitary facilities for workers. Garden City is, if idiosyncratic, a proposal to design a new way of living. But it is also a literary fantasy published ten years after Edward Bellamy's *Looking Backward* and three years after H.G. Wells' *The Time Machine;*[8] a utopian vision five years after Canon Barnett's *The Ideal City* (Meller 1979: 55–66); and one year after Charles Booth's report on the real poverty of east London. Thomas Hardy's evocations of an already extinct rural Wessex are another possible context. But Garden City remains a reaction to the deprivations which characterized parts of Britain's cities at the time, highlighted by Booth, and to the vicious circle in which declining rural life led to migration into towns and rent levels too low among residual populations to stimulate new building in villages.

Schoon says he warms to Howard 'because, like most journalists, he was a complete amateur' (Schoon 2001: 32). A website for Letchworth today states '[Howard] had no particular advantages of class, or special education, but at an early age he was sent away to school, where he received an early exposure to a more rural environment' (www.letchworthgardencity.net/heritage/index–3.htm). In 1871, aged 21, Howard emigrated to North America to become a farmer, but failed. Moving to Chicago, working as a law stenographer, he saw the rebuilding of the central business district after the 1871 fire and the construction of garden suburbs. Howard returned to England in 1876 and was associated with the introduction of the Remington typewriter. As Schoon observes, his talent was based not in a specialist expertise but in synthesizing elements of other progressive urban visions and turning them into a reasonably practicable proposal.

Peter Hall and Colin Ward note a change between Howard's first publication of his vision in 1898 and its republication as *Garden Cities of To-morrow* (1902): the replacement with a truncated version of a diagram of a Social City, on 66,000 acres, with a population of 250,000 and a ring of satellite districts separated by green land and joined by canals and an inter-municipal railway. They write that 'most readers [of *Garden Cities of To-morrow*] have failed to grasp the vital fact that Social City, not the individual isolated Garden City, was to be the physical realization of Howard's third magnet' (Hall and Ward 1998: 24, fig. 7.25). On that scale it is a dispersed, not a small, city.

Garden City was never, anyway, built as imagined. The immediate problem was raising the capital to buy land. A series of speaking engagements by Howard led to establishment of the Garden City Association in 1899, which became the Garden City Pioneer Company – supported by W.H. Lever and George Cadbury and (after initial doubts) by Bernard Shaw – then First Garden City Limited in 1903. When the first project was built at Letchworth, mainly designed in a uniform style influenced by the Arts and Crafts movement by architects Raymond Unwin and Barry Parker, it was funded by a share offer (Hall and Ward 1998: 34–35). This represents a more serious compromise than the truncation of a diagram. Howard had initially foreseen a purchase in trust to a small group of responsible people, the land reverting in time to common ownership in a city in which none owned freeholds. In another diagram, 'The Vanishing Point of Landlord's Rent' (Hall and Ward 1998: 27, fig. 8), Howard explains how, after repayment of the loans (from philanthropists) on which the land was bought, all rents would go to social welfare, including pensions and poverty relief. Hall and Ward summarize, 'each Garden City would be an exercise in local management and self-government' (Hall and Ward 1998: 28).

The Garden City is a project, it seems, for the kind of self-organizing, cooperative society envisaged by Peter Kropotkin (1976) in works such as *Mutual Aid* (first published in English in 1902). Yet, as Hall and Ward point out, there was from the outset a rupture between the intention and both the specific organization of the Garden City into a hierarchy of central authority and municipal departments and the role of a group of shareholders. Compromises also occurred in what was built at Letchworth. The architects planned a mix of workers' cottages around courtyards and middle-class homes with gardens. The Cheap Cottages Exhibition (1905) produced almost 120 homes for agricultural workers on strictly limited budgets, in concrete, weatherboard and other materials. Designs were exhibited in converted sheds near the station, 'previously used for housing unemployed labourers and road builders from London' (www.letchworthgardencity.net/heritage/index-5.htm). Elizabeth Wilson notes, however, that unskilled workers cycled to work from surrounding villages while Letchworth dwellers were middle-class radicals. She describes them as remembered by long-standing residents in the 1950s: 'vegetarians, dress reformers and eccentrics' (Wilson 1991: 102). Dress reform entailed the wearing of shorts and sandals – it is not clear whether with or without socks – by dwellers of both genders in summer. Wilson cites Ethel Henderson, who moved to Letchworth in 1905:

> Of course we were looked on as cranks … I suppose we were cranks, but I think we were very nice ones … Bare legs and sandals for both men and women soon became quite commonplace.
>
> (Wilson 1991: 103)

It is very 'home counties'. Wilson reads, still, a gender equality in the situation, as later at Hampstead Garden Suburb. But she concludes:

> The garden city was a sanitised utopia, with the obsessional, controlling perfectionism that characterised all utopias. The working classes were to be harmonised with and reconciled to the middle classes by becoming in effect a reduced copy of them.
>
> (Wilson 1991: 104)

This accords with Jane Jacob's scathing appraisal of Howard's 'city-destroying ideas' (in Barton 2000: 22) and may be a useful lens through which to look at Poundbury, a philanthropist-led project.

Howard's vision of a growing population at Letchworth did not materialize any more than his complementary idea of a migration out of London, yet Garden City informed English town and country planning, in contrast to urbanization in most European countries.[9] The garden city idea nonetheless has histories in other places, including North America.[10] Agnotti (1993: 122) notes a proposal for small, planned towns across post-revolutionary Russia, 'somewhat along the lines of the British Garden City'. The group of architects and planners responsible were known as the 'disurbanists'. Farès El-Dahdah writes of Brasilia as 'like the Garden cities of Unwin or Howard … circumscribed by a green belt, the purpose of which … is to impede the coalescence with other urban agglomerations' (El-Dahdah 2004: 50). Helen Meller remarks that 'the only elements which really made the transition to Europe were those relating to the achievement of high-quality housing for the workers, the creation of a "healthy" environment … and the decongestion of old-established cities by building new settlements to absorb excess population' (Meller 2001: 117). She cites a parallel in Zlín, Slovakia, a model city for 30,000 people. The social vision incipient in Howard's first proposal emerged, too, in other forms such as the satellite suburbs of Stockholm, Helsinki, Oslo, Paris, Rome and Berlin. As Agnotti summarizes:

> In most cases, there is social ownership of land, significant public planning at the metropolitan level, integration of new neighbourhoods with other parts of the metropolitan area, and the use of public policy to achieve some level of social and spatial diversity. These new neighbourhoods generally go beyond the small-scale Garden City villages that also dot the European landscape, and the first generation of British New Towns. They are decidedly urban, not suburban, in form and function.
>
> (Agnotti 1993: 232)

Agnotti cites the Stockholm districts of Farsta and Vallingby.

Perhaps, then, Howard's Garden City is a contribution to modernist planning of a kind which, in continental Europe, took a more urban form. If there is a traditionalism in Howard it is derived from Kropotkin's history of mutualism in village societies rather than from a specifically English model. But, as a modernist, Howard's dream was undermined by its distance from practical and common actualities. The lesson is, for Garden City as much as for post-war social engineering, that a new society cannot be designed.

## A BRAND NEW LETCHWORTH?

Before moving to a comparison with more explicitly self-organizing projects, I note a recent art project: *Brand New Letchworth*, organized by the London-based curating partnership B+B (Sarah Carrington and Sophie Hope). In a project briefing they state:

> Today, the radical ideals on which Letchworth was founded have become less relevant to current inhabitants. The bohemians and 'cranks' of the last century have been replaced by young professionals and commuters who are drawn to the town [sic] by its regular train services to London, twee cottages and leafy environs. Perhaps Letchworth has become a victim of its own success.
>
> (B+B 2005)

Carrington and Hope remark that Letchworth is characterized by excessive planning restrictions on new development and an 'overwhelming emphasis on preservation' (ibid.). Invited by the Place Arts Centre, B+B organized a series of community-based events in Letchworth's centenary year, 2003. A period of research included making international contacts, at Zlín for instance.

B+B set out to 'debate the future of a city that has lost many of its socialist founding principles' (ibid.) by testing attitudes to collective planning and asking how artists, perceived since the 1960s in some circles as incidental people able to connect to diverse social groups, might contribute to the process. Meetings were held with an estate agent, a shopkeeper, a resident, a retired planner and staff at Letchworth Garden City Heritage Foundation and the local Museum and Art Gallery (which houses archive material). A call to Letchworth residents was circulated by letterbox distribution for ideas as to how Letchworth might be in an imagined future. This produced drawings, written proposals, an extensive manifesto (eventually given to John Prescott, the Deputy Prime Minister) and models by both adults and children. Lay ideas for Letchworth were put in a visual form by invited artists and exhibited at the Place Arts Centre for a month. The project was publicized through posters and a newsletter in the style of a local political campaign. A series of workshops was hosted during the exhibition. One involved staff at a local bank brainstorming a Barclaysville as successor to the equally brand-conscious Bourneville. Other workshops involved A level students and young people, with a parallel event at Dunaujvaros, a socialist new town in Hungary built in 1950. There was an Internet site and workstation at the arts centre, through which anyone could contribute ideas. Among those received was a proposal to change one of the schools to a special-needs school, a request for a big sandpit and a plan to renovate old buildings as studio and gallery spaces for an arts community as a driver of economic regeneration. B+B concludes:

> *Brand New Letchworth* created a temporary platform for debate. In so doing, it also raised local expectations for change and led to questions of who might be responsible for effecting change. … By providing a critical look backwards and forwards at a specific town [sic] with its myriad of views and understandings of ideal ways of living, *Brand New Letchworth* attempted to shake up the comfortable commuter belt that Letchworth has become.
>
> (Project briefing 2005)

## RECLAIMING GROUND

To what extent has Garden City, as a tool for social change, been overtaken by events? Was it in any case flawed by contradictions, as between planning and self-organization, or an aspiration for land ownership and a debt to shareholders? I move now to a reconsideration of some of the issues and the relation of approaches to size.

In his diagram of three magnets, Howard cited 'Trespassers Beware' as an aspect of the countryside (Howard 1898 in LeGates and Stout 2003: 312); and beneath his diagram of the Garden City wrote the words 'Freedom and Cooperation'. As Hall and Ward argue, the Garden City was to be 'an exercise in local management and self-government' (Hall and Ward 1998: 28). It was envisaged that some people would build their own homes with capital loaned by friendly societies. Hall and Ward continue:

> It was a vision of anarchist co-operation, to be achieved without large-scale central state intervention. Not for nothing did Howard admire Kropotkin. Garden City would be realised through individual enterprise, wherein individualism and co-operation would be happily married.
>
> (Hall and Ward 1998: 28)

The self-build concept initially favoured by Howard has been taken up. In Lewisham in London, for instance, a group of self-build houses designed by Walter Segal reasserts the possibility for a non-speculative development in which families and individuals collaborate in the construction process. But there are more direct occupations, too, in squatting and squatter gardening.

Peter Lambourne Wilson writes of walking around New York's Lower East Side imagining gardens everywhere: 'Every vacant lot was a pocket garden. Most of the rooftops were blooming with flowers and vegetables, and dripping with ornamentals, giving each block the baroque tropical look of the Hanging Gardens of some Babylon-on-the-Hudson' (Wilson 1999: 7). Squatter gardens were, by then, disappearing under the pressure of real estate, but Wilson argues that 'Gardening will emerge as one of the major economic forces' of anti-globalization (Wilson 1999: 34).[11] The squatter gardens in the Lower East Side were spontaneous, beginning in 1973, supported by the City a year later. The means of that support, however, was a system of licences which limited access to legitimacy and gave the City power to de-sanction gardens when developers identified plots for their more speculative form of neighbourhood improvement (Ferguson 1999). John Wright argues that 'Community gardens are targets because they are liberated zones, areas free from consumption and mediation; at a time when the very idea of urban public space is under assault' (Wright 1999: 128); and Sarah Ferguson that 'More than green spaces, New York's gardens are microcosms of democracy, where people establish a sense of community and belonging to the land' (Ferguson 1999: 78). The perceived threat of squatter gardens was, then, that they were socially and culturally, as well as human-scale, models of sustainability outside the dictates of money and the administered world.

The sustainability of squatter shacks, or informal settlements, as housing for poor people was demonstrated by John Turner (1976). Today, some informal settlements are being legalized in north Africa and Latin America (Fernandes and Varley 1998). But the act of informal building is not confined to non-affluent countries. Skinningrove on the Cleveland coast in north-east England is an industrial village marooned by the closure of the pits and reduction of steel working, written off in the Durham County Plan. Men have coffee mornings, but they also have pigeon lofts – huts they build themselves from scrap wood. Most can no longer afford to keep pigeons and use the huts as a place to go during the day. Most huts have armchairs and a stove, the odd one a fish-smoking box. Many have stores of materials found on the beach and collections of shells or driftwood. This informal accumulation of huts is an interesting comparison to the parochial ordering of Poundbury or Arts and Crafts vernacularism of Letchworth. But while the link to locality is vital, it may not be size as such which matters most. Some informal settlements, after all, are large and sprawling but may be no less sustainable as a result. Perhaps a factor is the culture in which a model for settlement is located. Howard saw Garden City in a moral culture of improvement, as such, a top-down model, also informed by North American suburbia. A different tradition is suggested by self-build schemes and squatter gardens, closer to that of anarchism (by which, albeit at a literary level, Howard was influenced via Kropotkin). Anarchism has, at times, involved making settlements; but it also, like many recently founded eco-villages, entails the direct democracy of consensus decision making in place of both top-down planning (as in a representative democracy) and the divisive act of voting.

This can be fraught. In 1895, for instance, a notice appeared in the *Newcastle Daily Chronicle* seeking 40 acres for an anarchist land colony. William Key and Frank Kepper duly leased a site at Clousden Hill (Coates 2001: 203). In 1897, *Freedom* reported that 17 adults and two children were resident there, but under various pressures the colony divided into factions. Chris Coates records 'The attempt to reach consensus on the way ahead resulted in a seemingly endless round

*Figure 10.2* Poundbury, Regency-style terrace. (Source: courtesy of Malcolm Miles.)

of increasingly acrimonious meetings, with the only thing that everyone could agree on being to ask visitors to give a week's notice of their arrival' (Coates 2001: 205). It is a familiar story. Consensus decision making, nonetheless, is key to the self-organizing lives of many intentional communities and eco-villages which proclaim the slogan 'Another World is Possible'. ZEGG (Zentrum für Experimentelle Gesellschaftsgestaltung) at Belzig, Germany, is an example; home for 80 or so dwellers. This scale means that personal issues necessarily inflect community decisions and need to be faced openly. An eco-village affords none of the anonymity of a large city. ZEGG practises consensus decision making (and free love). Meetings can be long and sometimes difficult, but are key to social cohesion in their accommodation of differences, in every sense, between dwellers. The result, unlike Letchworth, is not a 'sanitised utopia' (Wilson 1991: 104) and ZEGG, like many other such settlements, demonstrates the viability of alternative means of living. The question, obviously enough, is how such processes can be mapped onto urban development in which far larger numbers of dwellers are present.

I do not see, still, why the social architectures of settlements such as ZEGG cannot be adapted to urban sites. The facing of difference and the time taken to work through decision making by dwellers of equal status (and provided with equal access to information), could be mapped onto local decision-making fora. This implies a shift away not only from the social engineering of post-war planning but also from the aestheticization of urban space in much current urban design and redevelopment. Jane Jacobs argues that '*a city cannot be a work of art* … Confusion between them [a city and a work of art] is, in part, why efforts at city design are so disappointing' (Jacobs 1969: 373); and that a city's public life is underpinned by an informal, non-planned use of spaces: 'The casual public sidewalk life of cities ties directly into other types of public life … Formal public organizations in cities require an informal public life underlying them, mediating between them and the privacy of people of the city' (Jacobs 1969: 57). The difficulty is simply, though it is not so simple in practice, one of scale. Yet if the rational-comprehensive model of planning was a means to apply professional expertise (and implicit ideology) to large cities, perhaps participatory and action planning are fitted to neighbourhood-level development. Perhaps the provision of infrastructure, which needs city-wide planning, need not be the basis for other kinds of decisions on the character and uses of the urban fabric. Oddly, the idealism of, say, the City Beautiful movement of the nineteenth century, with its insistence on visual harmony and vistas, is more closely allied to the planning of large cities than to small ones, despite a traditional association of beauty with objects small enough to be seen as single entities.

This leaves me wondering whether it is not so much that small is beautiful as that human scale enables diversity. That might be the lesson of a tradition of English radical practice from the Diggers to direct action (Sheehan 2003: 81–90), anti-roads protest (McKay 1996: 127–158) and occupation of a redundant brewery site in south London in 1996 (Schwartz and Schwartz 1998: 54–65). The squatter gardens of the Lower East Side provide a further precedent for dweller-led development which sets its own boundaries. But there is no reason a similar model could not operate in networked form across a large city. It is unlikely that aesthetic unity would be preserved but, although aesthetic unity underpins the, largely visual, appeal of small cities in certain areas of urban design, or for developers of urban villages, I would argue against confusing it with social or cultural cohesion. Actually, I do not think it matters what a settlement looks like other than in the eyes of those who occupy and fashion it according to their needs (which do not exclude but may endlessly redefine visual pleasure). In the end, I see the idealism of Garden City as its undermining. It is an idealism linked to small scale, but which could as easily have applied to a larger city (as in Howard's Social City). Jacobs criticizes Howard for ignoring the cultural complexities of city life and avoiding the politics of a city's daily operation (Barton 2000: 22). I would add that Howard's diagram is exactly that: a representation distanced from the actualities of occupation it describes *and which it in time prescribes*. The attitude of prescription, however, is tied to neither small nor large cities.

At Port Sunlight there is a sharp contrast between the half-timbered workers' homes and the white stone columns of the neo-classical art gallery. An obvious reading would be that the art gallery is for the educated middle classes and the houses for the deserving poor. Or, more likely, that the poor will be morally and socially improved by visits to the gallery. The assumption is, then, that *the poor cannot improve themselves*. Writing on Thamesmead in south-east London, Edward Robbins reads a 'deep suspicion of traditional working- and lower-class neighbourhood form and possibly unconscious and unstated distrust of the life it is assumed to produce' rooted in a tradition of reform and nineteenth-century socialism: 'From Dickens to Morris, Dore to Kingsley, the solution to the old city lay in the open spaces and single family homes of the middle classes' (Robbins, 1996: 289). The result is crushing:

> a deep distrust of working- and lower-class life phrased as a critique of aspects of the physical condition of the neighbourhood. The working and lower classes cannot be allowed to reproduce their old spatial patterns … Images of middle-class familialism and individualism replace images of the spaces of working-class solidarity and sociality.
>
> (Robbins 1996: 289–90)

Barton cites Gordon Cherry (1996) that the preferred model 'remained the decentralist tradition based on Howard's garden city' (in Barton 2000: 22). The outcome, parallel to the taming of the working class as a secondary middle class, is a suburbanization of the city. I do not think a vogue for small cities will do much to change this and may affirm it. What matters more, perhaps, for sustainability is the process of human-scale decision making.

## NOTES

1 Agnotti argues that 'only 20 per cent of the world's population lives in metropolitan areas, and only about 33 per cent in cities over 100,000' (Agnotti 1993: xv).
2 See Schoon (2001: 234–235) on the Millennium Village at Greenwich in south-east London.
3 In context of comment on Alice Coleman's *Utopia on Trial* (1984) and Alison Ravetz' idea of dispersed urban communities (1980). Wilson notes that Coleman was an adviser to Mrs Thatcher.
4 Schoon notes a letter to the *Observer* (21 May 2000) in which Richard Rogers dismisses Poundbury as Hardyesque nostalgia.
5 These include schemes at Mylor Bridge near Falmouth, Clyst Vale and Exminster in Exeter, Moorhaven on Dartmoor (previously a lunatic asylum for Plymouth), and St Clements Vean in Truro.
6 Republished in 1902 as *Garden Cities of Tomorrow*, London: Swan Sonnenschein.
7 Hall and Ward see three influences: the Crystal Palace in London, from the 1851 Great Exhibition; the glass-roofed arcades of central London, such as Burlington, and Leeds; and the winter gardens 'which were then just becoming a prominent feature of English seaside resorts' (Hall and Ward 1998: 21).
8 Bellamy writes of a future state of the world (in 2000) in which 'Every man, however solitary may seem his occupation, is a member of a vast industrial

partnership' in which all, regardless of ability, receive equal benefit in a system of state provision (in Carey 1999: 286). Wells writes of a society in which the middle classes have mutated into the Eloi, living above ground and playing in the sun, while the working classes have become an underground rabble, the Morlocks, who eat them. In *When the Sleep Wakes* (1899), over-population leads to huge, enclosed cities surrounded by industrialized agriculture, and literacy is dead. In *Anticipations* (1901), Wells foresees a state practising genetic selection (in Carey 1999: 367–72). Wells joined the Garden City Association after a talk by Howard at the Fabian Society (Hall and Ward 1998: 30).

9 A key example of urbanization is Cerdá's 1859 plan for the extension of Barcelona (Miles 2004). For comparison of English and French development strategies, in Kingstanding and Villeurbanne, see Meller (2001: 222–52).

10 Fishman writes of 'Cottage suburbs and garden cities … [which] seemed to extend amorphously over three hundred square miles on the monster map [of Los Angeles]' (Fishman 1987: 156).

11 Wilson explains that the term avant-gardening was derived from Dreamtime Village, Wisconsin, an international arts community.

# 11 Small cities for a small country

## Sustaining the cultural renaissance?

Graeme Evans and Jo Foord

This chapter[1] explores the implications for smaller cities of adopting culture-led regeneration strategies. It is suggested that there is a divergence between cultural planning for long-term sustainable urban cultural renaissance and culture-led makeovers which rely on externally oriented projects devised to draw in new visitors, residents and enterprises. Drawing on evidence from Sheffield, an industrial city in South Yorkshire, northern England, it is suggested that by thinking big, small cities have been seduced into entering a culture-led city competition in which the stakes are high and the prospects of success limited.

### CULTURE IN THE URBAN RENAISSANCE

Cities have been placed at the centre of government policy (if not the centre of power over resources[2]) in the UK with core objectives of promoting creative, competitive economies and enabling social inclusion and participation. For the first time in the history of urban policy the goal is to encourage people and businesses to return to, rather than leave, city centres. An 'urban renaissance' has been championed to counteract socio-economic deprivation and the sprawling car-dependent edge cities which emerged in the 1970s and 1980s (Department of Transport and the Regions 2000). Increased residential densities, mixed-use development and the reuse of brownfield sites are now commonplace in spatial plans for central city areas and they have been accompanied by calls for improved design standards, quality open space and an enhanced public realm (Commission for Architecture and the Built Environment 2004). This physical renaissance, combined with re-population and new economic investment, was initially focused on the metropolitan and core cities of the UK (Greater London Authority 2004; Core Cities Working Group 2004) but is now driving planning policy and practice in market towns and smaller settlements, as well as major planned urban development areas and new town extensions, notably the Thames Gateway (South East) and the Northern Way.

Embedded within this strategy to re-centre cities and towns is a parallel concern with promoting an urban *cultural* renaissance. Here there is a general desire to capture and renew all that is best about cultural activity in cities – the ability to bring strangers (domestic and immigrant) together; to provide platforms for cultural engagement and creativity; and to foster social inclusion (Department for Culture, Media and Sport 2000b, 2004a). Thus there is recognition that it is within cities that cultural expression is forged, not just in the established cultural institutions but also in the informal spaces and events of everyday life. This attention to culture as an 'urban way of life' brings the identities of cities, places and communities and their quality of life to the forefront (Evans and Shaw 2004; Matarasso 1997). The development of Local Cultural Strategies by UK local authorities linked to 'Best Value' performance indicators has encouraged the strategic use of cultural resources for the integrated development of neighbourhoods, cities and regions (Gilmore 2004).[3] Using a cultural planning approach (Evans 2001; Mercer 2003), the focus is not meant to be on cultural amenities per se nor the local cultural infrastructure, but rather on how strategic intervention into cultural activities can foster the development of a place, its communities and its economy. The aim is to encourage and support a cultural renaissance from within city communities.

However, other aspects of the cultural renaissance have had a higher profile, in particular, culture-led physical revitalization and creative industries-led economic revival. In Europe and North America these have become commonplace and they are increasingly included in the urban strategies of a wide range of developed and developing economies (Evans *et al.* 2005). Flagship cultural buildings, iconic cultural institutions, cultural and heritage quarters alongside cultural events, festivals and markets have all been used to kick-start both physical regeneration and visitor economies. Instead of cultural practice emerging from urban life, here it is being mobilized as the dominant driver of the post-industrial urban economy, pushing cultural activity and creative enterprises to the mainstream of the urban renaissance and inward investment (Evans 2001; Parkinson and Bianchini 1993). The cultural economy has thus begun to re-shape urban cultures as consumption cultures (Zukin and Smith Maguire 2004), even in former production districts which now serve as heritage locations combining residual design production with visitor activity, such as the Lace Market, Nottingham; Jewellery Quarter, Birmingham; and Clerkenwell crafts quarter, London (Evans 2004). Placed at the leading edge of the post-industrial economy, cultural assets have become a key resource in inter-city competition where specific cultural (including sporting) projects are used to deliver re-branding and economic development (Evans 2004; Garcia 2004a; Stevenson 2004). Strategies initially based on extracting the economic advantage of local arts and cultural activities have been replaced by initiatives encouraging the growth of cultural and creative clusters (Evans *et al.* 2005) mirroring current popular policy belief in the value-added of economic clustering and related social networking (Porter 1990, 1993). The perceived goals are economic diversification and employment creation as well as image enhancement, with further external benefits arising from the compact city and mixed-use development – namely crime reduction (natural surveillance, vitality), less car usage and therefore less pollution.

It has been argued that such economic and cultural transformations of cities are being led by three (overlapping) social groups. Indeed the competition between cities is increasingly played out through culture-led strategies aimed at attracting one or all of these groups. First, there is a new group of urban citizens (young and single or older and child-free) who are discovering the advantages of city living and engaging in high-consumption lifestyles (living near to work or live-work, networking and socializing in convivial settings, taking advantage of the city's entertainment and cultural activities, being part of a diverse community) (Zukin 1998). Second, there are day visitors and tourists who are being drawn to city destinations such as specialized retail and entertainment areas, night-time zones of bars, clubs and restaurants, as well as themed ethnic enclaves and heritage, or market-based tourist attractions (Shaw *et al.* 2004). Finally, there are those with 'creative talent', identified and popularized by Richard Florida (2002). This 'new social class', he claims, is not only reshaping how work is being undertaken but, through a flair for both bohemian ('boho') city living and entrepreneurialism, is also at the vanguard of transforming city spaces.

Much of the national and international research on documenting and evaluating the progress and impact of this urban cultural renaissance has focused on metropolitan and core cities (Evans *et al.* 2005). Strategic emphasis on gaining competitive advantage and attracting investment, new residents and visitors has prioritized major investments in high-profile infrastructure such as galleries, concert halls, sports and exhibition arenas, new public art and spaces (Garcia 2004b; Bailey *et al.* 2004), in branding places (Evans 2004), and in developing and promoting cultural or creative economic clusters (Mommaas 2004).This culture-led city competition favours metropolitan and regional centres where past investment in cultural institutions and arts education has resulted in a comparative over-representation of established cultural activity. This uneven pattern of cultural investment has been strengthened by the current tendency for new creative enterprises (for example in advertising, design and new media) to locate in particular metropolitan and regional centres (Bryson *et al.* 2004; Grabher 2001; Pratt 1999).

Nevertheless, there are examples of culture- and creative-led change in smaller urban centres that suggest that the cultural approach to an urban renaissance may be appropriate for smaller cities. Christopherson (2004) notes that many small towns and cities have specific resources that can be drawn on to build a cultural and creative economy, such as a local college or university, a health facility or other major public or private sector institution. Here the key is to draw on an existing pool of 'knowledge' workers as both audience and participants in new cultural activities and creative enterprises.

A small city, Sheffield, has also provided the first and one of the most influential examples of intervention aimed at changing place and economy through cultural production. Sheffield's Cultural Industries Quarter (CIQ) has been used as an exemplar in UK national policy and also internationally. Based on providing local authority- and European-funded work units and production space in a ring-fenced area of the city, the CIQ was aimed at retaining creative talent (especially in popular music, crafts and the visual arts) within the city's economy and so providing alternative forms of employment in a city that had had heavy losses from its steel industry in the 1970s and 1980s (Urban and Economic Development Group 1988). This also reflected a move to distance regional cities from the capital, London (Fisher and Owen 1991), using a combination of national and European regional structural funding. This was a strategy adopted by 'cities of culture' including Barcelona, Bilbao, Dublin and Glasgow. Resistance to the centre was a long-standing sentiment, as was active opposition to the trap of becoming what Borsay coined as the 'little Londons … importing their theatrical and musical performers from the metropolis … [and which] provided a blueprint for many other areas of provincial urban life' (Borsay 1989: 286–287). In Sheffield's case, maintaining a local presence with national reach has required upgrading its theatre quarter and appealing to contemporary drama and popular events alike (such as the televised World Snooker Championship held at the Crucible Theatre since 1977).

The initial driver for Sheffield was jobs and enterprise rather than a new cultural identity for the city – more a new-industrial than a post-industrial cultural renaissance. Sheffield had been an exemplar innovative city in an earlier age, through its specialist skilled steel craft and manufacturing base. Alfred Marshall had identified the benefits of agglomeration here (as well as in Sweden) as the foundation of his (creative) cluster and milieu theories which drive current creative class and local economic development initiatives today:

> The leadership in a special industry, which a district derives from an industrial atmosphere, such as that of Sheffield or Solingen, has shown more vitality than might have seemed probable in view of the incessant changes of technique. Yet history shows that a strong centre of specialised industry often attracts much new shrewd energy to supplement that of native origin, and is thus able to expand and maintain its lead.
>
> (Marshall 1919: 287)

Inspired by Sheffield's initial audacious intervention and influenced by aspirational toolkits (Landry 2000), or transfixed by blockbuster accounts of urban change (Florida 2002, 2005), other smaller urban centres throughout the UK have mobilized their local arts and creative/cultural sectors in their pursuit of regeneration and growth through cultural production and consumption.[4] Several have developed relatively successful niche interventions around one particular art form or a group of creative practices (Aitchinson and Evans 2003; Cooper 2004; Murray 2004; Wood and Taylor 2004).

However, there are pitfalls. The intense concern with inter-city competition, place-making and leveraging in middle-class populations and inward investment has tended to sideline the wider goals of holistic and socially inclusive cultural planning (Chatterton and Unsworth 2004). In the UK, a proposal to subsume Local Cultural Strategies with Community Plans has raised concerns among some policy analysts that, despite central government endorsement for a wide understanding of culture (Department for Culture, Media and Sport 2000a, 2000b, 2004b), this process of withdrawing from community-based cultural activity for its own sake will be accelerated (Gilmore 2004). Furthermore it may be that smaller cities and towns have been seduced into an unequal culture-led city competition in which they can never succeed. The competitive cultural advantage of metropolitan and regional cities leaves smaller cities at an irreconcilable disadvantage. If small cities focus their strategies and investment on the same types of externally-oriented projects used by metropolitan and regional centres they can fall short on some of the key elements which enable metropolitan success. Small and geographically peripheral cities often lack a sustained year-round critical mass of trade and cultural activity, generating the level of audiences, tourists and visitors that are required for regeneration to be viable (this is notwithstanding the short-term benefits of some highly specialized festivals or arts events in smaller towns and cities). Small cities can also find that externally focused projects are easily detached from local cultural identity, leading to increased social division and therefore undermining the sustainability of any regenerative effects (McCarthy 2005).

Given such health warnings and accumulating evidence which suggests that few culture-led regeneration projects fully meet all the set social and economic objectives (Evans and Shaw 2004; Evans 2005), it is surprising that many small cities and towns continue to pursue conventional 'creative' strategies. This may, in part, be due to the persuasive nature of claims for the universality and growth prospects underlying the new creative economy.

## CREATIVE SUCCESS FOR ALL?

In his 2005 book *Cities and the Creative Class*, Richard Florida argues that his original thesis on the rise of the creative class has been misunderstood. He suggests that he did not mean to imply that the creative class were an elite vanguard of the new urban economy and society (2005: 3–4). He goes on to argue that he chose 'creativity' as his key concept because every human being has creative potential and the creative society (and by proxy economy) is open to all. This notion that creativity is a limitless resource is central to the power of the culture/creativity-led transformation notion and why small towns and cities become seduced into believing that they too can use the cultural renaissance to transform their communities.

The clusters associated with creative class enclaves are also predominantly a male, white, under-thirty-year-old group, whose shelf life is limited in terms of establishing a residential community (Nathan 2005). In the case of Sheffield, for example:

> when people buy an apartment to live in – they've been there a year or two and are now at a point in their life where they want children or they want a garden and they move to the suburbs … Having a school in the city centre is a waste of time – it's not going to be needed – there are so few kids that live in the city centre and there *are* going to be so few – I don't see a change in people's attitudes to having children in the city centre. On a Friday and Saturday night, it's full of drunk young people. Is that the kind of environment you want to bring kids up in?
>
> (Interview: Local Estate Agent and Resident, Sheffield 2005)[5]

This is a fundamental challenge to the higher density, mixed-use model which is the basis of the government's Sustainable Communities vision (Office of the Deputy Prime Minister 2003). In mature inner-urban areas which have pioneered residential development through industrial re-use and mixed-use design, local amenities and schools do not follow the commercial developments (Aiesha and Evans 2006) and even where more enlightened developers plan school provision as part of their masterplans (as in New Islington, Manchester, by *Urban Splash*), local education authorities will not back these plans if it means building a school only to have it lie empty while a new community settles and grows.

In a comparative study of creative industries interventions worldwide (Evans *et al.* 2005), several US mid-west cities rejected the notion of this kind of creative class and of the creative industries associated with them, preferring to stick to the *cultural* industries and their link to local heritage and identity. For these small cities, regeneration and quality of life was for 'local people', not for an imported group of creatives in downtown lofts.[6]

## NEW GROWTH THEORY

A principle tenet of the creative economy thesis is that creativity, the ability to use human imagination to develop new or original ideas or objects, has different properties from any other input into the economy. In a challenge to neo-classical models of exogenous economic growth where inputs are used up in the process of production, the concept of a creative economy draws on 'new growth theory' (Cortright 2001; Romer 1993) and its analysis of 'ideas' as endogenous drivers of the knowledge economy. Essentially, it is argued that creativity produces a limitless supply of ideas and knowledge which can be shared, used and developed by more than one person, group, organization or firm at a time. Creative knowledge, in theory, can be accumulated without limit. However, the intangibility of creative knowledge, its transfer through tacit and explicit exchange, makes the capture of potential economic value uncertain. Yet this uncertainty also means that any number of innovative events, products or services can occur and there is always the possibility of unforeseen innovations.[7] It is also argued that the benefits of new ideas in the knowledge/creative economy flow to individuals or economic actors other than those who created the new knowledge. This 'spillover' is central to new growth theory, suggesting a necessary openness and an ability to draw inputs from dispersed fields. Though attempts are increasingly applied to restrict this process through intellectual property rights (Santagata 2002), Cortright (2001) argues that such mechanisms are likely to have negative effects by restricting the relatively free flow of ideas on which creative knowledge depends. This openness and uncertainty therefore means that creative knowledge remains undervalued until it becomes embedded in market-tested innovations. In some respects, the rankings or 'Creativity Indices' used by Richard Florida and others suggest a link between the presence of creative talent and the rate of innovations. The indices correlate selected location advantages with creative milieus. They also highlight socio-cultural divisions, spatial inequalities and gentrification effects. Highest scoring places on Florida's composite creativity index also score high in terms of inequality between the highest and lowest earners (e.g. in Boston, San Francisco, Silicon Valley, New York and London). In the UK, the creativity index has been tested in an analysis of regional cluster economies and was found to show little or no significant link between higher productivity and creativity, as defined, and no demonstrable link at all with innovation, despite the knowledge-based creative industries the creative cluster supposedly represents (Department of Trade and Industry 2004).

Nonetheless, 'new growth theory' proposes that there are limitless opportunities for the creative economy to produce an unpredictable number of new products, services and types of enterprise. Following this evolutionary approach, generating innovations from creative ideas is a process of trial and error. Furthermore, it is suggested that small interventions (by an individual, private enterprise, government body or institution) can produce disproportionate and lasting returns. A particular role for public institutions in cities is identified:

> As the world becomes more and more closely integrated, the feature that will increasingly differentiate one geographic area (city or country) from another will be the quality of public institutions. The most successful areas will be the ones with the most competent and effective mechanisms for supporting collective interests, especially in the production of new ideas.
>
> (Romer 1992: 89, cited in Cortright 2001)

In many senses, new growth theory which implicitly underpins much of the current claims for the 'creative economy' suggests a win–win scenario for small cities: exploitation of ideas in the creative economy is not limited by finite resources and, within an open and astute entrepreneurial/institutional/organizational context, successful intervention can take place within the unpredictable business of capturing creative knowledge for innovative products and services. Although large cities, especially metropolitan and regional cores, may have an historical advantage of cultural investment, a critical mass of knowledge and creative workers and the exposure to diversity identified by Jane Jacobs as the key to the production of 'new work' (Jacobs 1969), new growth theory implies that, although creative knowledge is neither evenly distributed across nor within nations nor equally accessible in every location, its unpredictability implies it can emerge anywhere where institutional openness coincides with the tacit exchange of creative knowledge. Big things can grow from small ideas in small cities. As Scott suggests, 'Provided that the right mix of entrepreneurial know-how creative energy, and public policy can be brought to bear on the relevant developmental issues, there is little reason why these cities cannot parlay their existing and latent cultural-products sectors into major global industries' (2000: 209). However Scott also sounds a note of warning: 'As the experience of many actual local economic development efforts over the 1980s demonstrates, it is in general not advisable to attempt to become a Silicon Valley when Silicon Valley exists elsewhere' (Scott 2000: 27). This warning is even more pertinent in the early twenty-first century as, in blind faith, metropolitan cities the world over develop major campus-based digital media cities and corridors – from Seoul to Singapore and from Barcelona to Berlin (Evans *et al.* 2005). Small cities have a tendency to try to follow in their footsteps, encouraged by their enterprise and regeneration agencies.

## Cultural renaissance in towns and small cities

Wood and Taylor (2004) appear to provide positive reinforcement of the new growth thesis in their account of Huddersfield's mobilization of culture for regeneration. A former woollen mill town, west of Sheffield, Yorkshire, set within a wider de-industrialized metropolitan area, they show how at a point of economic and political crisis a combination of new open leadership at the local authority, combined with risk-taking public sector and not-for profit local arts workers with plenty of creative knowledge, an economic development partnership including the local 'new' university and entrepreneurial bidding for European funding, resulted in a 'creative milieu' which aims to: 'initiate a wide range of diverse projects designed to find, stimulate, nurture, attract, harness, exploit, recycle, embed and keep creative and entrepreneurial talent, in order to rebuild the prosperity of the town' (p. 382). The Huddersfield Creative Town Initiative has delivered site-specific physical regeneration, with premises for creative enterprise and cultural activity, a festival-based visitor economy, clustering of creative enterprises, new jobs and a change of image for the town. Although the initiative has now ended, evidence suggests that the number of creative jobs, albeit from a low base, continues to grow and although no one body assumes responsibility for the creative milieu, several new organizations have taken forward cultural and creative projects, alone and in partnership with both the University and private sector. General lessons were drawn from the Huddersfield experience and a generic model to 'create a form of renewable urban energy' has been proposed (*Cycle of Urban Creativity*[8]) which, it is suggested, can be used to inform strategy development for the cultural renaissance in other small towns and cities. This developmental approach is thought to be particularly beneficial in towns and cities where there is little obvious 'tradition of creativity', economic uncertainty or a reluctance to participate. The authors do note, however, that the cultural renaissance is not overtly evident on the streets of Huddersfield. The socio-economic landscape changes slowly and this change has to be continually mapped and demonstrated. In line with evolutionary thinking, serendipity could have played a part in Huddersfield's success – 'There is an argument to be made that Huddersfield was simply in the right place at the right time' (p. 393).

A different picture emerges from McCarthy's (2005) review of the cultural renaissance in Dundee, East Scotland. This is a small river-port city that has experienced the annihilation of its staple industries ('jute, jam and journalism') and heavy losses of jobs in light manufacturing during the 1980s restructuring (see Chapter 8 this volume). Here, the cultural renaissance was centred on the development of a 'cultural quarter' anchored by a new building, the Dundee Contemporary Arts Centre, complementing the long-established Dundee Repertory Theatre, and bolstered by two successful computer games spin-off companies linked to the nearby Abertay University. An initial driver behind Dundee's cultural quarter strategy was a perceived need for the city to signal a presence within the national cultural sector: Dundee's cultural image within the Scottish (and UK's) cultural sector was poor despite a number of key local institutions (including 'the Rep' and Duncan of Jordanstone College of Art and Design). The identification of Dundee as a key hub in a growing Scottish digital media cluster boosted the rationale for a cultural quarter (Scottish Executive 2000a).

McCarthy suggests that it was a fervent desire for a TV- and newspaper-friendly image makeover that made the idea of a 'cultural quarter' attractive to the City Council. This rationale is ready-made rather than bespoke. A number of critical points are raised in the evaluation of this small city's use of a cultural quarter-led approach to the cultural renaissance. The strategy was neither home-grown nor locally embedded. It was the result of 'policy transfer' derived from experiences and applications elsewhere (in particular from Sheffield) and delivered through external consultants' reports and feasibility studies. Consequently it has been suggested that:

> While there are accepted examples of 'good practice' in culture-led regeneration, acceptance and application of such good practice would seem to be frequently based not on formal evaluation or analysis but largely on anecdotal grounds. The process of policy transfer may therefore be one of 'serial replication' rather than sensitive adaptations to context that take into account the peculiarities and specificities of local aims and circumstances.
>
> (McCarthy 2005: 287)

Such copycat tactics undermine any potential uniqueness of Dundee's cultural renaissance and therefore jeopardize the desired aim to change the image of both the city and its culture. The link between local cultural abilities, assets and desires and the development of ubiquitous cultural facilities – workspaces, galleries, rehearsal rooms, cafés, bars, restaurants and new residential spaces – is questioned. Dundee's 'cultural quarter' is unhooked from the wider city, especially its large socially-deprived population living in peripheral housing estates at some distance from the resources of the cultural quarter. McCarthy goes as far as to suggest that this strategy is likely to undermine inclusive indigenous cultural activity in its pursuit of external promotion, validation and audiences.

The argument has been made that 'hard branding' through cultural projects is becoming less sustainable as cities – metropolitan, regional and small cities alike – struggle to find and then maintain a 'cultural USP'.[9] Metropolitan and regional cities are becoming culturally homogeneous and are in danger of losing tourists, day visitors and new investment (Evans 2003). For Garcia (2004a), too much imitation in culture-led regeneration is problematic. Driven by economic rather than cultural agendas, they failed second-tier cities in their competitive struggle to rise up the global hierarchy. Policy diffusion to smaller cities is in danger of replicating these inherent problems.

Nevertheless, new growth theory suggests that creativity is universal if unpredictable. One element of Dundee's story is more in line with this analysis: the two internationally successful digital media firms grew as an unpredictable consequence of developments in the research and teaching activities of Abertay University where computing and creative arts merge in the development of games software, virtual reality and 3D-imaging. This creative success, which is transforming the cultural identity of Dundee, has had little to do with the designation of a cultural quarter and much to do with the tacit transfer of creative knowledge, its spillover and, once again, serendipitous timing. The small intervention that delivered significant results was first within the higher education institution; latterly it has been through financial support from Scottish Enterprise and in infrastructure development of both new and reclaimed premises. The triumvirate of university, boosterist city regime and symbolic or residual industrial heritage, is now a formula for city growth and for achieving the hope value promised through a combination of the cultural renaissance and the knowledge economy. Regional cities such as Birmingham and small cities such as Sheffield have navigated local and national political regimes (and regime change) over the past 20 years to reach their current position and renewed focus on central city development.

## Creative Sheffield: from productive culture to consumption culture

As already noted, the model from which many subsequent cultural-quarter strategies drew inspiration originated in the small 'large' city of Sheffield. Sheffield is the smallest of the English Core Cities outside of the metropolitan centre in the UK.[10] It is not the regional (capital) centre. This is Leeds, a city which has also undergone an externally-focused programme to deliver cultural renaissance and has been labelled 'boomtown' because of its rapid growth (Chatterton and Unsworth 2004). It also competes with former industrial cities such as Bradford, Doncaster and Rotherham, and lacks the natural and political advantages that cities such as Barcelona and Birmingham possess in their respective regions.

Benchmarking city performance has been one goal of the UK government's Core Cities initiative, in a similar way to the measurement of European cities against a norm as the qualifier for regional assistance. In this case those city-regions below 75 per cent of the European average GDP are eligible for support, and Sheffield and adjoining regions have been designated under Objective 1 of the European Regional Development Fund (ERDF). Sheffield, like other UK-assisted cities, is facing the move out of this ERDF zone as the benchmark is recalculated following the accession of poorer East European members and economic growth and returns from decades of public investment take effect. This means that the Euro gravy train will pull out of such city-regions and the dependency culture will be challenged. An acceleration of city redevelopment and private sector involvement encouraged by a boosterist council is one response to this scenario. Politically, this has required a conversion from urban new-left to centre-right policies, which is also consonant with national politics. Comparisons with the English Core

Cities therefore look to continental Europe as their reference points. This has seen Sheffield's economic performance compared with that of cities such as Stockholm and Helsinki (see for example the 2002 prospectus – *Creative Sheffield: Prospectus for a Distinctive European City in a Prosperous Region*). Since these are established national capitals and natural cultural capitals, such comparisons not only represent the 'small city thinking big', but also set up cities to fail. Competitive advantage cannot be manufactured (without continued public subsidy), whilst it is comparative advantage that small cities such as Sheffield might better build upon.

Sheffield has a population of 516,000 and some of the worst deprivation in the UK. The national Index of Multiple Deprivation in 2004 shows that 10 of its 29 wards fall within the lowest 10 per cent, that is 65,000 households of which 27 per cent are receiving Income Support. The city is sharply divided between two wards in the south-west, which are amongst the most affluent in the UK, the low-scoring wards and a number of extremely deprived neighbourhoods within higher-scoring wards. The proximity of wealth and poverty within the city boundaries is unusual (Sheffield City Council 2005). In recent years, in a bid to attract a new population, the expansion of higher education in the city has led to rapid increases in the number of students living in the city. In key city-centre areas (such as the Devonshire Quarter), one-third of the population is aged between 16 and 24.

The high concentration of students (45,000 between the two universities) is matched by a similar clustering of white-collar and professional workers in the city centre – two universities, the City Council, the Department of Trade & Industry HQ, the Regional Passenger Authority, teaching hospitals and law courts. This fuels the social divide between the city centre and the peripheral and rural hinterland, and is reflected in house values and the demand for school places in a small area west of the city centre. Thirty per cent of employment in the city is in public service (versus 25 per cent in the UK as a whole). The city's proximity to open space also defines its character, transport usage (a high car commuter traffic into and out of the city) and the limits to its urbanism: 'the Sheffield mentality – we are not London or Leeds – maybe Sheffield people like the character of the city – they don't want it to be a Leeds or a Manchester. We are like a big village' (Interview – Estate Agent and Local Resident, 2005, *Vivacity 2020* [5]).

Over 50 per cent of the population lives within 15 minutes of the open countryside, with one-third of the metropolitan area lying within the Peak District National Park itself. The city's topography – built on seven hills with five rivers – has defined both its industrial past and the contemporary urban renaissance (see Figure 11.1). Quality of life is a key factor and selling point in attracting and retaining students, new residents and employers. Sheffield has the potential to trade on its locational advantages, despite its secondary position in the regional city hierarchy. Safety and crime are an important measure of 'liveability' and city image. Sheffield claims to be one the safest cities in the UK (www.sheffield.gov.uk/facts-figures). This is the case for violent crime, but burglary and theft from vehicles is 50 per cent higher than the national average (Office of National

*Figure 11.1* View over Sheffield from University of Sheffield. (Source: courtesy of Graeme Evans and Jo Foord.)

Statistics 2001), an indicator perhaps of an economically divided community and problems of social cohesion.

## CULTURE CITY AND CITY OF SPORT

Though culture (including sport) was the means of Sheffield's early move into the urban renaissance, it was as a route to economic development rather than an end in itself. Moss (2002) notes that the motivations for intervention through culture were the sudden and heavy loss of employment in the city; the politics of localism including the fervent opposition to the Conservative central government; and the rising influence of claims for the economic contribution of the arts and creative enterprise. These came together at a time when there were significant amounts of national and European funding available for regional economic development.

Like continental European cities, large-scale public infrastructure projects were seen to lever private investment and engender public pride. This included public transport (South Yorks Supertram) and new sports facilities (Don Valley, Ponds Forge) in preparation for hosting the World Student Games in 1991. Neither projects achieved their hoped-for impact – passenger targets were not met (Lawless and Gore 1999) and the Student Games were a financial disaster for the city which carried the debt burden for several years. Local sports and community facilities, including four older local swimming pools, closed, effectively reducing access for groups most in need of public leisure facilities. Swimming participation in the city had been in decline, relative to the national trend, despite a fillip in 1991 which was short-lived and followed by a continued decline (Taylor 2001).

Positioning Sheffield as a Sport City continued in the 1990s. The city played host to some of the football matches for Euro96 (but with marginal economic benefits); over 300 sporting competitions have been held there since 1991 and it made a successful bid to locate a centre of the new UK Sports Institute in the city, exploiting its sports facilities and growing sports studies and science faculties in both of the city's universities (Sheffield and Sheffield Hallam). The national headquarters of the Institute was, however, located in Manchester. Sheffield featured in a popular film, *The Full Monty* (1997), which Edensor characterizes as 'nostalgically lamenting the passing of the industrial landscape and way of life' (Edensor and Kothari 2004: 39). A group of unemployed men strive amongst the 'abandoned forges and smelters, echoing the uselessness of male labour habituated to a working life in a steel industry which has drastically trimmed its workforce' (ibid.). The city seeks to put such images of the industrial past behind itself, often using the more positive rhetoric of the cultural renaissance, quality of life and the new (smokeless) economy. Industrial inheritance, where used, features as either thematic museums (Metalcraft Museum) or as residential re-use and backdrops for heritage tourism.

An out-of-town shopping centre at Meadowhall also reinforced the city centre's decline, as experienced elsewhere in the UK, encouraged by land-use liberalization and car-borne economic development exploiting motorway access and cheaper land opportunities. This opened the floodgate to an urban fringe landscape of sheds, mono-use multiplexes and shopping malls (Evans 1998). A decade of policies by the incoming New Labour government to reverse this spatial and policy drift has only now started to see town centres revive. This is evident in Sheffield and other city centres such as Manchester, by increasing population numbers and residential density.

Local public sector expertise in leveraging regional development funding also grew rapidly in the 1980s and 1990s, especially in the Cultural Industries Quarter (CIQ), starting with a music venue (Leadmills) and the provision of work units in collaboration with the Yorkshire Artspace Society. This was followed by training and exhibition space, all enabled by public sector funding. The CIQ, from its inception in the mid-1980s to date, was public-sector-led and dominated by not-for-profit activity. It became an extension of the service provision to the cultural sector. This has created a funding dependency which, Moss argues, has eroded the capacity for sustainable and creative development of both cultural production and cultural consumption in the city. 'Through its strongly interventionist approach Sheffield locked itself into a particularly narrow interventionist model from which it is now difficult to expand because of geographic and competitive disadvantage' (Moss 2002: 212).

In 1999, Sheffield did attempt to extend its range of activities through the building of a National Centre for Popular Music, a museum and interactive visitor attraction in an iconic building designed by Nigel Coates and funded by £17 million of National Lottery and European (ERDF) grant aid (see Figure 11.2).

*Figure 11.2* Lapdancing Club and Music Training College/Student Union Centre. (Source: courtesy of Graeme Evans and Jo Foord.)

Other regional cities had exploited this opportunity where 'form followed funding', using extensions of national collections to create cultural tourist venues such as the Imperial War Museum (Salford), Tate Gallery (Liverpool; St. Ives, Cornwall) and the Tower of London (Royal Armoury, Leeds).
However, again it was a radical move for a small city to attempt this transition with no history of national museum-based tourism. The National Centre closed within nine months due to poor attendance and loss of confidence in its limited content. This raises critical issues for small cities. A top-down approach to developing a visitor attraction is risky and requires an exceptional collection, 'experience' and investment. In the absence of other tourist attractions, an existing consumer culture, experience of curating national specialist collections and passing footfall, it is not surprising the project failed. The long-term consequences damaged Sheffield's image and capacity as a destination and raised the spectre of the 1991 Student Games. The building is now being used by the local Hallam University as its student union centre.

The CIQ, though beginning to attract a wider range of media and other employers, is still dominated by small and micro-enterprise with marginal profitability. Its history isolated it in an industrial landscape with little development of either the public realm or the mixture of uses and activities which have made other cultural quarters attractive to both new urban dwellers and visitors (Mommaas 2004). Organic development of related enterprises – cafés, independent bars, small retailers, service industries and professional activities related to the creative industries – is absent in the CIQ (this isolation is symbolized by the Workstation café and cinema which is only open part of the week). An exception is a lapdancing club situated between the music training college and ex-Pop Museum buildings (see Figure 11.2).

Attempts by the newly established CIQ Development Agency to promote diversification through the public–private partnership in the development of heritage buildings, new residential building and dual (rather than mixed) use, are evidence of its exit strategy, albeit still dependent on city resources. Whereas this has brought a new residential population to the CIQ, this is largely made up of students who are housed in mono-use developments with limited aesthetic appeal and whose spending power is both limited and drawn to subsidized specialized venues

*Figure 11.3* Student accommodation, Cultural Industries Quarter. (Source: courtesy of Graeme Evans and Jo Foord.)

within their higher education establishments. Adjoining retail property has lain empty for up to two years (see Figure 11.3).

For Moss, the long-standing culture of a highly controlling local state, looking to the objectives of funders, actively prevents the kind of small incremental creative enterprise which allows a change to take place. Likewise Dabinett (2004) is also critical of this legacy and questions current masterplanning approaches to Sheffield city centre 'quarters', which are artificial rather than vernacular. Residential property development (if not occupation) is booming in Sheffield as land is released for private-sector investment. However, the residential and activity mix is not necessarily creating the Creative Sheffield identified in the Core City Prospectus (*Creative Sheffield* 2004). 'Studentification' (Smith 2004) produces a transient population and limited market demand for (creative) goods and services. Even the Devonshire Quarter, a small central area north of the CIQ identified as a vibrant 'mixed-use' area is, in reality, dominated by new residential developments, which have seen new property bought by investors for lease to students, graduates and young tenants (see Figure 11.4 and note 5). Non-residential amenities and ground-floor commercial and leisure uses for these North American-style apartment blocks have been absent or under-used, leaving the area less than vibrant. This reality is lost on the extreme hype used by the Single Regeneration Budget (SRB) Community Participation Newsletter (*City Talk* May 2005): 'The Devonshire Quarter, renowned for smart city living and innovative retail establishments, is one of the most cosmopolitan places to shop in England.'

New growth theory assumes that creativity can emerge in unexpected places as long as there is an openness to the institutional culture. However, in Sheffield, the legacy of a centralized, public funding-dependent local state and its current overly enthusiastic boosterism and welcoming of residential property developers appears to be stifling the environments in which creativity can thrive. As Drake (2003) found in his interviews with creative producers, it is the old industrial spaces where manufacturing still hangs on that provide inspiration and connection to the creative process. Despite the *Creative Sheffield* strap-line and new projects for an E-Campus, Advance Manufacturing Park and the City Centre

*Figure 11.4* Devonshire Green – West One Plaza. (Source: courtesy of Graeme Evans and Jo Foord.)

Masterplan, a familiar but flawed top-down rather than bottom-up approach is being replicated.

## CONCLUSION

Although it was heralded as a model of culture-led urban renaissance, Sheffield, with its CIQ, sports and new Central City 'Quarters' Plan is struggling to deliver the changes perceived necessary to attract and retain the new urban populations and visitors who are seen to be central to the consumption-dominated cultural renaissance of metropolitan and regional cities. Sheffield's experience, like that of Dundee, suggests that the promises of an externally-driven 'new economy' and 'creativity' are not easily delivered. Strategies largely focused on external audiences, markets and investors in a competition with cities higher up the urban hierarchy, make high-profile initiatives for cultural change in small cities a precarious business. If successful, the dangers are of exacerbating existing social divisions within the small city community; and if not, the loss of local amenities can make the opportunities for cultural activity, with all its attendant benefits for social inclusion (skills, health and education), far more difficult for those communities most in need. The promise of this kind of new 'creative' economy may indeed be the 'Emperor's New Clothes' (Pratt 2005). For all cities, small and large, a sustainable cultural renaissance is more likely to emerge from within city communities where a variety of cultural amenities, activities and providers respond to and challenge local contexts. This does not mean becoming isolated and inward-looking. The cultural renaissance thrives on stimulation from beyond local borders.

However, as in all vibrant cities, linkages, networks and flows of ideas need to be more than one way. A cultural renaissance is not a quick fix. It does mean quality cultural opportunities for everyone – to be entertained as well as enlightened, to create for leisure and pleasure as well as enterprise, to consume culture and to participate in its everyday production.

## NOTES

1. After Richard Rogers (1997) *Cities for a Small Planet*, and Richard Rogers and Anne Power (2000) *Cities for a Small Country*.
2. Between 80 per cent and 90 per cent of local government financial resources are controlled or distributed centrally, including business property taxes (UBR) and regeneration programmes (including European), or are ring-fenced according to central government formulae, e.g. education.
3. Instigated by the Government Culture Ministry (Department for Culture, Media and Sport 2000b) *Creating Opportunities*, these are voluntary although most local authorities have completed or are preparing their own Local Cultural Strategies (LCS). In future these may well be merged with more wide-reaching Community Plans where the significance of culture may be minimized.
4. See the many creative and cultural strategies of numerous smaller centres, for example *Creative Plymouth* (www.plymouth.gov.uk).
5. This research is being undertaken by the Cities Institute in collaboration with University College London under the EPSRC *Vivacity 2020* programme (vivacity2020.org.uk). The project 'Generation of Diversity' is investigating mixed-use development in central city neighbourhoods.
6. In most cases this reflects a genuine desire to mobilize local assets for endogenous growth. However there is some anecdotal evidence that this rejection is driven by intolerance to the diversity and openness that Florida espouses, especially to gay communities.
7. Evolutionary economics is a growing area of debate. See Hodgson (2002).
8. This proposed generic model has five key elements identifying points in the urban creativity cycle – ideas generating capacity; ideas into practice; networking and circulating; platforms for delivery; markets and audiences. Each element may require input and support. Some cities may be good at one element and not others, therefore unbalancing the cycle.
9. Unique Selling Point.
10. UK government statistics rank 'large cities' as those with populations of over 250,000. The 14 largest used in the government's 'Urban Audit' averaged 444,000, placing Sheffield as the smallest of the large cities. In contrast, Sheffield's population of 516,000 is equivalent to only two of London's 33 boroughs whilst London's regional population is set to grow by a greater number than Sheffield as a whole over the next 15 years.

# 12 Creative small cities

## Cityscapes, power and the arts

Gordon Waitt

World cities are often heralded as cultural centres, foci of creativity and the arts (Hall 1998; Florida 2002). Metropolitan centres such as Paris, London, Sydney and Washington, DC, appear to dominate geographies of creativity and 'cultural industries'.[1] This lead is often attributed to their concentration of cultural infrastructure, large audiences, the operation of agglomeration economics and the desire of many people working in the arts to live in the 'buzz' of cosmopolitan suburbs (Scott 2000; Gibson *et al.* 2002). Florida's (2002) work has been influential amongst economic policy makers and moves to harness the creativity of cities. As Gibson and Klocker (2004) note, his ideas have resonated particularly strongly amongst Australian decision makers.[2] This thinking would suggest that, when compared to the creative nexus operating through the agglomeration of film, music, writing and painting in world cities, small cities lacking clusters of creativity are always 'less' (Gibson and Klocker 2005). Yet, as Rodger's (2005) study of small cities in non-metropolitan Victoria, Australia, illustrates, creativity is not an innate attribute of people living in large Western cities. Smallness itself would not appear to work against the creativity of people. Indeed, smallness itself may by regarded by creative workers, creative businesses and decision makers as being appealing, particularly in the context of sustainable futures. If so, why have recent regional and city policy makers interested in boosting economic indicators through cultural industries almost consistently ignored how cultural production operates in small cities?

To explore why decision makers often ignore cultural industry production in small cities, this chapter turns to the concept of cityscapes. Drawing on Olwig's (2002) ideas about landscape, cityscapes are conceptualized as always more than a way of seeing, an ideology, a place image or dreamwork. Cityscapes are also grounded in the material, given they are conceptualized as produced through and embedded in everyday life. Cityscapes are continually made and remade through everyday actions. Cityscapes are reflections of social practices. Therefore, beneath the groundworks and dreamworks of cityscapes lie very different social relationships. Cityscapes are sites of struggle. Power, and how it is exercised, is therefore at the heart of cityscapes. Drawing on Foucauldian theories of power, my focus is on the contested quality of the transformation of landscape through the polity of the creative economy. My aim is to try and make sense of how particular ideas of creativity have been folded into the neo-liberal discourses of governance that shape how the arts are manifested within the creative economy cityscapes of small cities. To provide empirical evidence, I draw on a specific Australian example; the city of Wollongong, Illawarra, New South Wales, Australia.[3] Here, neo-liberal discourses of creativity are strong amongst policy makers within a corporate-style city planning. Moreover, this is a city positioned as seemingly lacking creativity according to Florida's 'creative class index'.[4]

This chapter therefore focuses on the transformations of the cityscape of Wollongong through the embrace of neo-liberal discourses of creativity. In doing so, this chapter explores four questions often raised in regard to the contribution of the arts in revitaliszng small cities. (1) Why have arts become repositioned as a key motor of neo-liberal urban rejuvenation discourse? (2) What are the assumptions and meanings that redefine the role of art as part of the creative industries? (3) What are the social implications of urban revitalization policies that are reliant upon folding together the 'cultural' and the 'economic' through the arts? (4) And, finally, how does smallness operate with artistic creativity?

## CITYSCAPES AS CULTURAL LANDSCAPES OF GROUNDWORK AND DREAMWORK

How can a discussion of creativity in small cities benefit from the recent elaborations of cultural landscape as both groundwork and dreamwork – or, to appropriate Creswell (2003), the geographic *practices* that defined the material, representational and affective landscapes within which we live? Most importantly, these theoretical discussions of landscape provide an important departure point, conceptualizing the geography of creativity in terms of place-competitiveness in an increasingly globalizing economy. Florida's (2002) *The Rise of the Creative Class*, the most recent example of this type of thinking, tends to result in a geography that highlights the concentration of creativity into world cities and tends to ignore how cultural production operates in smaller cities or non-metropolitan areas. For Florida, creativity underpins the economic performance of large cities through the concentration of his so-called 'creative class', who work in agglomerations of creative industries, such as music, film, television and publishing. Creative people are themselves 'hip' cultural consumers; helping sustain creative industries through purchasing designer foods and homewares; attending concerts and galleries; and eating out at fashionable cafés and restaurants. Moreover, creative people are mobile. They actively seek out 'cool' places – expressed through a vibrant street life, architecture and arts precinct. Consequently, city quarters can be refashioned to accommodate the needs and wants of the 'creative class'. Planning for 'creative futures', small cities are provided with, paradoxically, rather uncreative blueprints derived from replicating the 'creative' buzz and facilities found in Western metropolitan centres of creativity (see Ryan 1992; Landry 2000). Plans to reposition de-industrialiing small cities within local, regional, national and global economies are triggered by 'place wars' and norms of creativity found in the world cities of Paris, London, New York and Sydney.

Thinking about the role of arts in creative industries through the lens of landscape theory draws attention to the relationships established between the city, policy and social justice. Cityscapes require creative industry policies to be reconceptualized in the context of everyday place-making activities. Thinking cityscapes as everyday practice draws attention to how they are continually fashioned through social webs that operate across different geographical scales simultaneously. Global cities are not self-contained hubs of creativity. As Gibson and Connell (2004) demonstrate for indigenous popular music in Australia, metropolitan centres rely upon creativity occurring far beyond the city limits. In another sense, cityscapes materialize the needs, ideas and laws of the people who make them. Creative industries policy is, then, just one of many ongoing transformations to cityscapes. But these policies are particularly significant as their transformative potential is taken in order to 'improve' economic futures. Yet, how the dreamwork of policy makers' creative city becomes grounded is always contingent upon the very different relationships that exist between the cityscape, the arts, the municipal authority and residents. There is always the potential for struggle in how the arts should be folded into creative cityscapes. At the same time, this is often a struggle for social justice. Therefore, a conceptualization of cityscape able to examine, respond to and intervene in the transformation of cityscapes must include a theory of power.

In a Foucauldian approach, power is not conceptualized as a stable, pre-existing influence possessed by an individual or agency, but is negotiated through discourse. Neither is power conceptualized as repressive. Instead, following Foucault (1980), power is everywhere and can be productive. This Foucauldian approach suggests that both planners and residents are caught up in the circulation of power through negotiating discursive norms. In this case, municipal authorities and residents are negotiating a range of social positions regarding the value of arts in a 'creative Wollongong'. Consultants' reports, town plans, official websites and image campaigns are important mechanisms that circulate official discourses about the value of the arts. Equally, some residents express their agency and ideas by lobbying council. It is through investigating how policy makers and residents, in their everyday lives, weigh up, act and respond to the discursive norms presented by the impact of creativity to economic performance of the city, that the conceptual tool of cityscapes can help provide insights to the role of the arts in the 'revitalization' of Wollongong.

## Wollongong: cityscapes of steel

As a city, Wollongong has ambiguous qualities. On the one hand, located on the east coast of Australia, 80 km south of Sydney, Wollongong is now part of greater metropolitan Sydney, with a population approaching 5 million. On the other, it is a city of around 250,000 people, separated from Sydney not only by the Royal National Park but also by indigenous and settler-Australian histories and geographies. In part, Wollongong remains a small city as urban growth is constrained by the Illawarra Escarpment. Spread along 30 km of coastline, contemporary metropolitan suburbs are former settler settlements with a mixed legacy of estates, coal mining, steel, resorts, shipping and fishing that now function, in part, as dormitory suburbs for Sydney. Once the homeland of the Wadi Wadi and Thurawal people, settler-Australians have since made this coastline their own, marginalizing Aboriginal people both physically and socially. Indigenous and settler-Australians continue to sustain their identities and sense of belonging through place-making practices of home, leisure and work (Hagan and Wells 1997).

Steel, not the arts, underpins the geographical practices of both past and present cityscapes in Wollongong. Steelmaking remains an integral part of the everyday lives of many people living in the city and the dominant place image for many non-residents. Cityscapes of steel (and coal) date from the 1920s following the building of a new harbour and establishment of a nexus of manufacturing industries around Port Kembla. Throughout the 1960s, Broken Hill Proprietary (BHP) positioned Wollongong in the geographical imagination of the Commonwealth as an industrial 'heartland', the Sheffield of the South (Eklund 2002). Between 1971 and 1981, BHP, or the 'Big Australian' as it was then known, employed over 20,000 people in Wollongong (Castles 1997). Since the 1980s, like many smaller manufacturing cities in older industrialized economies, Wollongong has not been immune to the emergence of a global steel economy and new production technologies (Haughton 1990). The cityscape was transformed. Between 1981 and 1996, BHP downsized to around 6,000 employees, escalating local unemployment rates (Haughton 1989). Yet, at the same time, productivity per employee increased. Consequently, steel remains a primary source of employment and attribute of the cityscape. Nevertheless, a place image and city identity once built on the strengths of steel became a source of ridicule. In local, State and Commonwealth planning circles the higher-than-national unemployment figures became a source of concern. Wollongong's place image was tarnished. In the national and regional imaginary, Wollongong was a 'rust city', or worse, a 'problem city', its cityscape 'obsolete' (Watson 1990).

Since the 1980s, the municipal authorities have therefore sought a new cityscape, one grounded in alternative geographical practices. Through the Steel Cities Program and other job-creation schemes, attempts were made to diversify into new sectors, including information technology and tourism. The Illawarra Technology Corporation is one example of diversification into information technologies. Less successful was the attempt employing Australian comedians Norman Gunstan and Grahame Bond, 'Aunty Jack', to position Wollongong as the 'Leisure Coast'. Undoubtedly, the coastal escarpment provides tourist spectacle. In regional terms, however, the legacy of steel meant that Wollongong as the Leisure Coast could never compete with place images of the Gold Coast or the Sunshine Coast in Queensland. Instead, the most notable transformations of the cityscape occurred through growth of the University of Wollongong, the Area Health Service and the Institute of Technology. These transformations are reflected in the growth of employment in community services (see Table 12.1).

*Table 12.1* Wollongong Employment Profile. The percentage of people employed in selective industrial sectors as a proportion of the total number of people employed in Wollongong, 2001.

| *Industrial sector* | |
|---|---|
| *Extractive industries* | |
| Coal mining | 1.2% (884) |
| *Health and community services* | |
| Education | 9.5% (6861) |
| Health services | 7.4% (5371) |
| Community services | 2.7% (1912) |
| *Creative industries* | |
| Printing and publishing | 0.7% (534) |
| Motion picture, radio and television | 0.4% (327) |
| Libraries, museums and the arts | 0.6% (438) |
| Cultural and recreational services, undefined: including music and theatre productions | 0.05% (41) |
| Retail and support services | 6.4% (4617) |
| *Manufacturing industries* | |
| Metal product manufacturing | 8.0% (5730) |
| Machine and equipment manufacturing | 1.8% (1316) |
| Total employment | 100% (71867) |

Source: Australian Bureau of Statistics 2001.

## WOLLONGONG: CITYSCAPES OF ARTISTIC CREATIVITY

In the 1990s, creativity became a buzzword that was added to neo-liberalism's lexicon of economic revitalization (that already included free trade, privatization, user-pays and global competitiveness). In Australia, the role of the arts in the competitiveness of the national economy in the emerging post-industrial world economy was enshrined in the 1994 *Creative Nation* policy of the Keating Federal Government, and the 1995 *For Arts Sake – A Fair Go* policy of the Howard Federal Government. According to these documents, the future economic prosperity of Australia lay in the cultural industries. The Australian government was not alone in aligning the arts to a knowledge-based economy that is internationally competitive. Positioning the arts as central to successful economies was also apparent in international reports (for example UNESCO's *Our Creative Diversity Report* (World Commission on Culture and Development 1995) and *In from the Margins* (European Task Force on Culture and Development 1997) and in case studies from other countries including Canada (Toronto Culture 2003), the USA (Hannigan 1998) and Singapore (Chang 2000)). National, regional and metropolitan economic futures lay in the talents and creativity of individuals. The creative industries were positioned as holding the potential to boost civic pride, establish social capital, enhance a sense of place and personal well-being and help to transform abandoned industrial precincts into heritage quarters. Postmodern futures were driven through the talents of a new 'creative class'.

In 1998, Wollongong City Council embraced the discourse of creativity, making cultural planning a central responsibility. Wollongong City Council commissioned a report to enable the preparation of a city-wide Cultural Plan: *Point of Take Off: Cultural Policy Framework and Cultural Plan 1998–2003* (Australian Street Company 1998a, 1998b). In Australia, this was novel. Cultural planning guidelines had previously never been a core interest of local government. Indeed, Wollongong municipal council was one of the first Australian authorities to develop cultural planning guidelines which, since 2002, have become formally required in New South Wales to comply with a Department of Local Government Circular (No. 02/33). Designed from an approach that prioritized public consultation, these documents depict the multi-faceted possibilities of arts in the city's future. The conventional role of the arts in facilitating 'belonging' through fostering social relationships that could help rebuild communities and create social tolerance was outlined (Mills and Brown 2004); innovative ideas were also presented for how the arts could be employed to create a local skills base for future creative businesses as well as more orthodox ideas of refashioning the city's identity (Australian Street Company 1998a). Wollongong municipal authority responded positively to a vision of the arts, understood in terms of their designation as part of a creative postmodern economy and socially diverse, tolerant and inclusive society. One immediate outcome of the first report was the branding of Wollongong as the 'City of Diversity'. Recognition was given to the value of cultural diversity in sustaining a creative economy in Wollongong. Moreover, a cultural approach to planning was given a clear priority.

In 1999, momentum for valuing the creativity of the arts as an integral part of Wollongong's social and economic futures continued to gain strength. A Cultural Audit was commissioned (Guppy and Associates and National Economics 2000), funded through a partnership between the City Council and Illawarra Regional Development Board. Visionaries behind this audit included the then Lord Mayor, David Campbell, and coordinator of cultural services, Barbra Wheeler. Results of this audit confirmed three key dimensions that defined the value of art as part of city planning: the economic, the personal and the institutional. Confirmation was given to the growing economic value of art as part of the cultural industries in both the national and regional economy. In 1998, the cultural industries contributed \$A129.4 million (2.7 per cent of total industrial output) to the Wollongong economy. In 1996, Illawarra residents spent around \$A204 million on cultural industries events, products and services (Guppy and Associates and National Economics 2000). Under the heading 'community cultures', arguments were presented for the personal value of creative activity. Projects were outlined that demonstrated positive outcomes, in terms of 'belonging', amongst socially marginalized people from the ability to record their experiences through projects such as Destination Port Kembla and Bellambi Safe Streets. Respondents valued the opportunity to counter their sense of loss in a cityscape that is seemingly constantly changing, as well as the recognition of social diversity and the expression of difference. Finally, under the theme of city image and cultural tourism, substantiation was given to the value of art in re-fashioning Wollongong's place image. Art, as part of the creative economy, could be employed to sustain ideas of Wollongong having 'energetic city spaces' (Guppy and Associates and National Economics 2000). The audit emphasized the conjunction of cultural policy addressing social justice issues such as 'belonging' with a creative industries agenda of folding the local arts into the economy.

One explicit outcome of the audit was funding for a new project targeting local artists. The Council became directly involved in a number of partnerships that prioritized local artistic talent as part of the creative economy. Illawarra artists were seemingly being repositioned as key drivers of a local creative economy. The Council became actively involved in helping local artists establish networks through the creation of the web page, Create Illawarra (http://www.createillawarra.com). A cultural broker or creative intermediary was employed for two years through a partnership established between Wollongong City Council, the Illawarra Regional Development Board and the Department of State and Regional Development. The broker's role was to establish networks to incorporate local artists into the regional economy. Partnerships were to be established between local artists and business, metropolitan, state and federal government agencies.

Council funding was also approved for Film Illawarra. An initiative of the Creative Arts Faculty of the University of Wollongong, this company is a partnership between the University, the NSW Department of State and Regional Development (DSRD), the Federal Government (Department of Transport and Regional Services) and with members of the Southern Councils Group, including Wollongong City Council. Proximity to Sydney and diverse cityscapes that include steel plants, beaches and rainforests makes Wollongong marketable to film makers (http://www.filmillawarra.org.au/index.htm). Through practices of Film Illawarra, cityscapes of Wollongong now include television and movie crews. Since its inception in 1999, Film Illawarra is estimated to have attracted about $A1 million to the region (Sharpe 2004). Films have also provided indirect tourism promotion and employment opportunities for local actors and film professionals as well as traineeships. Unquantifiable social value has also accrued to those local residents who enjoy watching their everyday places translated into stories on film and television.

Moreover, the practices of Circus Monoxide, since its council-sponsored relocation to Wollongong, have transformed an abandoned industrial warehouse into the headquarters and training space of a national performing company. This building is now the site of masterclasses run by international circus experts and educators. The skills of this professional company have inspired creative activities and enhanced local social capital. Two new amateur companies have emerged: Elixir Youth Circus and Wild Older Women's Circus (Circus WOW). The value to the women who perform in Circus WOW is not solely in terms of learning new skills and entertaining audiences, but also in creating a safe space for women. This performance space provides a forum in which to raise particular issues that concern them. Simultaneously, the Cultural Services Division of Wollongong City Council was increasing the visibility of the arts in cityscapes of Wollongong. A number of arts projects designed to enhance a sense of personal well-being and belonging have been implemented with socially marginalized groups in the suburbs of Berkeley and Cringila, as well as the Graffiti Wall for young people in the city centre (Buckland 2002).

At one level, policies acknowledging the importance of enhancing the arts in the cityscape are unremarkable. In the nineteenth century, 'The Arts' building in the coal mining settlement of the northern Illawarra provided a focus for education and maintaining a sense of community. Particularly since the 1950s, enrichment of the arts has been ongoing through processes of immigration; the population now comprises 29 nationalities, 24 languages and 22 religions. For many migrants, establishing and maintaining their ethnic identity relies upon sustaining particular cultural practices. Indeed, there are over 95 ethnic community/cultural associations (Australia Street Company 1998a); some of these support artistic practices including local bands, choirs and folk-dancing groups.

At another level, cultural planning in Wollongong is revolutionary. The plan disrupted conventional norms held by many Wollongong residents and the municipal authority. Surveys suggest that 'the arts' are normally associated with a 'night out' in Sydney (Australia Street Company 1998a). At the civic level, the leisure cityscape dominated by sports activities suggests a lack of appreciation for the creative arts. One consultant points out that: 'Arts-related activity is widely perceived to be 'soft' (feminine) and without significant commercial value' (Australian Street 1998b: 12). The culture/economy and masculine/feminine binaries are clearly illustrated in the ideas and practices of the Council. In 1998, the local government area contained only 32 'arts-based' facilities; this includes a City Library, Youth Centre, Performing Arts Centre and City Arts Gallery. In comparison, there were over 300 indoor, outdoor and water sports facilities; the discrepancy in the number of sports to creative arts facilities would appear to support an assertion that voluntary art-based associations have not received the same level of financial support and civic recognition as sporting facilities. Indeed, since the 1930s, sports have played in integral role in fashioning Wollongong's civic pride and identity, particularly surf-lifesaving and rugby league (Kelly 1997). More recently, civic identity and sports have been maintained through corporatized sports, specifically rugby league (St George-Illawarra) and basketball (The Hawkes). The heteronormative masculinity performed on amateur and professional sports fields easily translates into a civic rhetoric of competition, mateship, strength, performance and skill. In Wollongong, related myths of sport, steel and masculinity appear to have become complexly intertwined with civic pride and identity. Sustaining this masculine civic pride has been the apparent priority given by the municipal authority to funding and promoting cultural activities performed on sports-fields over those on canvas or stage.

At the civic level, there is also little evidence of previous municipal decision making that made any connections between creativity, economy and social justice. On the contrary, the Council appears to have actively *prevented* the role of the arts in these practices. For example, during the 1980s, strong partnerships between some local artists and trade unions meant that some miners, steel-plant operators and wharfies had been encouraged to think about art not only in terms of design and aesthetic, but also as an important political mechanism to express social injustice. The mutual benefits for artists and trades union members were acknowledged through the Australia Council's Art and Working Life Program (Australian Council of Trade Unions 1983; Cassidy 1983). In Wollongong, Redback Graphics is one example of the outcome of this exchange. In the 1980s, Redback Graphics heightened the social profile of Native Title, youth unemployment, exploitative working conditions and the threats of nuclear technologies. Along with the writer Julianne Schultz (1985), the film producer Mary Callaghan (1982) and playwrights of Theatre South and the Bread and Circus Theatre, this company also addressed the sense of placelessness generated through the demise of steel. Redback Graphics' inclusion at the Los Angeles Olympic Games Arts Festival brought international acclaim, yet this recognition was not celebrated locally. Instead, such an international profile was seemingly regarded as bringing shame on Wollongong: social injustices should apparently be silenced rather than profiled overseas. In 1985, the City Council therefore exerted pressure on this company to leave, a force applied by refusing to renew Redback Graphics' lease of council-owned property. Local or community art was seemingly deemed troublesome, if not militant.

At this time, a conservative municipal authority only sought the kudos that, potentially, art could bring to the cityscape through the work of internationally eminent artists. Consequently, civic support turned the city's engineering and industries' recommendation of organizing a competition to commission a sculpture for a public monument to honour the steel industry. In 1979, Ken Unsworth's 'Nike, The Steel Industry Sculpture', won this competition. Nike, seemingly, caught the judges' imagination. Perhaps the splendour of ancient Greek mythology could be bestowed upon Wollongong through Nike's references to the winged goddess of victory. Alternatively, Nike also referred to an intercontinental ballistic missile – which, for some, potentially glorified Wollongong as a city on the cutting edge of future energy, power and technologies (Vlandys 1999). Certainly, this role was envisaged by the Wollongong City Gallery Director, Tony Bond, who is reported to have said: 'Ken Unsworth's sculpture Nike, accelerating out of the earth like a space rocket taking man to new visions, will carry the name Wollongong around the world' (*Illawarra Mercury* 1979: 2). The symbolic quality of the sculpture was put to work by the City Council in place promotions as an insignia to pitch Wollongong as a winning location.

## WOLLONGONG: CITYSCAPES OF SCIENTIFIC CREATIVITY

Despite these achievements, city-wide cultural planning that incorporates ideas of Wollongong residents is seemingly no longer the driver of socio-economic futures in Wollongong. The durability of expressions of the arts in the cityscape through municipal authority initiatives appears to be losing strength, seemingly following the departure of the Lord Major, David Campbell, and of the Director of Planning, Mike Mouritz. Instead, following the conventional language of civic boosterism, definitions of culture have narrowed to 'good taste' (and of the arts to the aesthetic). Policies turn instead to science and technology. Hijacked by neo-liberalism's discourse of competitiveness, the arts are redeployed for institutional gain (of both the Council and University) to strengthen the external image of the city rather than the personal lives of residents. Questions of social justice emerge in the practices and place images of the public–private partnerships behind the Wollongong Image Campaign, 1999–2004, The Innovation Campus and The Creative Energy Centre.

In 1999, the Wollongong Image Campaign was conducted by marketing consultants commissioned specifically by the Council to undertake place-image making. Employing conventional ideas of civic boosterism, they designated a brand: the 'City of Innovation'. This place image is allegedly inspired by the strengths of the manufacturing and tertiary education sectors (Wollongong City Council 2004). The brand was adopted immediately. With the jettisoning of the slogan 'City of Diversity' went the ideas of the Cultural Plan that repositioned the arts at the centre of municipal planning. A city of innovation appeared to redraw familiar boundaries within the municipal authority that repositioned the arts as entertainment or promotional images and always separate from science and the economy.

The Image Campaign ran from 1999 to 2004 and received annual Council funding of around $A500,000. Priorities of the general manager in charge of this place-image campaign were directed to commerce rather than cultural services. The pitching of a city relied upon the quango, Image Wollongong, joining an existing network of business interests, Illawarra Alliance. Cultural planning was relatively absent or disempowered in Image Wollongong networks. Rather than focusing upon maintaining and enhancing creative skills of people already living in the city, the emphasis of this campaign was to sell Wollongong to potential businesses, students, gentrifiers, tourists and 'sea-changers' (those people seeking to leave metropolitan Sydney is search of a coastal 'village' atmosphere) as an innovative place. Consequently, although a number of arts festivals received Image Campaign funds, including Wine Wollongong, the Short Sited Film Festival, Mt Kembla Heritage Fair, the Illawarra Folk Festival and Viva La Gong (a ten-day arts festival), these events were only valued as an opportunity to showcase and promote the city to potential tourists, residents and investors (see Bradley and Hall, this volume). In place-marketing campaigns to brand a city, priority is given to aesthetics. Images derived from arts festivals became an integral part of advertising strategies that are deployed to signify 'good taste' and a 'happening', entertaining and cosmopolitan city atmosphere. Viva La Gong unquestionably established a vibrant atmosphere in the city centre.

The branding of Wollongong can be interpreted in terms of what Kotler *et al.* (1993) term 'place-wars'; the frenzy of place-marketing between cities to attract increasingly mobile capital and people. In the place-marketing of a city, allegedly unique attributes of a city's identity are lifted out of their social context and edited back together, depriving them of their initial meaning. As Shields (1991) points out, place-marketing always trades in oversimplifications, stereotypes and labelling. The brand logic of marketing places generates sameness and blandness through an indifference to complex social realities. Within networks sustained by the Image Campaign, arts festivals were valued primarily for their symbolism and aesthetic qualities. Not surprisingly, media images of the dreamscape of the new 'innovative' city would appear to express a very classed perspective through the affluent lifestyles and tastes portrayed in the re-branding campaign (of course, it is also the work of the place image to create the illusion there is no social injustice, corruption, violence, pollution or unemployment).

The Wollongong Innovation Campus is an example of groundwork that is implementing this latest branding. This campus is reliant upon a well-resourced partnership between the University of Wollongong, the Premier's Department of the New South Wales State Government, the Illawarra Institute of Technology and Wollongong City Council. The State Government has committed $A24 million. The proposal is to establish the campus with 'a world class Research, Development and Technology Campus' and the intention that the University will form strategic alliances with the business and research section' (http//www.uow.edu.au/about/community/innovation.html). The web page is quite certain about the synergies that will be created between research and the business of information technologies and telecommunications. Less conviction, however, is expressed about the potential links between creative arts and the film industry.

Following the removal of a film school from the proposal, some members of the University Faculty of Creative Arts are not happy with the low level of consultation. They have been left with the impression that the Faculty has been locked out of the decision-making process. Frustration about being marginalized within the planning process resulted in one faculty member commenting that: 'We [the Faculty of Creative Arts] will only be called in at the end, to hang the pictures on the wall'. As this response suggests, the value of art amongst key decision makers shaping this campus is narrowly framed by the aesthetic. The role of the arts in social justice is absent. Indeed, the site accommodating the Innovation Campus was once a hostel for steel workers and their families. The value of creative arts in highlighting the ongoing process of place-making and belonging was apparently dismissed as unimportant on the Innovation Campus.

Indeed, the proposed Creative Energy Centre demonstrates how the creativity of the arts is being prevented from becoming a more prominent part of the cityscape. Located in the city centre, a two-storey council-owned building has been the proposed focus for the arts as creative industries since 1998. Following Florida's (2002) argument of vibrant streetscapes to attract his 'creative class' and the lead of British cities such as Dundee, Scotland, the concept of a 'creative incubator hub' is used in planning documents to describe the concentration of people associated with small-scale, innovative 'start-up businesses' in the graphics, multimedia, sound and film industries (see Lloyd *et al.* this volume). The concept is to provide facilities often required by the creative class to start businesses. The creative hub is the focus of folding the arts into a creative economy. The Creative Energy Centre was the planned focal point of the city centre, renamed the 'Cultural Precinct'.

Despite successive consultancy reports prepared for Wollongong City Council that highlight the potential of a creative hub, the municipal authority has not acted on this proposal. Clearly, in the City of Innovation it is not creativity that has become unfashionable; nor is it neo-liberal arguments that Wollongong is engaged in place-wars and requires an image and industries that are competitive in the global economy. Equally, resistance to the logic of a cultural quarter and creative hub did not stem from fear of zoning-out local residents. There is no evidence to suggest that planners had considered arguments that inducing creativity through creating urban quarters may only act to fracture the city along socio-economic lines (Bell and Jayne 2004). The in-migration, property speculation, rising rents and subsequent displacement of lower-income residents may operate via processes of revitalization operating through a creative hub and Cultural Precinct. Nor are municipal authorities versed in critiques of the creative economy that suggest that those employed in creative artwork can experience variable rates of pay, long periods of unemployment, variable rates of job satisfaction and high levels of job insecurity (du Gay and Pryke 2002; Gibson 2003; Hesmondhalgh 2002). Instead, despite audiences illustrating the growing employment and economic wealth generated by creative industries in Australia, many key players in both local business and the council still do not appear to appreciate the potential roles of the arts in the economic revitalization of the city. For economic decision makers in Wollongong, discourses of the creative class appear to imply too big a departure from the norm, too radical. The arts remain positioned, amongst key decision makers, in the realms of entertainment and culture (ethnicity) or as a 'soft' option. Whilst the discourse of creativity has been well received by economic policy makers in Wollongong, the creativity of the arts is slipping from their agendas. Present policy makers appear afraid of departing from norms and accepting risks posed by embracing the arts in a small city still producing steel.

## CONCLUSIONS

This chapter reveals that Wollongong is not lacking creativity per se. Instead, through investigating Wollongong as a cultural cityscape, I have argued that the cityscape of creativity reflects the interests and ambitions of key municipal decision makers; people who have certain conceptions of creativity. Wollongong, as the City of Innovation, is a place where science and technology have an apparent monopoly on creativity. In contrast, the arts in the creative industries remain undervalued in the cityscape within local governance through discourses that continue to frame ideas about economy/culture and science/arts in narrow, elite and hierarchical terms. This underestimation persists despite a wealth of artistic talent, the existence of a cultural infrastructure and consultancy reports that outline innovative city-wide cultural plans.

Such insights help provide some assistance to addressing the four questions posed at the start of the chapter regarding the trend towards the arts in urban planning in small cities. It is clear, then, that the ascendance of the arts in the geographies of de-industrialization of small cities can be explained as part of the embrace of cultural planning by local, regional and national policy makers. Creativity of the arts became a 'tool' for addressing not only social justice, but also urban renewal and environmentally sustainable economic regeneration. Creative people are positioned as driving future economic growth through the potential synergies with film, media and information technologies.

Assumptions about the value of the arts in small-city planning are contested. Initially in Wollongong, creativity in the arts was valued in terms of building social tolerance and community, enhancing skills and increasing employment opportunities. These ideas became embedded into the cityscape through the presence of artworks that have increasingly become part of the everyday interactions of residents. Yet, since 1999, the Council appears less convinced of the arts as offering a (creative) economic future. A civic identity founded in the highly masculine activities of steel production and team sports seemingly still operates to position the arts as 'soft', 'feminine' and 'radical'. Therefore, it is perhaps unsurprising that priority appears once again to have been given to economic futures reliant upon the creativity of 'hard' sciences. Creativity in the arts becomes restricted to a limited role in place-pitching rather than place-making. The arts as a signifier of 'good taste' or as entertainment are particularly valued to help sell a specific place image to prospective investors or scientists.

Social implications arising from folding culture into economy would therefore appear to arise from the process of integration. On the one hand, folding the arts into *place-marketing* always runs the risk of imposing unconvincing ideas on a city (such as 'innovation' or the 'cultural precinct or quarter'), excluding long-term residents from the refashioned and commodified cityscapes and implementing urban policies oriented towards tourists, gentrifiers and sea-changers rather than addressing questions of social polarization and exclusion. On the other hand, a *place-making* approach to the arts values creativity not in terms of making a place competitive internationally but in terms of sustaining the everyday lives of residents. In this cultural planning approach, emphasis is given to public consultation, the transfer of skills and knowledge and social justice issues of belonging, including sense of community and sense of place.

Through the theoretical lens of cityscapes, small cities are shown not to lack creativity in the arts per se. Instead, in adopting this approach and drawing attention to the discursive structures underpinning the groundwork and dreamwork of economic policy makers, I have argued that it is conventional economic policy thinking that masks the creativity that is always present in people living in small cities. Ideas of small cities lacking creativity are framed through ideas embedded in particular classed ideas of creativity and the arts. Ideas about creative people and cities are imposed from elsewhere, usually New York, London or Paris. Moreover, the idea of creativity has become an essential attribute to succeed in neo-liberal economic policy discourses. This suggests that creativity follows a blueprint provided by consultants which is essential to compete successfully in the heightening place-wars of the global economy. Yet, these ideas do not take into consideration the geographical practices that underpin the everyday lives of people living in small cities. Smallness does not inhibit artistic creativity. Instead, artistic creativity is inhibited by the ability of some policy makers to depart from the conventions that govern the arts as separate from the economy, and neo-liberal policies that fold the arts into the economic in terms of place promotion.

## NOTES

1 This term refers to that sector of the economy fashioned by creativity. The Australian Bureau of Statistics (ABS) defines the cultural industries as comprising printing and publishing; film, video, radio and television; libraries and museums, music and theatre production; and retail and support services to these activities. The cultural industries integrate the arts, media, technology and telecommunications sectors from the stage of creativity, through production and distribution to marketing and consumption. Artists, craftspeople, dancers, film-makers and writers were thus repositioned as integral to the economic vitality of a city, region or nation.
2 To assist the development of Florida's creative class theory, he quantified the creativity of cities through employing statistical modelling and measures of a 'creative class index'. The index is composed of aggregate scores from a number of sub-indices; 'gayness', social diversity, 'talent', patent registrations and employment in creative industry sectors. In 2002, National Economics (2002) employed Florida's techniques to calculate the 'creativity' of each Australian region for the Australian Local Government Association's *State of the Regions 2002* annual report. Not surprisingly, the aggregate scores suggest that Sydney and Melbourne are Australia's only internationally competitive creative centres.
3 The empirical material for this chapter is drawn from a range of sources, primarily: reports commissioned by Wollongong City Council; the Australian Bureau of Statistics; and the Illawarra Regional Information Service; and semi-structured interviews conducted in December 2004 with employees of Wollongong City Council, Film Illawarra, The University of Wollongong and Illawarra Ethnic Communities Council.
4 Following the neo-liberal doctrine of place-competitiveness, the message sent to the Illawarra, the region in which Wollongong acts as the local metropolitan centre, was that it lacks a creative cluster of people to compete globally (Cutler 2003). This region scored high only on the ratio of the regional level of freelance artists to the Australian average (the 'bohemian index') (National Economics 2002). Whilst acknowledging that the spatial pattern of creativity

outside of Sydney and Melbourne strongly reflected other indicators of socio-economic marginalization (including lower educational attainment, higher unemployment and larger Aboriginal populations), solutions were presented in terms of market-based, competitive planning. In fostering creativity, the municipal authorities of Wollongong would appear to have much to do to succeed in a competitive world economy.

# 13 Rethinking small places – urban and cultural creativity

## Examples from Sweden, the USA and Bosnia-Herzegovina

Tom Fleming, Lia Ghilardi and
Nancy K. Napier

Over the past 25 years, the disappearance of local manufacturing industries and periodic crises in government and finance have increasingly made culture and the broader *creative economy* the business of cities and the basis of their tourist attractions and their unique, competitive edge (Zukin 1995). During the 1990s, in particular, notions such as the network society, the experience economy, creative cities and globalization were used to define new modes of production and consumption within the 'new economy' (Kelly 1998) and emphasis was put on the interplay between the economy and culture, as well as on creating crossovers between media and technologies (Amin and Thrift 2005). The result for urban policies was that, rather than selling just goods or services, even small cities began to mobilize tourism, the retail trade, architecture, event management and the entertainment and heritage industries, as well as the media and the wider creative industries, in order to produce and sell 'experiences' (Pine and Gilmore 1999).

This is true today of large cities but even more so for small or 'second tier' cities such as those presented in this chapter.[1] Essentially, the more outside competition these cities confront, the smoother must their operations be in order to harness their internal resources or their 'creative capital'. In this scenario, a dynamic and creative-led urban policy becomes part of the image of a city and acts as a catalyst for its symbolic economy (Verwijnen and Lehtovuori 1999). Thus, tourism, culture and the creative industries – the fastest-growing industries in Europe – play an important role in the urban image-creation processes, providing a major rationale for the aestheticization of city landscapes and the creation of new urban identities. This understanding of the complex and often contradictory nature of urban space is explored in Henri Lefebvre's notion of 'the production of space':

> Space is permeated with social relations; it is not only supported by social relations but it is also producing and produced by social relations (1991: 286).

To this he also adds that every society in history has shaped a distinctive social space that meets its intertwined requirements for economic production and social reproduction. This social notion of space – applied to the current concerns of small cities – will form the basis for the analysis of the three case studies presented in this chapter. In particular, we will argue that while new narratives of regeneration, urban culture and heritage have been employed in the conversion of these places into post-industrial, knowledge-oriented creative hubs, the process of implementing this regeneration has not been unproblematic. As more small cities compete (using similar mechanisms) in (re)producing and promoting themselves to attract a globally mobile middle class and other forms of flexible capital, their ability to create uniqueness diminishes, the economic benefits turn out to be short-lived and the 'creativity potential' is depleted. In *The City and the Grassroots*, Manuel Castells (1983: 314) has noted that the problem with this new 'tendential' urban meaning is that it creates the spatial and cultural separation of people from their product and from their history. The implication here is that these policies can have an effect not just on the urban form but also on governance and, ultimately, on social justice.

This chapter draws together case studies from three contrasting regions – southern Sweden, Idaho USA, and Bosnia-Herzegovina – to explore the divergent approaches to urban development policies and the broader 'creative economy approach' that smaller, 'second tier' cities adopt, according to different levels of political and economic stability, cultural integration and metropolitan aspiration (see Table 13.1). The three case-study cities – Malmö, Boise and Tuzla – are each going through a process of re-thinking their position in the urban hierarchy through a focus on creativity, new forms of governance and partnership – each with varying levels of success. These responses are provided, variously, against a context of political upheaval, cultural inferiority, low levels of economic capital, bureaucracy and relative prosperity. The contrasts between the case studies are marked, yet the ambition to steer each of these cities towards international competitiveness and local harmony – based upon innovative approaches to cultural planning and regeneration – reveals what these very different 'smaller cities' have in common.

## LEARNING TO BE CREATIVE THROUGH REGENERATION: THE CASE OF MALMÖ

With a population of 270,000 inhabitants, Malmö is Sweden's third largest city and the commercial centre of southern Sweden. It is a cosmopolitan and multicultural city where high-tech and knowledge-intensive activities are slowly replacing the old, traditional industrial structure that had given it its 'working-class' character since the 1960s. In particular, the integration of the Öresund region, brought about by the link with Copenhagen, plus other major infrastructural investments, is putting the city on the map along with the most advanced European 'second tier' centres such as Rotterdam or Lille.

The expansion of Malmö University College (with a student population of 21,000) and the development of the Western Harbour area, with housing and innovative workspaces such as the incubator MINC, has brought an atmosphere of youthful creative energy to the city. Cafés, open spaces, galleries and shopping

*Table 13.1* A brief introduction to the case-study cities

| |
|---|
| **Malmö, southern Sweden** is a progressive and forward-thinking city, with an increasingly qualified workforce, where high-tech and knowledge-intensive activities are replacing the old, traditional industrial structure. The expansion of Malmö University College, the opening of the Öresund fixed link between Malmö and Copenhagen, in neighbouring Denmark, and the regeneration of the Western Harbour symbolize how the city is aggressively repositioning itself as a player in the global urban competition game. An additional ingredient in this process of identity building is the (still ongoing) implementation of a number of housing and regeneration projects in the area of the Western Harbour. However, as the Malmö section will show, this process of identity rebuilding – from an industrial centre to a sophisticated, high-tech, new-economy-driven hub – poses a number of challenges which even a relatively successful city such as Malmö cannot ignore. |
| **Boise, Idaho** is by many measures a most unlikely place for urban creativity and the strategic pursuit of a creative agenda – it is relatively small, remote and, for many people even in the US, little known; it is prosperous and increasingly so; and it is located in a region not recognized for its creativity and the progressiveness this implies. And yet Boise is developing into a small urban creative city on the move. Unlike many cities, Boise is facing no serious economic crisis that pushes it towards embracing a creative-led agenda. Unemployment is typically below four per cent, housing starts and price increases are among the fastest growing in the US, and the city and state continue to rank high in top business, recreation and retirement reports. However, key people in business, arts, government and education are increasingly realizing the importance of creativity as a crucial part of maintaining the city's quality of life while also preserving a strong economic base. This is in some part linked to the city's long tradition of entrepreneurship: the city's key companies (many of which have been on the Fortune 500 list of the country's largest) stemmed from the pioneering efforts of a few hard charging individuals. Thus, the notion of moving forward, trying new ideas and accepting failure, has a precedent. |
| In **Bosnia-Herzegovina**, small cities are struggling to 'break into' international networks and to undertake calculated risks that advance their status and profile. Key strategic considerations often centre on re-building civic identity and intercultural pride from the 'bottom up', with memories from the 1990s war too painful and divisive to allow a more progressive international focus to take hold. For some, building an international role and profile is seen as a relative luxury; yet for many it is increasingly recognized as a necessary move if cities are to shed their internal parochialism and its dangerous consequences and climb inter-regional and international urban hierarchies. Vital is the contribution of relatively young leaders and intermediaries, less implicated in the war, more open to the cultural and creative opportunities of inter-city connectivity, and critically aware of the positive impact this 'openness' will bring to the political process and community. This section will introduce the ways the northern Bosnian city of **Tuzla** is basing its future harmony and success on a broad approach to 'cultural democracy', where every project, service or intervention is positioned as part of a wider cultural plan to raise the profile of the city to the 'outside world' while connecting locally to the aspirations of a local population which is for the most part keen to move on from the torment of recent history. |

areas are dotted all around the city and the overall feel is that of a compact, lively place. What was once a brash, blue-collar city, is now in the process of becoming an acknowledged centre for information technology and biotechnology. This transition, however, does not appear to have been an easy one to achieve for a city that still carries the burden of recent industrial failures on its shoulders. The 1990s recession hit Malmö harder than any other city in Sweden, with 27,000 jobs disappearing during a period of three years in the mid-1990s. As a port and business city, with roots going back to the Middle Ages, Malmö seems to have a peculiar 'genetic code' that makes it prone to constant swings between periods of growth and prosperity and decades of relative decline.

From an urban sociology perspective, one could argue that this is a cycle often observed in medium-size European port cities, and here the names of Hamburg, Bilbao, Rotterdam, Bristol and Glasgow come to mind. These are all cities that had a great history of commercial and urban development and then hit the rocks of recession in the 1970s. But, since then, they have all been able, to a certain degree, to reinvent themselves. This successful 'remaking' of smaller cities is often underpinned by a focus on the resilience, creativity and cultural mix of these urban centres. In the case of Glasgow, it was its internationalism as well as its urban and cultural heterogeneity that made it possible for the city to successfully begin – in the early 1990s – to reverse the cycle of decline which, until then, seemed intractable.

Similarly, early commercial development brought cultural variety to the city of Rotterdam, so much so that it was precisely this successful cosmopolitan mix of skills and potential that saved the city in the early 1980s from the decline of the shipbuilding industry and encouraged the planners and policymakers to design a new regeneration plan (the Binnenstadsplan). This, in turn, was the first step towards the successful creation of a new identity for the city. This notion of urban and social mix, along with the importance for cities of having a creative milieu, are increasingly seen by urban commentators as central in the making of successful and competitive cities. In particular, the argument put forward by economic development experts such as Allen Scott (1998), Jeremy Rifkin (2000) and, more recently, Richard Florida (2002), is that today's economy is fundamentally a 'creative' economy.

In his study of what makes cities and regions grow and prosper, Florida observes that, rather than being exclusively driven by companies, economic growth is occurring in places that are tolerant, diverse and open to creativity, mainly because these are the places where creative people of all types want to live. Scientists, engineers, architects, designers and artists are all part of a new creative global class that cities need to nurture in order to be able to compete internationally (Florida 2002). So, by extension, Florida's message is that development policies need to be aware of the benefits of creating an environment in which tolerance of different lifestyles and a good quality of life for everybody living in a particular place go hand in hand.

In the Swedish panorama of old industrial cities attempting to recover new functions, Malmö seems to have done better than other cities. From its thirteenth-century origins, rooted in the herring trade, to the splendours of its Danish period and the great developments of the nineteenth century, characterized by the expansion of the textile industry and the shipyards, Malmö is a city open to external influences, proud of its achievements and not afraid of taking risks.

Now that its once-thriving docks are being redeveloped into both University

Island and the new housing and business district of the Western Harbour, Malmö is learning to develop new ways of dealing with planning and governance. This new approach is the result of a process of '*learning by doing*' which has its roots in recent attempts by the city to implement large regeneration projects using the tools of masterplanning and welfare policies – typical of an old utopian Swedish tradition – despite the 'discovery' that these tools don't work any more in an environment in which cultural diversity, fragmentation of interests and new forms of democratic representation dominate.

## Learning to be creative

> The big issue for Malmö is recreating its identity. The image of an old industrial city belongs to the past; we now have to come up with a major project for the future, something indicative of a shared vision.

With these words, back in 1999, the City Director, Ilmar Reepalu, articulated Malmö's aspiration to join Copenhagen in the new Öresund region while at the same time sending out a message that the city needed to be bold in its thinking about the future. The response of the city came in the shape of the Draft Comprehensive Plan published less than a year after Reepalu's speech. The Plan singled out a number of mainly infrastructural projects for implementation – such as the expansion of the University and the transformation of the Western Harbour – which in the minds of the planners were to give tangible benefits to the whole of the city's economy. However, these projects ran into difficulty as soon as they started, because of a number of social and economic reasons, the most important of which was the underestimation by planners and policy makers alike of the social changes the city was undergoing at the time.

The late 1990s was a period of great upheaval for the city and while refugees from Kosovo and other European war zones were increasingly choosing to settle in the city – putting a strain on Sweden's legendary liberal ideals of social and housing policy – unemployment (particularly in the public sector) was rising as the spatial segregation between the new middle classes and the old inhabitants increased. The launch of the regeneration of the Western Harbour (spearheaded by the 2001 Housing Expo Bo01) came right in the middle of these changes and, with mounting controversy in the press about the ability of regeneration to deliver benefits for the whole of the community and not just for the new middle classes, it looked as if, for a time, the success of the plan was in danger.

The root of the problem was later identified as the lack of support given by the local community to this big regeneration plan. In particular, the language adopted in the marketing literature for the Western Harbour prompted community leaders to criticize those in charge of the delivery of the plan for bypassing the basic principles of democratic accountability.[2] This reaction can only be understood by looking at the past 40 years of Malmö's history, whereby since the 1960s (a time of great urban expansion for both the inner city and the suburbs), economic growth and urban expansion were achieved through carefully nurtured relations between the political establishment (mainly the Social Democratic Party), the public sector, the banks and the construction industry. Thus, while in the past such transformations were the result of a shared, top-down, carefully planned long-term vision, today they seem to happen more as an urgent, 'ad hoc', response to perceived outside threats.

For any small city to go from an essentially industrial economy to a 'creativity and knowledge-driven' mode of production and consumption, all in little more than a decade, would be a challenge, but for Malmö the stakes are even higher. There are two reasons for this. The first, as mentioned earlier, is the 'historic baggage' of welfare and democratic accountability to which the city is still tied and the second is the rapidly changing social, economic and cultural environment in which policy makers were and are trying to operate.

With more than a quarter of the population having foreign roots, Malmö is the most multi-ethnic city in Sweden. Here, diversity also extends to a highly visible variety of lifestyles, political and ethical allegiances, consumption patterns and sexual orientations. This richly diverse environment requires constant renegotiations of trust along with redefinitions of legitimacy by local government. This is the task that local politicians and policy makers alike are learning to perform and the testing ground is the completion by 2010 of the cultural and housing redevelopment of the Western Harbour. The redevelopment will involve turning 160 hectares of harbour-front brownfield land into a fully developed new, ecologically sustainable, 'neighbourhood' of 10,000 inhabitants. Malmö's 'City of Tomorrow' is made up of the old Swedish ingredients of meticulous planning and a quality of housing and urban design of such standard that the 'new neighbourhood' has already been heralded as a prototype for the new European urbanism.[3]

So far, the city appears to have turned what could have been a 'high risk' development into a successful tool to attract interest and investment from Sweden and elsewhere, but what is interesting here is to look at the management and governance components of this success. In particular, it is worth noting a precondition, which is that, over the past ten years, Malmö City Council has increasingly assumed responsibility for urban and economic growth. This 'decentralization' of responsibility was implemented through the establishment, in 1996, of a 'Districts' Reform'. The Reform involved the division of Malmö into ten city districts with their own councils and administrations. The result was a much leaner government with the ability to respond quickly and flexibly to local needs. This new 'culture of flexibility' prepared the ground for the establishment of the public–private partnership that is currently presiding over the development of the Western Harbour. Although the lead organization is the City of Malmö (backed by the Swedish government), this partnership includes private development companies (a total of 13 developers), the university, business leaders and residents' associations. The partnership is a flexible mechanism that is allowing the city to shorten the time span considerably between planning and implementation and it is a strategic mechanism capable of delivering on issues ranging from the infrastructure for the creative economy to education and training and housing.

Examples of the projects implemented recently through the partnership include the creation of MINC, an incubator for high-tech start-ups, now expanding into a key support mechanism for local design companies, and the creation of University Island (the new university campus which specializes in new media and communication). These two projects were achieved in parallel with the creation of 500 ecologically-sustainable residential units, which are currently rented or sold to occupants working in creative jobs in the Western Harbour. Though implementation has turned out to be quite a steep learning curve for the city, this level of regeneration and risk-taking in a 'small town' such as Malmö is still unprecedented in Sweden and especially as far as 'governance' is concerned. Here, too, Malmö is experimenting with new mechanisms aimed at ensuring democratic participation and a higher degree of transparency in the decision-making process in relation to large regeneration programmes. One such mechanism is the City Planning Forum set up two years ago. The Forum is a permanent place where the planning department can hold exhibitions, meetings and seminars on the subject of Malmö's urban developments (especially those in the Western Harbour). This is a strategic tool put in place not only for the dissemination of information to the general public, but also as a way of inducing a collaborative approach to the design and planning of the areas in need of transformation.[4]

In conclusion, the key lesson from this example is that 'little to lose', upstart, ex-industrial cities such as Malmö are performing a 'trailblazer' role in Sweden by challenging traditional, top-down approaches to urban and cultural development. In this case, a mix of risk-taking, flexible management, cross-disciplinary work and democratic participation combine to establish a broader understanding of the city's cultural resources that put Malmö on the map of cities to visit, invest and live in. Finally, Malmö's quintessentially 'adaptive' quality has opened the way to experiments in governance which other cities in Sweden and elsewhere may wish to follow.

## THE PIONEER MODEL: MOVING TOWARD THE CREATIVE ECONOMY IN BOISE

Boise, Idaho, is a most unlikely place for a creative-led approach to urban development – it is relatively small, remote, little-known and growing very quickly. Malmö is a more 'obvious' location for creative-oriented growth: it is closer to the metropolitan core of the continent, it is more diverse and it has a long history of industrial development. And yet, for several reasons, Boise is developing into a small urban creative city on the move, with several characteristics similar to Malmö. For example, Richard Florida's (2002) rankings have shown Boise, Idaho, to be one of the top US cities that are attractive to creative people (and Malmö is an increasingly promising propositon for aspiring creative people in Sweden). With such confirmation of an already strong position, Boise's decision makers and urban strategists recognize that a valuable intellectual market will be available only if the city grows its economic strength and remains an attractive place in which to live: creativity is thus an investment in future competitiveness and prosperity. This section explores how Boise is investing in its future by adapting approaches that contributed to its very existence: it was once a pioneer city on the edge of the American West, a place where only the most adventurous settled and invested. Today, new pioneers are seeking to establish a creative city for the future.

## Conditions defining Boise

Boise and Idaho are known for odd reasons or not known at all. Potatoes, racists and 'where's that?' are the most common image challenges. It is remote – Boise is the state's (only significant) population centre, with nearly 400,000 in a 50 by 30-mile valley. It has been know for being the 'potato state' (as the major supplier of McDonald's French fried potatoes), as a (former) neo-Nazi haven, until 2000 when the Aryan Nations lost a major legal battle and left the area, and for having 'no image', since people confuse it with states having similar sounding names (for example, Iowa, Ohio). The negative (or lack of) images of Idaho and Boise are fast being replaced by one of a high quality of life, a good business environment and an attractive retirement option. In the last few years, Boise has received much positive attention in a variety of press outlets as a place to do business, retire, or enjoy recreation. For example, it has attracted a collection of large firms (for example, Micron Technology) and start-ups and the state's largest university (more than 18,000 students).

However, remoteness continues to scare off potential employees and employers ('why would I move to the end of the world?') and makes some aspects of business life (international travel) slightly more challenging. The CEO of one of the city's top software firms – small, but with clients worldwide – recently commented that as his firm grows, it becomes harder to recruit. Without a large cluster of software firms in the city, high-tech experts are reluctant to move from areas like Silicon Valley, where their career options are broader. However, remoteness in Idaho does have its positive sides. Because of the lack of amenities, original settlers – and modern-day pioneers as well – have created a full spectrum of cultural, economic and social fabric in the city. In addition, having major corporations based in the city means that many managers have come from more cosmopolitan settings, and hence expect access to cultural and related activities. The city has its own philharmonic, ballet, modern dance and opera companies; it has two professional theatre groups – a nationally recognized summer Shakespeare Festival and a highly regarded contemporary theatre. While the city lacks the range of options of a metropolis like Seattle or San Francisco, it has cheap and accessible parking, relatively low event ticket prices and high-quality performances.

Today, Boise sits at the state's political, (newly developing) educational and business centre. It hosts the state capital, state agencies, branch offices of several federal agencies and even headquarters of selected federal agencies, such as the National Interagency Fire Centre,which coordinates all forest fire efforts for the country as a whole. Boise State University leads the state in the rate of growth of student numbers, with lead programmes in engineering, business and public affairs. Finally, as introduced above, the business centre of the state is undoubtedly in Boise. Many firms, like Micron Technology, have their headquarters in Boise, or their major divisions, such as Hewlett-Packard. The actual number of spin-offs from the larger technology firms is unknown but estimated to be about 300. Such conditions lead to questions of why and how such a small urban community should move towards a more robust, creativity-based economy. The 'why' is born out of a lack of complacency and a realization that for the city to grow beyond its small-city status it requires a broader range of creative assets and approaches; the 'how' is through an emerging model that differs from any other non-East-Coast US city and has more in common with cities in Europe: the pioneer model.

## The Pioneer Model of Small Urban Development

Compared with the US, cities in Europe have led the wave of interest in and development of creative industries and creative communities. Only in the past few years have pockets of creative-led urban strategies begun to emerge in the mid- and western-US. Much of the initial deliberate efforts have come in the eastern part of the US, especially in New England, the south-east (for example, Georgia and Florida) and south (for example, Memphis, Tennessee and Louisiana). In the UK/Europe and eastern US, efforts appear to be primarily driven or led by key champions, often government or political officials. For example, the Department for Culture, Media and Sport in the UK spearheaded early work on creative industries; and the mayors of cities in the UK (for example, Huddersfield) and Europe (for example, Freiburg, Germany), helped lead work to distinguish their cities. Likewise, governors of Maine, Vermont and Louisiana have made the creative economy and creative industries a key focus for economic development. On the other hand, a top-down, champion-driven model, common in Europe and the eastern part of the US, may not always be appropriate or desirable, especially in the western part of the US. The 'pioneer' metaphor helps to illustrate an alternative model of how creative economies and creative industries may evolve, at least in one part of the 'Wild West'.

The notion of pioneers is fundamental to American culture and identity. It suggests migration of a group of nameless adventurers, who see themselves as highly independent and set out (mostly) together on a journey to a generally unknown destination (and they shape it as they go), where they use external guides (like experts on creative industries). Boise's pioneers are just beginning the trek. Despite no clearly identified pathway, signs already exist to suggest some of the outcomes for Boise, especially ways to blend different disciplines to support high-tech, the arts and educational endeavours. For example, an informal group has formed (TekKlatsch) that brings together people from business, government and education to share information and ideas about how to build the high-technology industries within the region. In particular, there is growing interest in creating a Media Centre that would house high-tech, arts, educational and meeting points for people from a whole range of disciplines. In addition, state-level effort has begun to find ways to bring film industry production to the state as well as ways to support the budding local, indigenous, independent film industry: the acclaimed 2004 film *Napoleon Dynamite* is a case in point.

The pioneer model is at an early stage of development, yet it is already clear that there are at least four reasons why the model may be appropriate for a remote, small urban community like Boise, Idaho. First, a lack of clear 'leaders' or champion(s) *can* be an advantage because it allows creative community development to emerge organically, permitting unexpected pathways and options to arise. Second, with no clear champions on the horizon in the city, a group of nearly invisible or quiet 'guides and translators' is appearing. These people are quietly guiding groups to move in similar directions and are helping 'translate' knowledge from outside sources for usefulness to the Boise community. They have formed no coordinated group, have no official name, but rather 'nudge' quietly in their respective areas. A third reason for the Pioneer Model's success is the notion of being able to 'find new veins', in this case creativity – a renewable and distributed resource. Finally, the model seems to work because the pioneers are willing to create pockets of change or 'forts' as the community develops. The quiet guides are creating formal and informal discussion groups, websites with access to information and experts, and advocacy teams to work towards building certain industries.

Boise, Idaho, a remote city in what many (still) consider to be the frontier of the US, may offer a model for how smaller creative cities develop over time. The mix of location, people and history has blended to yield a more pioneering (bottom-up) potential model. To succeed, several factors are critical. First, eventually, the community will need a champion(s) to offer cohesive vision. In addition, the model allows opportunities for many trails or directions. Yet, eventually, the community must decide whether it can (or should) sustain all of the 'good ideas' or creative industries or whether it should focus on a few. At present, emerging areas range from biosciences to film to media centres to high-tech to agribusiness and beyond. Will the state, the community, or organizations want to support all (in time, money and other resources)? Such questions and longer-term monitoring may reveal the Pioneer Model as an emerging alternative approach to creative community development and the revisioning of cities at the margins.

## Revisioning the scarred city: cultural democracy, leadership and the young voice of Tuzla

Tuzla, in north-east Bosnia-Herzegovina, provides an example of how a relatively small, undistinguished city in a marginal and stricken part of the world, can invigorate change that is innovative, empowering and distinctive. Through undertaking a 'cultural democracy' approach to city planning, Tuzla has managed to transform its decaying industrial heritage into a new economy asset; it has translated a negative identity based on this industrial heritage into a progressive forward-looking sense of place; and it has used the shock and pain of war to galvanize citizens into building cohesively for the future good of the city. Tuzla remains relatively poor, isolated and negatively received/portrayed; but the city has set in motion a process of change that is gradually transforming the physical and psychological landscape to configure a new city that is youthful, bold, international, innovative and increasingly intercultural. This process of change is led by a genuinely visionary mayor and supported by a population that has shown ambition, tolerance and huge appetite for a better and more connected city. The increased engagement and commitment of the city's younger citizens has been vital.

## There's no avoiding salt

Tuzla is Bosnia-Herzegovina's fourth largest city, with a population of 165,000.[5] It is one of Europe's oldest continuously populated settlements and can attribute its growth to coal and salt extraction, which has established the city as a centre for heavy industry, with a strong presence of complementary chemical and energy production.[6] It is this industrial city image that dominates notions of Tuzla, plus the tragic events of recent history, when in 1995, 72 young people were massacred by a Serb mortar attack. It is very much a scarred city: physically and psychologically scarred by industry and war.[7] For example, there is no escaping salt: the name Tuzla is derived from the history of salt extraction;[8] the city has a 'Salt Square' and, until recently, the old salt mines were subsiding, sinking parts of the city by more than 10 metres, scarring the landscape and city identity, leading to regular salt-water flooding and prompting a range of disincentives to invest in or think positively about the city.

However, there is more to Tuzla than salt and war. For example, the city has a strong cultural heritage. It is the birthplace of possibly the greatest artist in Bosnia, Ismet Mujezinovic, and home to one of its foremost living writers, Semezdin Mehmedinovic. It has Bosnia and Herzegovina's oldest theatre. It is also a university city, with a relatively young population for a country losing a high proportion of its youth through international migration. It is the energy, resourcefulness and vision of Tuzla's young people that provides the focus for change in the city: their capacity to reinterpret the past – the salt, the war – to create new identities and opportunities is crucial to future prosperity, competitiveness and, quite possibly, peace.

Unlike a range of other small cities in marginal, often fraught locations, Tuzla is grappling with change as a positive challenge. Three key themes are vital to constructing new positive opportunities, ideas, identities and infrastructure:

- building cultural democracy: engaging, listening, willingly accountable;
- maximizing the value of a past that won't go away: from salt and war come positive ways forward;
- developing new markets and a new voice: youth, ideas and the creative industries.

Together, these themes contribute to a nascent toolkit for Tuzla that has attributes transferable to other small, struggling, even marginal cities. Indeed, Tuzla teaches us that a city can never be totally reinvented, for the past cannot be destroyed; it shows us that a genuine engagement with local people can lead to both innovation and stronger, more creative communities; and it warns us that without allowing younger and diverse voices to take a lead, the city will cease to be a city.

## Building Cultural Democracy

In few other countries in Europe is cultural policy and engagement more important than in Bosnia and Herzegovina. Issues of culture and identity are both the cause and the solution to its problems. Landry (2002) explains that they are the cause, because cultural arguments were used to divide the country and to turn the different groups against each other in an 'orgy of destruction'; and they are the solution because culture might be able to bring people back together again through initiating cultural programmes and activity that increase mutual understanding. In Tuzla, a re-engagement with culture has been at the forefront of public policy rationale, with almost every policy action and its causative challenge or question understood in terms of its cultural impact or meaning: 'what will this do *for* Tuzla?'; 'how can we build a more open Tuzla?'; 'what are our disadvantages and how can we turn them into advantages?'. Crucial here has been the role of the Mayor and his office, asking difficult questions and seeking culture-based solutions. Mayor Imamovic represents a new type of politician in Bosnia-Herzegovina. In a country previously crippled by corruption, bureaucracy and ethnic-based fragmentation, Mayor Imamovic has built a persuasive vision for Tuzla based on the phrase: 'No one is as smart as everyone together'. Whilst sounding a little idealistic – especially in a recent war zone – there is genuine substance to this phrase, a substance developed through a practical engagement with local residents.

Put simply, in 2000 the Mayor introduced a two-year process of in-depth public consultation as the first stage in his *cultural democracy programme.* This was driven by two main agendas: to find out the needs and aspirations of local people; and to ensure that future corresponding actions have a relevance to, and ownership for, local people as the basis for civic pride. The consultation was and continues to be based upon in-depth engagement with over 40 local communes and voluntary organizations; the creation of a Mayoral Advisory Council of 25 unelected advisers that cut across religious, ethnic and gender boundaries; a six-monthly city-wide survey of public opinion; and an ongoing process of civic engagement, triggering informal debate and targeting themes that progress new ways of thinking about the city and its potential.

However, conviction, persuasiveness and a determination to build consensus around genuine reform rather than stasis, are insufficient to exact change by themselves. This is because a city such as Tuzla needs more than collaboration and strength of character if it is to progress as a growing internationally focused city. It requires good ideas. This is where the most significant strength of Mayor Imamovic lies: his ability to stimulate ideas from the population and to identify those ideas that will work to transform the city from economic decline and painful memories to an outward-looking city with real reasons for renewed civic pride. This requires an assured approach to the delivery of essential services; the development of a mix of high-profile projects that genuinely 'raise the game' of the city; an ongoing and deeply embedded relationship with local people – especially young people; and an outwardness, fostered through nascent relationships with other cities. Tuzla has now cultivated international links with cities as diverse as Götheburg and Bologna, Osijek and Ravenna. The city has gained membership of several pan-European and international networks and thus actions in the city are seen by local people as actions within a network of cities, impelling Tuzla to be more bold and creative than perhaps a provincial approach might produce.

## Maximizing the value of the past: The Salt Lake

Tuzla remains a city famous for salt. However, today the city is famous for a salt lake rather than a declining industry that was literally and psychologically sinking and undermining the city. Mayor Imamovic, through his consultation processes, identified a series of options for the area of Tuzla most damaged by salt-water flooding – an area adjacent to the city centre. In short, the solution agreed was both simple and brilliant: to capture the water through the construction of a salt lake that, in turn, acts as a resort in the heart of the city. The newly formed Lake Pannonia is a masterstroke of civic planning and cultural democracy: forged through a positive translation of local connections to salt, providing employment for workers who previously mined the salt, introducing a new unique leisure opportunity for people from across the region and beyond, and establishing an iconic feature that affirms the transformation of Tuzla into a progressive and innovating city.

No other city has a salt lake and surrounding beach that provides coastal features despite the great distance from the sea. Over 300,000 people visit the shores of the lake every year, bathing, sunbathing, using the lake as a base from which to visit other attractions such as the Peace Flame Centre. At least 300,000 people are changing their perception of Tuzla. Indeed, an Austrian Bank Director was so impressed by the lake and what it signified that he established the Bank's national headquarters in Tuzla. The Mayor and his growing networks of city visionaries have shown great understanding of the importance of effective branding and positioning with increasing inter-city competition, while realizing that it is crucial for high-quality, effective and convincing policy and action to underpin that brand. What's more, the city is no longer sinking.

Further initiatives are underway that re-engage with the past to establish a contemporary, outward-focused identity for the city. They include the Peace Flame – a collaborative international programme intent on maintaining dialogue and nurturing positive outlooks from the memories of atrocity and hate. This is physically expressed through the Peace Flame Centre, close to the salt lake, which operates as a type of community and arts centre, undertaking initiatives aimed at bringing local people together through creative practice. In addition, the Mayor has championed a range of cultural and creative initiatives and programmes, including an annual literary festival and a series of high quality public art and public realm schemes. These have included a streetscape re-fashioned to represent 'symbols of humanity' (such as Nelson Mandela, Shakespeare and The Beatles). The Mayor often seeks support from cities abroad – requesting that a piece of art be donated or access be made available to the cast. A focus on international connection is coupled with a focus on translating the pain of war into a positive affirmation of humanity with a revitalized, inclusive, tolerant and connected Tuzla being the primary aim.

Yet, despite the considerable success of the cultural democracy approach of Mayor Immamovic, it is clear that Tuzla's future prosperity and peace depend on a more sophisticated engagement with an even wider range of voices and a more focused and nuanced approach to determining opportunities in the 'new economy'. *A prosperous and inclusive society can't be built on cultural gestures and iconic developments only*: for the young people to stay (and they continue to leave in large numbers), feel engaged and become economically active, will require the development of new home-grown markets for creative products and experiences.

Moreover, by collecting and negotiating around the production and consumption of new cultural forms, young people will be able to explore their identities without recourse to remember the war or even salt as the primary drivers of their identities: new cultural markets will bring a new type of cohesiveness that is spared the pain and embarrassment of the past. There is a growing realization – borne out through connections with other cities – that for Tuzla to be competitive and progressive, it needs to develop its own new, distinctive cultural products and services, and it would benefit most from undertaking these activities on a commercial and entrepreneurial basis: developing new local creative industries markets and connecting these emergent taste communities to a huge potential market across the Bosnian diaspora and beyond.

## Ongoing challenges

Tuzla is far from a fast-growing, well-connected and respected international city. It remains very much a city on the margins of a wider European context: unemployment is high; many people are leaving the city (while many displaced people still arrive); negative self-identity is commonplace; post-war hatred and suspicion remain; and the cultural economy is far weaker than state-developed cultural programmes. In addition, culture remains too overtly politicized (often on ethnic terms); piracy is stifling commercial potential in the music industry; there is a severe lack of legal expertise to coordinate change; and the learning and skills sector is too didactic and narrow, discouraging students from thinking laterally and creatively. However, Tuzla *is* a model for other cities blighted by war and industrial decline because it is confronting these and many more issues in a direct and far from complacent manner.

John Kao (1996) defines creativity as:

> The entire process by which ideas are generated, developed, and transformed into value. It encompasses what people commonly mean by innovation and entrepreneurship … (I)t connotes both the art of giving birth to new ideas and the discipline of shaping and developing those ideas to the stage of realised value.

Tuzla has embraced this broad definition of creativity to begin to address a range of city development challenges. Through history, cities have thrived where the management of creativity and innovation is paramount (Hall 1998). In contemporary society, Florida (2002) and Landry (2000) highlight how cities are increasingly operating as providers of cultural currency, as key determinants in generating and attracting creative talent. Tuzla is currently grappling to establish the climate and ecology for the development and release of its creative talent, which in turn will be followed by the attraction and retention of creative talent. It is extending its organizational capacity (such as through networks and fora), engaging deeply to engender participation and ideas generation (a cultural democracy), introducing catalysing projects (such as the salt lake) and placing its bets firmly on its strongest card (the talent, energy and aspiration of young people working with new content and technology).

According to an UN survey, 62 per cent of young people now want to leave Bosnia and Herzegovina and most want to leave forever.[9] The best and the brightest are finding it easiest to leave. Without a direct commitment to building creative opportunities at a local level while ensuring the city is connected internationally, cities such as Tuzla will continue to lose the creative talent on which their future depends. In her groundbreaking book *Towards Cosmopolis*, Leonie Sandercock (1998) notes the focus for decision-makers in cities, regions and countries is how they can 'organize hope', 'negotiate fears' and 'mediate memories'. The youth-focused cultural democracy embraced in Tuzla is helping to arrest trends in youth departures, giving them a powerful, hope-provoking stake in the revisioning of their particular small but very significant city.

## CONCLUSION: THREE CITIES, COMMON CHALLENGES

This chapter has explored how three very different small cities are engaging with creativity agendas in pursuit of stronger and more distinctive identities, greater connectivity, higher levels of creative entrepreneurship and governance and thus more competitive economies. Each city is faced by divergent sets of challenges and opportunities. For example, Malmö is moving from an atmosphere of constant crisis, bureaucracy and top-down policy making to implement more adaptive structures that deal on a project-by-project basis to reorientate the image of the city, transform its physical landscape and build on its existing and incoming creative population. Boise is responding to a history of pioneering development to focus on the new creative pioneers – networks of forward-thinking individuals unsettled by the 'comfort zone' of a successful economy and increasingly skilled workforce: there is a recognition that for Boise to adapt productively and for it to match its increasing size in dynamism, it needs to raise the stakes and build a strong and distinctive creative economy. Tuzla continues to struggle with a fraught political and economic context and is in a painful transition from conflict and economic decline; yet the city is proactively seeking new ways to adapt old problems into new economy assets, to galvanize and engage a population that lost its way and in many cases moved away, and to move forward, emboldened through a direct embrace with its creative potential.

Though very different, each city has recognized the value of exploring its creative potential as a way to climb urban hierarchies while retaining existing strengths – whether through a direct development approach to the creative industries or through a creative approach to governance and place-building. This is not an 'empty gesture' approach to creativity where the creativity of a city is proclaimed as little more than an exercise in civic boosterism. This is a brave and risky engagement with often disenfranchised populations, inflexible and dying businesses, or at times complacent and insular politicians. Each city has recognized that continuing to decline, to stagnate or even – in Boise's case – to sit pretty, is not an option given the ruthlessness of contemporary inter-city competition, where investment and people are increasingly mobile and unflinchingly promiscuous until they find and settle in progressive, innovative, inclusive places with a strong mix of commercial and cultural opportunities and strong distinctive identities. The small cities of today that face this critical reality face-on are likely to be the big cities – in approach and reputation, if not size – of tomorrow.

## NOTES

1 For a definition of second tier cities see: Ann Markusen *et al.* (eds.) (1999) *Second Tier Cities:Rapid Growth Beyond the Metropolis*, Minnesota: University of Minnesota Press.
2 See City of Tomorrow Brochure, 1999.
3 See City of Tomorrow Brochure, 1999.
4 For an analysis of the City Planning Forum's role in the regeneration of the Western Harbour, see case study: Building a Collaborative Design Approach for the Development of the Western Harbour, in the dissertation by Maria Lundgren (2004) Malmö University, Department of Media and Communication.
5 Tuzla is the seat of the Tuzla Canton, which is a canton of the Federation of Bosnia and Herzegovina, as well as of Tuzla Municipality, which is one of the 13 municipalities that together constitute the Tuzla Canton. Administratively, Tuzla is divided into 39 mjesne zajednice (local districts).
6 For example, in AD 950, the Byzantine historian and cezar Constantine Porfirogenet mentions the existence of Tuzla's saltwater springs and settlements surrounding them.

7 During and after the war, over one million people left the country, around 25 per cent of the total population, including a large proportion of the professional classes – the intellectual and knowledge infrastructure of the country has been decimated. Over 300,000 people were killed – the majority Bosnians. Forty per cent of the population remains displaced – creating immense problems of community building (see Landry 2002).
8 It is derived from the Turkish word *tuz* meaning salt.
9 See Landry 2002.

# Part 4

# Identity, lifestyle and forms of sociability

# 14 Caudan

## Domesticating the global waterfront

Tim Edensor

Waterfront developments have become a banal element in strategies to regenerate, reinvent and re-image the increasingly post-industrial port cities of the West. Numerous motivations lie behind these projects, including the imperative to attract free-floating global capital, renovate dilapidated infrastructures, provide upgraded public (or semi-public) spaces for inhabitants, support flagship developments that contribute to the changing place-images of cities, assert distinct versions of heritage, house key creative and new media industries and vary the mix of consumption and leisure provision, drawing in tourists and shoppers. And these developments are devised by numerous alliances between investors, entrepreneurs, heritage managers, retailers, bureaucrats, regional, national and supranational government, typically comprising a mix of public and private partnerships.

Waterfront revitalization projects first emerged in North America in the 1960s and were subsequently taken up in Europe and Australia, later spreading to Asian, African and Latin American cities. This is a global expansion that, whilst hitherto confined to advanced countries, is now starting to impact upon developing countries as they seek to revive port cities in different contexts. These diverse venues now include cities such as Dar es Salaam and Zanzibar in Tanzania, Mombassa and Lamu in Kenya (Breen and Rigby 1996; Hoyle 2001, 2002) and the subject of this chapter, Caudan, Port Louis in Mauritius (see Simone, this volume). The globalization of waterfront development has been identified as producing homogeneous 'construction standards … organizational methods, spatial typologies and architectural forms, thus generating a monotonous sense of déjà vu' (Bruttomesso 2001: 49). This is undoubtedly the case and I will acknowledge those serial characteristics and the exclusions generated by waterfront development in Caudan. However, it is also vital to explore the particular translations through which formulae for regeneration take shape, the distinct political and social pressures that impact upon the production of space, the local opportunities and constraints that emerge and the unpredictable ways in which local people use and understand these new urban realms. Generalized assumptions about cultural imperialism and homogeneity tend to neglect the specific contexts in which waterfront developments occur, yet they provide an excellent terrain to explore processes of 'glocalization', the ongoing, often surprising ways that distinct local social and cultural processes becomes enmeshed with global forces and flows. After a brief depiction of Caudan, I will discuss its regulatory and exclusive effects, before exploring how the site has opened up commercial and architectural opportunities that belie preconceptions about corporate hegemony. I will then discuss how the development offers new possibilities for public conduct and social interaction that can potentially transcend the everyday conventions of Mauritian society.

With a population of almost 130,000, Port Louis is a bustling seaport that occupies an important strategic position in the Indian Ocean, the centre for exporting and importing goods to and from Mauritius and a port of passage for all manner of vessels. The capital city of the island, it is replete with administrative buildings, a large and vibrant public market, central business and retail establishments, a racecourse and numerous street traders. It is a synthesis of prosperity and poverty, with its gleaming office towers and shanty areas, and a blend of the global postmodern and the modern in its architectural expression and popular culture. Yet despite being the venue for several modernist and postmodernist tower blocks and office complexes, the overall characteristics of the city are shaped by the ad hoc housing and commercial developments, large and small, that have been superimposed on the planned colonial city and by the fervent activity that takes place in its central streets. In this context, the new waterfront development of Caudan seems incongruous, providing a striking contrast with the crowded city streets, modern and vernacular architectural forms and visual and spatial disorder. One inhabitant referred to the development as a 'mini Dubai', associating the space with luxury and the ambitious transformation that has been wrought in that Middle Eastern state, wholly at variance with the crowded bus stations, urban noise and cluttered landscape of the rest of the city.

Businesses and governments in Mauritius continually scan the world for potentially beneficial global trends for, having few evident advantages or plentiful resources, the country must continually jockey to establish a position that enables it to participate in the world economy. During colonial times – until 1968 – sugar was entirely central to the economy and with sugar cane covering the island, it is still overwhelmingly the prevalent agricultural crop and an economic mainstay. Since independence, however, the island has been highly successful in developing an upmarket tourist industry and a significant export processing zone in which hundreds of clothing factories manufacture branded goods for Western corporations. Formerly extremely poor, Mauritius is commonly identified as a middle-income country and is often used as a regional exemplar of good development strategy, although this assignation can disguise persistently entrenched forms of economic and social inequality. Under conditions of speeded-up globalization, however, this success story remains tenuous. The clothing sector is threatened with the demise of the multifibre agreement and the world sugar price has fallen dramatically. Recent attempts to develop an IT sector are very modest, yet although economic links to the West may be weakening, Mauritius plays a prominent role in SADC (the Southern Africa Development Community) and other regional associations so that the interpenetration of Mauritian and regional economies of South Africa, Madagascar and other southern African countries are developing.

Until recently, despite the increased wealth of Mauritius, a consumer culture comparable with Western contexts has been slow to develop. Part of the ongoing effort to globalize the Mauritian economy, to increase consumer spending and tourist provision, has been to create new spaces of consumption that supplement the local stores and markets. Accordingly, in the past decade, a spate of new retail spaces has emerged across the island, including large supermarkets, shopping centres, specialized boutiques and other complexes. The waterfront project at Caudan is part of this pattern of diversification and, like many new developments in Mauritius, including those in tourism and clothing manufacture, it has been funded by the desire of wealthy sugar estate owners to extend their economic interests. Money originating from the profits of the sugar industry and the sale of sugar estates has been invested in the private Mauritius Commercial Bank, the financial power behind the limited company that has invested in and developed the Caudan project. The sugar barons come from the Franco-Mauritian community comprising about one per cent of the population but owning most of the wealth. Fifty-one of the top 100 chief executives in Mauritius are Franco-Mauritians and five Franco-Mauritian families control 18 of the top 50 companies in the country.

At one level, the fact that this vast wealth persists to fund projects such as Caudan seems obscene. Originating from a plantation system originally dependent on imported African slaves and later indentured Indian labourers, further supplemented in 1834 by the £2 million (a huge sum at the time) given to the planters as compensation for the outlawing of slavery by the British colonial rulers, the continuation of a semi-colonial, exploitative sugar plantation system producing vast disparities in wealth has endured. However, as I will show, despite these origins of the Caudan project, the scheme offers liberatory possibilities for commercial interests and social interaction.

The ambitious scheme to develop the Port Louis waterfront was initially driven solely by private interests in the mid-1990s, but later on, after some reticence, the government decided that it, too, would invest in the project. Accordingly, there are two distinct areas of the waterfront, one privately organized and operated by Caudan Development and the other regulated by the state. This has ensured that the environs are not wholly characterized by a themed and seamless design, as I will demonstrate.

Although on a wholly different scale, the inspiration and model for Caudan has been the huge Victoria and Alfred project in Cape Town, South Africa, rather than any of the numerous waterfront projects that litter the cities of the West, although this wholly private-funded waterfront, cited as the 'top tourist attraction in South Africa' (http://www.waterfront.co.za), was itself inspired by American, British and Australian developments. Managers and designers of the Cape Town development have advised and provided expertise on new waterfront projects in Nigeria, Gabon, Seychelles, Greece and Cyprus as well as Mauritius. This highlights how a global network of groups and individuals involved in waterfront development has been established and how, through regular contact and visits, the latest trends in waterfront development and management are monitored by regional experts and chief players.

Built upon an old port area of about 42,000 square metres, at first glance Caudan offers a range of familiar features, part of an array of ingredients belonging to a global formula and seemingly unconnected to local urban conditions in the city in which they occur: maritime heritage, diverse outlets for shopping, eating and leisure, a highly-regulated environment patrolled by security guards and maintained by cleaners, predictable aesthetics, an abundance of office space, street furniture and pseudo-public spaces. Within the complex there are thousands of square metres of office space, a casino, two hotels, 15 fast-food outlets, four restaurants, two cinemas, large car parks, about 30 clothes boutiques, eight arts and gift shops, ten jewellery stores and a large craft market with some 30 stalls.

As at many waterfronts, there is a conscious attempt to retain and broadcast specific forms of heritage. Some original dock buildings have been renovated, notably the nautical observatory and former gunpowder store, which now accommodates administrative offices, and the casino, converted from an old storehouse. The Blue Penny Museum, named after a valuable Mauritian postage stamp, is housed in what was the former port administration building before it was demolished and rebuilt in its original style. And as elsewhere, there are forms of elaborate paving and the maintenance of cobbles, the highlighting of bollards, street-lamps and iron railings, with some fixtures painted in bright turquoise.

There are, again like many similar spaces, simulacra and quotations from globally familiar elements of popular culture, referents that herald an assertion of global belonging within a local context. There are murals depicting the cultural icons of Laurel and Hardy, Marilyn Monroe and Indian film-star Shashi Kapoor, and the casino has appended to it a large model of a galleon with an imposing lion as the figurehead, jutting out into the central promenade. Such forms of signage refer to wider spaces and cultures, cosmopolitan icons and Western notions of glamour embodied in Hollywood and Bollywood films and pop music, yet they hardly overwhelm the site.

The government-owned area is quite different in atmosphere and function. A marked contrast between the prestigious, stone-clad buildings of the private area is afforded by the more vernacular, generic design of the two-storey structures that house more downmarket businesses. The public area contains cheaper retail outlets and a pub, the occasionally rowdy Keg and Marlin, where loud music plays. A large promenade area connecting two sites at either end of the harbour is replete with a small wheeled train, a playground, exhibition spaces and a stage where promotional, official and public events occur. The public area is regarded with disdain by the private company for, besides attracting a broader spectrum of people who are not necessarily concerned with shopping and are suspected of increasing the possibility of crime and disorder, the more modest architectural style is held to clash with the smooth surfaces of the larger private area, thwarting the attempt to give a unitary sense of design and ambience to the waterfront as a whole, forging the ‘recomposition’ of docklands to provide a new sense of coherence between previously disparate realms of function and appearance (Bruttomesso 2001: 40).

## A SPACE OF REGULATION

Undoubtedly a highly-regulated and somewhat exclusive space, like many waterfronts (Marshall 2001), Caudan is marked as separate from the rest of Port Louis. Tourists can be bussed in and remain untouched by what is going on across the road for, in contradistinction to the seething heterogeneity there, there has been a 'topographic tidying' of this part of the port (Bell and Jayne 2004: 251). In keeping with the values of its middle-class patrons, Caudan prioritizes 'visual aesthetics, artificiality, consumption, comfort and control' (Stevens and Dovey 2004: 352–353), seemingly constructing an over-determined, 'purified', 'single-purpose' space dedicated to consumption (Sibley 1988).

Surplus branded clothes manufactured in the export processing zone are sold across Mauritius at a host of retail sites (Edensor and Kothari 2004) and in Port Louis, besides formal retail spaces, such clothes are sold informally, by street market traders and pavement sellers. Next to the underpass beneath the busy road that separates the waterfront from the rest of the city, numerous traders peddle shirts, belts, bags and mobile phone facia, marking a profound difference between the city and its more sedate waterfront. Needless to say, such activities are barred from Caudan, where a more intensive regulation of preferred retail activities is mobilized, even though the wide expanses of the waterfront would be ideal territory for such traders. The small scale of the development and few access points means regulation is facilitated so as to restrict the range of activities that can be performed in Caudan, and this is further sustained by the exclusive prices and by the internalization of new norms of public behaviour that embody which activities are and are not appropriate in these new spatial settings. Formally, Caudan offers forms of leisure that feed on the urban desires of the largely well-heeled clientele and it channels them 'into consumption and carefully managed forms of play' (Stevens and Dovey 2004: 352), organizing social relations and social practices by mediating them through a 'predetermined, generic and predictable palette of images, perceptions and opportunities for action' (ibid. 361).

Caudan's aesthetic and sensual regulation also bolsters its preferred identity against dissonance, disruption and challenge. Themed spaces like this depend upon a vigilant resistance to disruptive features and the ongoing maintenance of design codes – the careful pruning of trees and shrubs, the removal of clutter and the upkeep of smooth surfaces. These measures are evidently designed to assist the unhindered mobility and undistracted consumption-oriented activities of shoppers, but they also aim to foster the reproduction of a particular sensual realm that conveys a relaxed atmosphere and unambivalent meaning.

Yet despite the surface impression that Caudan is designed and managed according to well-worn, globally formulaic regulatory techniques that extinguish diversity and improvization, the meanings and uses of spaces such as this can never be controlled totally and may be read and used in different ways at variance to official and preferred encodings. There certainly are forms of class-based exclusions that inscribe the desirability of middle-class lifestyle on the city and that etch a distinction between Caudan and poorer urban areas in ways that mirror those of other waterfront developments that effectively exclude large numbers of a city's inhabitants (for instance, Waitt's 2004 example of the physical and symbolic barriers erected at Pyrmont-Ultimo in Sydney). Yet the modes of public conduct inculcated at spaces such as Caudan are fairly recent, reflecting the aspirations of a would-be cosmopolitan Mauritian middle class. It is difficult to envisage the emergence of spatially resistant practices akin to Western activities, for there are no Mauritian traditions of skateboarding, busking and street theatre, youth subcultures or counter-cultures that mark their presence in public space. However, as mentioned, the publicly and privately managed areas of Caudan are organized according to different aesthetic and moral criteria, with the former more loosely maintained and less thematically ordered; as we will see, drunken antics, rowdiness and the less-kempt appearance of the public area can disrupt the carefully controlled look and ambience of the private zone.

Moreover, as Atkinson (2006) observes, albeit in a different context, smoothed-over, gentrified waterfront areas frequently depend upon the disappearance of the industries that produced them in the first place, for heavy industry and activity disrupts illusions of serenity and respectability. The docks in Port Louis remain a vibrantly busy setting for all manner of ships – indeed the port's traffic has expanded with the signing of a number of transshipment agreements with major shipping lines – and so the non-gentrified is always close at hand, including dockland activities such as piracy, smuggling and prostitution.

## THE PRODUCTION OF OPPORTUNITIES

I have suggested above that, despite initial appearances, Caudan cannot simply be characterized by an over-controlled, homogeneous importation of Western waterside developments, with the conformist standards of behaviour and a limited range of practices suitable to a single-purpose space. Accordingly, I now reflect upon the distinct architectural, touristic and commercial opportunities stimulated by the project.

### Architectural innovation

It is often the case that gentrified developments rely upon the production of highly similar built forms and, whilst the renovation and reconstruction of the old dockside buildings appear similar to the stylish regeneration of waterfronts across the world, the larger buildings accommodating offices, malls and hotels are unique in style. These designs have been depicted as embodying 'an architectural balance that combines the very particular features of the existing dockyard architecture with the classical elements of traditional Mauritian architecture' (Giraud 1998: 14), and they were described to me by a managing director of the site as an example of Mauritian postmodern style, singularly unique in its blend of international modernism and local design; principally 'colonial creole' – an innovative architectural form symptomatic of the hybrid cultural effusions that emerge through global processes. Caudan is thus a small area of architectural innovation in which the old dock buildings and media-inspired facades are complemented by a form of contemporary architecture. Rather than constituting a kind of ubiquitous non-place (Augé 1995), the resonances of distinct historic features lie alongside the production of new expressions of Mauritian-ness that reconfigure a sense of place that refers to past and present, home and elsewhere.

## Strengthening and Diversifying the Tourist Product

The development at Caudan is intended to contribute to a widening of tourism in Mauritius, to produce a 'second level of tourism' that re-develops the rather reductive product presently on offer. In a context where global competition to attract wealthy tourists intensifies, it is felt by many that the tourist product in Mauritius – the provision of high-end resort holidays at the numerous luxury hotels that line the coast – is somewhat limited in appeal. In recent years, a range of modest themed, zoological, shopping, sporting and other leisure spaces have been developed to lure tourists away from these resort enclaves and also attract middle-class Mauritians. This has met with only limited success, for most resort hotels are primarily concerned with keeping their residents captive, ensuring that they spend a higher proportion of their holiday income at in-house retail outlets, restaurants, bars and other attractions. Moreover, many hotels organize their own highly planned tours that restrict interaction with spaces outside the enclave. This partly testifies to the corporate power of the transnational businesses who own the resorts and are able to thwart initiatives to disperse tourist revenue more equitably and tourism across a wider spatial range. Accordingly, it is likely to be the more independently minded tourists who venture beyond the compound and the organized tour to Caudan. Nevertheless, there is some potential for tourist expenditure to be more equitably distributed if tourists venture further into spaces such as Caudan. Because of the stringent regulation of the waterfront, unlike the environs of the city across the road, tourists may 'safely' experience the 'other' in a place designed for gazing, ambling and shopping (see Edensor 2005b; Edensor and Kothari 2004) without epistemological, aesthetic or social disruption to their perusal of the 'exotic'.

## Selling Mauritian Products

It is often asserted (and is evident, for instance, in critiques of Cape Town's Victoria and Alfred Dock) that waterfronts, along with festival marketplaces and malls, produce a serial homogeneity, offering a bland concoction of the same corporate outlets the world over. Yet startlingly, the vast majority of goods and services on offer at Caudan are Mauritian in origin and not imported global brands. There are a few French outlets and a Pizza Hut, but little else. A few of the 30 retail outlets distributed across the site sell well-known branded items (for instance, Diesel and Billabong) that are manufactured in Mauritian factories for overseas corporations, but many sell locally-made clothes exclusively made for the Mauritian market. Accordingly, Caudan provides an opportunity for Mauritian entrepreneurs and helps to promote a local fashion scene. In recent years, several chic clothing outlets have emerged in response to the demands of middle-class Mauritian youth who have become dissatisfied with the often ubiquitous and somewhat homogeneous designs of the widely available corporate clothing that is manufactured locally for the high streets of Europe and the USA. Whilst these creolized clothes are designed as 'street wear' and 'club wear' and borrow from Western notions about what is fashionable, they are designed and made in Mauritius and are more innovative and idiosyncratic than their corporate counterparts. Caudan is therefore a retail environment that appears redolent of the malls and festival marketplaces of the West but actually promotes local businesses and an indigenous fashion scene.

A similar point might be made about the fastfood outlets which, rather than the ubiquitous burger bars and pizza parlours found elsewhere, offer the distinctive Mauritian melange of Indian, Chinese, Creole and French cuisine available throughout the island. The limits of Mauritian entrepreneurial manufacture are, however, evident in the craft market where the lack of any substantial craft tradition in Mauritius means that most skilfully wrought items are imported from regional sources in Africa and India. Yet even here, the market peddles regional crafts rather than Western goods. The small scale of the project, local ownership and desires of Mauritian consumers have thus far fostered a peculiarly non-corporate retail and service provision which contrasts with numerous other waterfront developments and larger retail spaces where the serial homogeneity of corporate outlets predominate.

## Broadening the Social

Besides the opportunities identified above, I want to consider the ways in which the development at Caudan has facilitated a more valuable service: the production of new social formations and practices. Despite its emphasis on activities oriented around consumption, Caudan has become a space for social gathering and mingling, for consumers and non-consumers alike. This is important in a city such as Port Louis for, like most other areas in Mauritius, there is no great tradition of a public, urban street culture in which people seek out conviviality or spectacle, the chance of marvellous encounters or informal social participation, although public beaches serve as somewhat liminal spaces of play and some inter-ethnic mingling. In the capital city, there are few places, apart from the racetrack, where large numbers congregate, for Mauritian life remains circumscribed by ethnic and religious allegiance. Social groups are primarily divided according to the incomparable categories identified as 'Hindu', 'Tamil', 'Muslim', 'Chinese', 'Creole' (the descendants of African slaves) and the aforementioned 'Franco-Mauritians', and there are few shared spaces and occasions that do not reflect these identifications.

As Eriksen (1998) has argued, there are inter-communal entanglements in work and other social practices but these remain limited and interaction between ethnic groups remains cautious amongst all classes. There has not developed, as yet, a public culture that can be termed 'Mauritian'. For instance, much television programming is organized according to the requirements of different communities.

There are soap operas and Bollywood films from India and French and African programmes for the different communal constituencies, but a dearth of offerings that serve all Mauritians irrespective of their ethnic and religious groups. Moreover, until recently, the football league clubs in Mauritius were formed, owned and supported by members of specific communities, a situation that produced antagonism yet provided crowds. Now that league clubs have been forced to reconstitute themselves on a regional, rather than a communal basis, support has evaporated (Edensor and Augustin 2001; Edensor and Kooduruth 2004), revealing the limits of state-led policies to enforce communal desegregation. There also remains a continual jockeying in political bodies, state organizations and business between communal groups for influence and representation, yet precious little public space for inter-ethnic mingling.

The dearth of inter-communal spaces in the city – spaces that might ameliorate some of the hardened, prejudicial attitudes towards others resulting from voluntary segregation – has been alleviated by the development of Caudan, which instantly became a new kind of public space, untainted by historical ownership or inhabitation, unclaimed by specific groups as *their* territory. All ages, ethnicities and religions mingle in Caudan, which is regarded as safe and, moreover, as a venue for ethnically-blind lifestyle pursuits, such as movie-going, gambling, dining out, drinking and shopping that offer social alternatives to the leisure and social activities that typically remain confined within families. This inter-communal mingling remains tenuous, for it was mentioned to me that the security guards have to maintain a delicate, 'softly softly' approach to policing to minimize the potential for inter-group conflict to emerge. Particularly since the devastation that resulted from riots in 1999, tensions have bubbled under the surface of much public life, yet Caudan provides a more inclusive space than is usual in Mauritius, notwithstanding the barriers to lower-class participation which limit this to a degree.

In addition, there is in Port Louis, and in Mauritius more generally, a dearth of venues for a range of social activities that usually take place in a public context, for they have been delimited by the spatial structuring of social practice and the cultural conventions that restrict social interaction. Most evidently, there are no suitable places where people may simply stroll, alone or in small or large groups. The ordinary pleasures of 'people watching' and being part of a crowd and engaging in what Erving Goffman (1959) has called 'fancy milling', with its connotations of showing off, parading one's clothes or group affiliation, have been greatly limited. Yet in Caudan, there are public places in which to meet. In this highly rationalized space there are numerous sites for 'hanging out' in solitude or with others and as a consequence, in the early evening and at lunchtime, the air resounds with cries of greeting as people meet each other publicly rather than covertly.

Caudan also serves as a venue for youth to meet, as a space to congregate and play and, more importantly, as a site for romantic encounters. Free from prying neighbours, couples can meet, whether at secluded points or in spaces in which there is a collective clustering of young couples kissing and talking. This existence of a more temporally and spatially expansive venue for romance means that couples are not forced to meet furtively in uncomfortable, private or crowded places and thus the development of a youth culture featuring overt boy–girl encounters may evolve. Escaping the surveillance of parents and neighbours, young people may engage in a more intimate relationship with each other, a form of rebellion that might seem minor from a Western perspective, but transgresses normative Mauritian social codes. Besides providing a context for youthful encounters, Caudan also offers a space in which others may stroll without fear. For instance, many Muslims wander around the site in large or small groups, a characteristic not seen in other public spaces, for Caudan is perceived as a space free from religious intolerance. In addition, this is a space that is not so inscribed as masculine as many of the highly-gendered public spaces in Mauritius for here, women may move around without fear of abuse and without being tracked by other family members, as it is assumed that they are under little threat.

Additionally, Caudan provides lifestyle choices for Mauritians that have not, until recently, been available. The extensive food court, along with other, pricier restaurants and the Keg and Marlin pub, have introduced the possibilities of a dining culture where people may go out to eat and drink in the evening, an introduction that is not widely available elsewhere in Mauritius outside the tourist complexes (few restaurants remain open in the evening, with food largely being consumed in private, domestic settings). Without minimizing the potential impact of the rise of the 'global culture-ideology of consumerism' (Sklair 1995: 16) or a social context in which the citizen-consumer is king, the development of Caudan has nevertheless enriched and diversified opportunities for leisure and consumption and provided more spaces for communal participation and, potentially, a basis for wider democratic involvement and the development of a sense of belonging to a shared culture. For instance, it might be argued that the Keg and Marlin, a pub which sailors and other less middle-class customers attend, provides a convivial venue for alcohol-fuelled loud talk and expressive behaviour, modes of conduct that transgress the ordinary constraints of familial and communal social practice, as well as the middle-class values and codes of behaviour which prevail in most waterfront venues.

Caudan is characterized by a mixed temporal use. The site becomes crowded at lunchtime, tends to be otherwise very quiet during weekdays, but again becomes thronged during early summer evenings and at weekends. In this regard, the waterfront has expanded the range of times when Mauritians might engage in social activity outside the domestic sphere. It is a common assertion that Port Louis 'died after 4 pm', indicating the lack of any substantial, public-based night-time culture in the capital city and across the island. Indeed, if one walks the streets of the city in the evening, there are few restaurants and bars open for business outside the Chinatown area, for social activity tends to be confined to the private sphere. The availability of cinemas, restaurants and bars at Caudan facilitates the development of a night-time economy, augmenting the paltry mix of leisure opportunities available after dark and again expanding the range of opportunities available for leisure and social mixing. Those visiting the waterfront in the evening are unlikely to be going on to a restaurant, pub, concert or club elsewhere and so Caudan tends to be the centre of the evening economy, especially at weekends, bringing with it an associated liveliness that is occasionally the source of concern, as worries grow about rowdiness and the minor presence of prostitutes who move across from their regular haunts on the other side of the boulevard.

## A GLOCAL SPACE?

In several ways, Caudan is an exemplary site of 'glocalization' (Robertson 1992), transcultural hybridity and cultural syncretism. It should be clear from the above that, whilst the site superficially resembles some of the generic characteristics of other Western waterfronts, festival markets and themed spaces in its design codes, forms of regulation, mix of facilities and preferred activities, these characteristics cannot pre-determine the use of place. The emergence of Caudan suggests that rather than producing an area of homogeneous conformity, a tedious blandscape, or a worn-out copy of numerous other Western waterfronts, it is an area of complexity, with a varied mix of functions, public and private spaces, people and practices (Bruttomesso 2001). As I have outlined, it has an idiosyncratic aesthetic and architectural texture, offers a largely local array of goods and services and is a venue for particular forms of social practice which have emerged out of a distinct, local context. Since the inspiration for the development comes from South Africa, this also gives the lie to notions that global flows are simply part of a process of Westernization or 'cultural imperialism' (Tomlinson 1999), for they are more diffuse and multi-directional. At Caudan, globally influential trends in urban development have been translated and adapted in ways pertinent to their social and cultural Mauritian context. This is not exactly a space in which practices resistant to global forms of power are articulated, not a site of heroic resistance to globalization, but neither is it a space of crushing homogeneity and the stifling of local cultures.

There is no doubt that consumer-led development has a tendency to exacerbate the distinctions between classes and in Mauritius this is no different, for the site marks itself as an exclusive realm of high-end consumption in relation to other shopping areas and wider urban space in Port Louis and it is a parading ground for the status-conscious and the fashionably dressed, those who aspire to an imaginary cosmopolitan lifestyle. Moreover, surveillant regulation and the expense of services and goods in Caudan deter the less well-off from inhabiting its shops and restaurants. Nevertheless, the best-laid plans do not necessarily eventuate in the outcomes envisaged by corporate planners and managers and there is no global blueprint that can be predictably applied everywhere, for social and cultural context are apt to shape the use of new developments. Accordingly, the small waterfront development at Caudan is also a space of opportunity for architectural experimentation and tourist exploration. More importantly, it is not dominated by the concerns of global capital as elsewhere. Neo-liberal development strategies have been focused on the imperative to attract the production and retail outlets of large global corporations, yet Caudan is a space largely bereft of such enterprises. It is instead full of small Mauritian retail businesses. And what might be regarded elsewhere as an over-regulated, over-coded, desensualized space that enforces social conformity to specified behaviours and deters social practices and disruptive people, is in fact a space in which inhabitants of Port Louis can obtain some release from dense urban living. Here, new ways of urban living and social interaction are being fostered that seem to be diminishing the exclusive activities and rigid identities mobilized around ethnicity and religion. It would appear that here the glocal is somewhat liberating for an everyday urban realm has the potential to become refashioned by new social practices and identity formations.

# 15 De-centring metropolitan youth identities

## Boundaries, difference and sense of place

Gordon Waitt, Tim Hewitt and Ellen Kraly

This chapter examines the place-based meanings, practices and feelings of teenagers within a small North American city, Townsville.[1] Young persons in the interstices of childhood and adulthood are a somewhat neglected group in social and cultural geographies. This chapter has its roots in helping to address this disregard for the geographies of young people (for overviews see Holloway and Valentine 2000; Matthews 2003). Applying Doreen Massey's (1993: 1998) conceptual lens of 'progressive place-making', the chapter investigates the relational connections among place, youthful identities, reputations and social change. Our focus is on how youthful identities are performed in and through everyday lives of city places beyond home, school and church. We report on ethnographic research of a group of 17-year-olds who live in the inner city and attend a high school in a socially and economically deprived neighbourhood in the north-eastern region of the United States.[2] In this task, attention is given to how smallness intersects with establishing and maintaining youth identities by drawing on and developing comparisons with other research.

Economic and social uncertainties in this small city have been generated through the processes of economic restructuring. Economic stability and social sameness in the demonstrably global–local nexus of this inner city have been replaced by economic instability and social difference. These uncertainties have stripped away many of the resources for identities that were once accessible in this city. In the context of the uncertainties that exist in this inner-city context, we pay particular attention to young people's construction of themselves and other people through practices and discourses of inclusion and exclusion from public space. We demonstrate how teenagers are defining themselves through practices that rely upon materializing in space clearly bordered by lines of differences, demarcated by morality, gender, ethnicity, age, fashions and interests. We identify three interrelated dimensions of identities used by different groups of teenagers to stake out their own territory: (1) changing localities, (2) markers of difference and (3) 'cool' or special places. In doing so, we make visible the social processes that constitute boundaries and the production of difference.

In this small city we demonstrate how actions and ideas of particular groups of teenagers are aligned along border-making spatial practices that hem their movements into particular places. Increased by economic demise, crime and a personal, heightened sense of ethnic and class difference, smallness not only operates against everyday life choices and personal anonymity but also enhances an increased familiarity with how, where and when different youth identities are practised in public outdoor spaces. We argue that, for young people, movement through, and gathering at, specific inner-city public spaces is a highly complex, reflexive and strategic action.

## BACKGROUND: THE CITY OF TOWNSVILLE

Townsville is not just small but a shrinking multicultural city, in the north-eastern region of the USA. In 2000, Townsville had a population of just below 100,000. Multiculturalism is a legacy of nineteenth-century early-twentieth-century prosperity, particularly from spinning mills and textile production. Immigrants drawn to work in these industries came from southern, eastern and central Europe and also included African Americans from the southern United States. However, from 1970, the total population in the surrounding metropolitan region fell in absolute terms by over ten per cent. The population of Townsville declined by one third. Everyday lives intersect not only with the marked seasonal changes of this region but the implications arising from the demise of its former manufacturing base. In the last quarter of the twentieth century, approximately 70 manufacturing firms closed or moved, resulting in the loss of nearly 10,000 jobs. Over the same period the number of retail trade companies declined by over one-half and wholesale trade companies by a third. In step with economic decline and population loss, there has been an increase in reports of organized crime, illegal drug trafficking (in particular crack and cocaine) and incidences of drug-related violence and guns being carried to school by young people.

Retail, industrial and domestic cityscapes materialize these statistics. Storefronts in the former city-centre retail hub stand empty. Surrounding industrial sites have been vacated or reduced to rubble. To revitalize the inner city, like entrepreneurial urban policies elsewhere, Townsville's authorities have recently turned to the symbolic economy. Mirroring public-private funded urban revitalization plans internationally, a new downtown business precinct has been zoned. In the case of Townsville, grounding new jobs relies upon convincing investors of a future prosperity within a dreamscape framed by past industrial prosperity signified by canals and industrial architecture.

Houses in the inner-city suburbs have also become empty in unison with the abandoning of industrial buildings. Housing stock is old – over half of the houses in the city were built prior to World War II. There is no evidence of gentrification. Instead, vacancy of residential units has nearly tripled in the past three decades. Present rates are well over ten per cent. Abandoned and dilapidated housing is a characteristic of many inner-city neighbourhoods. If it were not for international and federal refugee resettlement programmes the economic demise, the population decline and the housing vacancy rates would have been even greater. This flow of refugees has also increased the ethnic diversity of Townsville to include people from Bosnia-Herzegovina, the former Soviet Union and areas throughout

Indo-China. An ethnic diversity is now reflected in the cityscape through businesses, social clubs, churches and mosques.

Within this context of economic restructuring and socio-spatial polarization, young people living in Townsville are increasingly constrained in their use of, and mobility through, public space by an increasing number of city ordinances to control youth activities. In certain streets, skateboarding is banned. Curfews exist on entry into public parks. Since the 1990s, the local print media follows conventions that *all* young people on the street are demons and therefore an automatic source of threat to the moral order through their assumed criminal activity, drug dealing or gang behaviour (see Breitbart 1998; Valentine 1996). Municipal agencies in Townsville have responded by funding programmes designed to keep young people off the streets and in 'safer' places – including the Boys' and Girls' Club. This facility offers a number of year-round and specialized programmes in sports, first aid courses, drug and alcohol awareness and adventure travel. Supervised by adults, youth training is provided not only in competencies and responsibilities of future adulthood, but also in acceptable behaviours and experiences for young people. Whilst such programmes play a crucial service, they often rely on linear models of child development that conceptualize young people as always less than adult; and rely upon Western idealized notions of young people as 'incompetent', 'innocent', and dependent upon their parents (Gergen *et al.* 1990). As this chapter illustrates, the lived experience of young people is neither the Western imagining of childhood nor the one imposed on them by grown-ups. Instead, young people are adept at creating their own identities in and through space, resisting and negotiating both identities imposed on them by older people and the conventional values of city spaces.

## CONCEPTUALIZING INNER CITIES AND YOUTH IDENTITIES

Since the 1980s, the socio-spatial patterns of many inner cities have been reconfigured through processes of gentrification and the declining industrial and fiscal tax base of cities. These have often been theorized as the emergence of the post-industrial, post-modern or post-Fordist city underpinned by a movement to global capitalism or post-Fordist accumulation (Davis 1990; Harvey 1989a; Knox 1987; Soja 1989). Marcuse (1989) outlined how American cities were being materially and symbolically bounded into five types of intricately linked residential quarters: the dominating city, the gentrified city, the suburban city, the tenement city and the abandoned city. City authorities often fostered this division through new urban politics. Informed by an entrepreneurial rather than a welfare agenda, city authorities have embraced policies designed to foster economic growth through programmes of place-promotion, public-private partnerships and embracing initiatives to support creative and cultural industries.

Mapping Marcuse's residential quarters onto our Townsville suggests that his classification of the 'abandoned (inner) city' has relevance in thinking about it as a social space. The flows and interconnections of people that comprise our study appear to include individuals often effectively imprisoned by either their lack of mobility or their lack of control over their mobility. Our city location is characterized by recently arrived refugees, empty housing stock, poor households, contracting job opportunities and a level of violence that is of consistent concern to residents and city leaders. As Marcuse points out, the abandoned city is an urban space of social uncertainty, a place of violence, crime and ethnic tension. Equally, Marcuse's abandoned city is not conceptualized as a seamless, bounded area, defined by drawing an enclosing line that separates it from the rest of the world. Instead, the internal differences and conflicts of the abandoned city are conceptualized as a result of how the multiple social practices, flows and connections that comprise this place are intricately woven together across geographical scales from the local to the global. Whilst abandoned by certain industries, the inner city remains the place in and through which young people establish their identities.

Massey's (1993) conceptual lens of a 'progressive sense of place' offers a number of important points to help conceptualize the social process of identity making occurring within Marcuse's abandoned city. Most importantly, a progressive sense of place resists thinking in terms of an essential identity or a pre-given identity among young persons fixed by either their age or place of residence. Rather, identities are conceptualized as always multiple and becoming through the actions and discursive practices that constantly operate to establish the parameters of identification by which a person is known. Hence, becoming a young person is not a necessary or given relationship, but spatially and temporally contingent. Furthermore, becoming a young person is embedded within relational frameworks of power and what Massey termed 'the politics of difference'. In this understanding we have to go beyond identity as discursively constituted and examine the ways those very identities are embedded within and through material conditions.

In thinking space as practice we have to go beyond conceptualizing the inner city and identity as separate realms. Instead, identity and place are conceptualized as embedded within one another. In other words, the attributes of a de-industrialized inner city cannot be conceptualized as delimiting a fixed identity for young people. Young people living within the inner city abandoned by many businesses are not conceptualized as playing out essentialized meanings of fully-formed teenagers. The link that often sutures socially marginalized places to a negative (essentialistic) identity of young people is ruptured. Rather, multiple identities can inhabit a space. And, a young person's identities are thought of as progressively formed through actions, encounters and experiences negotiated in and through place. Inner-city places are rethought as fluid spaces open to constant uncertainties that allow opportunities for undoing the prescriptive connections between places and identities. The uncertainties of inner-city space enable identities to be conceptualized as ongoing and changing. For some, the uncertainties may result in a redrawing of imagined and material boundaries. For others, inner-city spaces hold the potential for people to negotiate and challenge the categorizations by which they have come to be known. In thinking about the quest for an identity through space, Massey offers a way of re-conceiving the inner city, not only in terms of the risks and insecurities associated with abandonment but also the pleasures and securities associated with home. In our thinking, the inner city is a place of ongoing possibilities rather than simple abandonment.

Finally, an important means by which a progressive sense of place reworks the interpretation of place is the emphasis on everyday knowledges and practices. Everyday lives are conceptualized as a complex interplay between experiences, understandings and actions within concrete geographical spaces. Everyday lives only continue as long as people continue to negotiate material spaces through the everyday practical activities of dwelling within and moving through a particular place. However, each person's route through a particular place, their special locations within it and the connections they make between places by foot, car, phone, email or imagination will vary. The social processes that underpin everyday lives are therefore intricately tied up with how people constitute home, safety, danger and pleasure. The everyday level seems very appropriate for thinking about how young people are actively involved in simultaneously making their identities and places through negotiating the complex interplay between meanings, practices and concrete spaces.

## CHANGING LOCALITIES: 'LIFE ON THE MARGINS'

Economic restructuring and the 'shrinking' of Townsville are interpreted by young people as having a negative impact upon the quality of their lives. For many, the lack of future job prospects and the deterioration of public and private spaces, since they were born, are read as illustrative of their 'life on the margins'. For example, Emma tells of her reason for wanting to leave in terms of employment:

> I am definitely not staying here. It has nothing to do with the people. I don't think there is much to do. There really isn't. I don't really look at places to hang out, it's more about the job. If I can't find a part-time job, how am I going to find a full-time job?

Marginalization is thus understood to operate though lack of opportunities to secure financial and domestic independence through paid employment.

How the inner-city streetscape has deteriorated also operates to undermine some teenagers' sense of belonging. The appearance of redundant industrial sites, abandoned houses and garbage in front yards became a source of anxiety. Moreover, the uncertainties and sense of powerlessness generated through violence and crime further erased some respondents' sense of pride in the locality. Emma recalled how one of her few special places, a local park where she could meet with friends, was 'erased' after a shooting:

> There used to be a nice park on Jupiter Street, and then there was a huge fight. And one guy was shot. Like, three years back, I think, or four years back, and after that no one goes there. They don't even link the swings up for kids during the summer and stuff like that. Cos its just like, whatever happened, like, made

> everybody stay away from it. But it was a nice place to hang out. I was there during the fight, but I left for home. I never came back again.
>
> (Emma)

Our respondents were also very aware of the lack of commercially-owned entertainment facilities for young people elsewhere, particularly in comparison with larger metropolitan centres. This consciousness spawned a sense of missing out through the type of entertainment facilities available in Townsville. Whilst not wishing to reiterate the negative reputation of the city, Luke also admitted to its inadequacies: 'It's not a terrible place to be, but it's not the best'. The expectation is that a greater range of youth services, both publicly and privately operated, should be available. The apparent lack of some services may arise from the desires of teenagers being more globalized than those locally associated with the older blue-collar workers – such as a choice of coffee houses, internet cafés or amusement arcades. For example, one teenager laments the absence of space to hold gigs for local rock bands:

> We go to the movies 'cause that's the only place in Townsville to go 'cause they closed down Mickey's, which is a coffee house. I don't know why they did that. And there's also like a community coffee house where they have like bands come and play. But that's closing down. I don't know why. So they are taking away everywhere for us to go, so I am leaving. I'm getting out of Townsville.
>
> (Zack)

Townsville does not perform well in the context of changing notions of what a place needs to have to be worthwhile for young people. There is no place for gigs or a designated 'leisure quarter'; respondents spoke of travelling to other cities to participate in entertainment spaces. Even Townsville's one mall is not spoken about favourably. Townsville's mall is small in comparison with those in neighbouring cities and has no indoor amusement park or ice-skating rink. Thus many young Townsville residents feel as if they are missing out. Jessie is typical of how many teenagers view the inadequacies of the private retail landscape: 'I hang out at the mall mostly and then Snowton [a nearby large regional city] is pretty good, you have got a bigger mall, shopping and indoor rides'. Smallness becomes equated with the city lacking choices for young people, such as the carousel ride in the mall. For many teenagers, this is just one symbol of Townsville's inadequacy. Moreover, smallness also operates to reduce not only the number of settings but also the size and diversity of the public and peer audiences, to which teenagers can display their styles and self-presentations. For Jessie, a trip to a neighbouring regional centre not only provides a space free from parental jurisdiction, but also an important place for being noticed as well as observing the latest fashions.

Within American youth cultures, teenagers in Townsville positioned themselves as increasingly marginalized through their limited choices of places to go and spend time with friends. Yet, this increasing lack of choices for young people cannot be ascribed to smallness. Katz (1998) points out how, in New York City, young people's choices are also being eroded. Economic restructuring, guided by neo-liberalism, in this world city has also resulted in fewer outdoor spaces where young people can 'hang out' and, simultaneously, reduced employment perspective for many working-class young people, notably African American men.

## CONSTITUTING BOUNDARIES AS MARKERS OF DIFFERENCE

The city centre is a focal point for young people in Townsville. Living 'downtown' means our respondents must constantly negotiate not only encounters with prostitution, dealers and ethnic diversity but also teenagers from other neighbourhoods. Like the teenagers in Toon's (2000) study for a small city in the United Kingdom, the public spaces of 'downtown' Townsville appear to appeal to most young people as a place to 'hang out'. The inner-city precinct seemingly provides an opportunity for many young people not only to feel free from parental jurisdiction but also to be in a place of visibility, excitement, interest, action, chance encounters and the unexpected.

In Townsville, the teenagers Tim spoke to are very aware of the inner city's sex workers, drug houses, violent crimes and meeting places for different groups of young people. 'Innocence' is not an attribute of our respondents. Respondents are also familiar with the designated territories of each group. Young people's identities are premised, in part, upon a sense of spatial opposition to these social difference. Self-identification is often operationalized through opposition to particular groups of people associated with a particular locality; prostitutes, drug dealers or young people associated with a particular style. These persons are often constituted as 'bad', using negative terms such as 'hookers', 'shady people' and 'gangs'. Struggling for self-definition, many teenagers are actively making territorial boundaries to actively distinguish themselves from these groups of people, a differentiation that is often underpinned by moral judgements and racialized politics.

For example, Sue tells how drug dealers operate from specific streets, and positions them as operating beyond the law:

> Like, you got your different, how are you going to put it, okay, you got your different drug territories, and when you try to sell in someone else's territory, you know there are going to have repercussions. You know what I am saying. Like you start a fight with the wrong person, you know that person got different people are going to do something to you and your people [are] going to do something to them. I stick to my own, I don't want to be out there hanging with the people like that, I ain't going to lose my life over something stupid.
>
> (Sue)

For Sue, spatial exclusion operates through the perceived risks of defying marked boundaries and entering zones positioned as 'someone else's territory'. This respondent actively creates boundaries as she negotiates her movements through space to clearly differentiate herself from, and reduce her exposure to, the perceived risks linked to the illegality of drug cultures. In Sue's case, her self-definition is limited to distinguishing herself from the moral panics associated with illegal drugs. Extended family still plays a crucial role in mapping parts of her neighbourhood as off-limits through the construction of difference. Adult constructions of what it means to be deviant are refracted in Sue's movements.

> When I was young my mother was like, well you're not allowed down at Elm Street, like a lot of drugs and stuff … Or now she is like, I don't want you hanging out certain parts of Elm Street. Like I don't want you going to Pine Street. It's like you can't go there.

'Safety' of the streets is a pressing issue for both Sue and her mother. Potential physical and verbal harassment underpins one key reason why some young women in particular are actively involved in both establishing and maintaining imagined boundaries around certain streets. Women's perceived fear is in a large part attributable to how public streets are not only moralized but also gendered. As noted for other cities (see Watt and Stenson 1998), how women negotiate public space in Townsville is often based on their concerns about male behaviour. For example, Lisa tells about her anxiety arising from passing a group of male drug dealers on a particular street corner:

> There is a street where there is a lot of gang members. Sometime I walk home from work and it makes you feel uncomfortable with all those people standing there on one little corner just staring at you. And that is how it is … Drugs, they sell a lot of drugs and stuff and guns. And you may have went out with [dated] one of those persons hanging out on the corner, and stuff, and they may not like you and try to threaten you and scare you away and stuff. So, I try to walk around from that place. It is a little mini market on the corner and [drug dealers] just stand there the whole day. I don't know how you could stand in that one spot, pretty boring, but everyday they [would] be out there, just standing there. Some more come, some leave, I know it sounds stupid, but that is what they do.

Lisa illustrates how this is an urban streetscape highly regulated by a male gaze that disciplines this young women's life. Superimposed on moral boundaries are divisions of gender imposing constraints on the production of femininity. This policing is pointed out by numerous authors (see Bartky 1990; Holland *et al.* 1994). In this case, the respondent tries to avoid walking past a particular street-corner to evade the policing of her femininity.

Imagining, marking and making boundaries is essential to establish and maintain a self-definition separate from identities constituted as different and deviant, particularly in the comparatively close geographical proximity of social differences within the inner city and the relative immobility of its residents.

Underscoring the need for materialized boundaries are women's expressed fears and insecurities regarding with how public streets are not only gendered but also shaped, in part, by the morality of drug cultures and prostitution. Concerns about women's lack of safety on streets, especially after dark, and a fear of crime, particularly linked to drug dealing, translated into a heightened awareness amongst female respondents of how streets act as boundary markers for particular identities. Crossing the dividing line into an apparently lawless world, young people are understood to become increasingly vulnerable to physical violence, assault and verbal abuse. Deviance and physical acts of violence fixed certain street territories for use only by sex workers and drug dealers. The inner city of Townsville is understood as highly spatially segmented through the activities and discourses of people who live and work there. The struggle for self-definition is, in part, established and maintained through spatial opposition to social difference found in the inner city. Smallness seems to help fix and generate familiarity with the demarcation of territorial boundaries.

## MAKING 'COOL' OR SPECIAL PLACES TO 'HANG OUT'

In these spatially fragmented, uncertain and increasingly constrained social circumstances of the inner city of Townsville, spaces that provide opportunities to be with friends who share similar values, interests or ethnicity, become constituted as particularly 'cool' or special places where teenagers can spend their free time. Here, Loader (1996) argues, young people have the potential to be creative makers of both uses and meanings of public spaces and their own identities. According to Ruddick (1998), in these places they can explore, establish and assert a self-identity in a social context in which they feel free from parental jurisdiction and other forms of adult authority.

In Townsville, the special places most often referred to as those in which to hang out during daylight hours were streets, parks, the mall, house parties, sports facilities and nightclubs. This list confirms findings for other cities (see Loader 1996; Toon 2000). Denied access to other amenities as a consequence of exclusion by age or finance, these places are constituted as special, not only offering experiences of safety, in the same way that respondents generally constitute their home, but also opportunities to be free from parental, guardian or other forms of adult supervision. Emphasizing the importance of safety, one female teenager, when talking about a park, evoked the idea of a 'comfort zone'. Friends who share particular identities play a profound role in achieving these feelings of comfort through opportunities to talk or play games. For example, Zack emphasized the importance of friendships in a public park becoming both cool and safe:

> Like chilling with my friends, it [is] like I don't care, we don't have to do anything, we can just sit there and talk about nothing and its fun. Just like, nobody fighting, nobody arguing, just sitting there talking about anything, any random thing that comes up.

Luke reiterates the importance of gathering and withdrawing with friends in the process of making a 'cool' place to hang out with friends. Such gatherings are not specific to youth cultures of Townsville. Leidberg (1995) and Toon (2000) found similar types of gatherings, which they interpret as operating to promote group identity and help constitute particular subjectivities in public space. Regardless of the intended use of the built environment, gatherings of young people have the potential to appropriate and use the street in creative and unexpected ways to give meaning to their social identities. Luke suggests how the potential for creative use is always present: 'Sometimes, wherever you are, wherever we go, we [group of friends] make it fun'.

Skaters are just one example from Townsville of how the gathering of a young male sub culture both genders and fashions public places in creative ways. However, access to the town centre for skaters is increasingly difficult as a result of exclusionary policing strategies. In Townsville, as in other Western cities, laws criminalizing certain activities and presences, including skaters and being in public parks after dusk, have been introduced in an attempt to constrain the spatial and temporal limits of young people who gather without adult supervision (see Davis 1990; Brietbart 1995). Clearly, the visible presence of youth skateboarding around the civic centre is not deemed appropriate when gatherings of young people in public spaces are understood as a threat to the social order. Teenagers in Townsville are very aware of being under increased adult surveillance. Some resent adults' suspicion of them. One male respondent spoke of the exclusionary policing tactics selectively deployed against skaters outside the County Building:

> We skate at the County Building 'cause we can jump off the steps … Ah, the only place that we would actually have to worry about the officials is the County Building, downtown. But other than that everywhere else is legal place to skate
>
> (Luke)

Luke's comments highlight that rather than gatherings of skaters being prevented by police removal strategies, skating at the County Building took on additional value and meaning in youth resistance to authority. With the increasing number of curfews on young people, skateboarding outside this civic building became particularly meaningful as an attempt to reclaim city space.

Despite the philosophy of public services in Townsville being aimed at eliminating the presence of young people gathering on streets and in parks, many teenagers talked about being attracted to these places when they are bored. Respondents tell how, through the activities of 'hanging out' on streets, they are actively and creatively constituting their identities premised upon a sense of imagined boundaries and spatial opposition to particular groups associated with specific streets. For example, Tim, who spends much of his free time hanging out on the street, demonstrates how he actively designates his identity through being able to distinguish a local neighbourhood associated with family and friends. In this territory he could establish and maintain his identities in relative safety. In

Townsville, knowing who and where your friends are relies upon the ways of differentiating identities and places through various practices that make 'claims' of ownership on the street through various modes of self-display such as playing music, dressing in a particular fashion and driving fast cars. Again, the simultaneous practices of getting noticed and admiring peers that are important in the process of self-definition among young people are not unique to Townsville (see Hebdige 1997).

In Townsville, respondents frequently spoke about the ways in which they are able to distinguish their social and geographical worlds of belonging through the boundaries of music, fashion and fast cars. For example, Emma thought about music and the streets in the following way:

> I don't know, too many kids are, like, out and walking all the time and people, and I don't know, you walk by if you had, like, whatever, the radio on during the summer, whatever, and they don't like your music they'll be shouting at you, and they're usually out on the street. So, I try staying away from those streets. I would go up to Sycamore, then I would go on Willow.

Emma is actively aware of how musicscapes act to exclude. She has experienced how expressions of distinctive identities are mobilized, materialized and spatialized through practices of her peers walking the street. She clearly speaks of her fears of being verbally 'hassled' whilst walking the streets by peers she may encounter who share different musical styles. Emma has designed a complex route to avoid certain streets to avoid confrontation.

Clothes, as a mode of self-display, can also help materialize this sense of spatial opposition to particular peer groups. When Zack discusses streets he tends to avoid, clothes operate as a marker of subjectivities:

> It's always a sombre feeling, like when you get brought down a couple of levels, like you have always got to look behind you to see what is going on. Cause, if someone doesn't like you or whatever clothes you have on, they'll try to fight you or whatever. So, I just try to stay away from that area … Yeah, I'm like you can have your place, I can have mine … Like, I am cool with everybody, but if you mess with me and cross the line, then I am going to get upset. But, I am basically cool with everyone.

Zack's narrative illustrates how clothes may also operate as a mechanism of self-presentation and their role in boundary demarcation. Zack is particularly sensitive about border crossings. Apparently, for him, violence is an everyday act of boundary maintenance at the lines drawn between different groups of teenagers. Clothes apparently act as one mechanism of collective appropriation and demarcation of inner-city public space.

Displaying and driving fast cars also seems an important practice, particularly for some male teenagers to maintain their masculine identities and delineate an identifiable geographical territory. Driving practices associated with display, illegality, danger, speed and skill intersect in the marking of boundaries. Fast driving becomes a symbolic act, closing-off and distancing certain streets from the adult world of authority. Mark's narrative illustrates these points:

> Like you know the cars, and anything though. You know, they got to look cool, and all. In the summer they do it [teenagers put their vehicles on public display], tons and hundreds just go around showing off, along Locust Street. Well, they race you know, Yeah, and oh, the cops! They're just trying to lay the law down on that. … People get in a lot of trouble they get tickets. … Cause, you go down Locust Street on a Wednesday night, you know in summer, that's all you see, is like, all you see, is like, motorcycles and cars, you know racing and stuff. Last summer, this guy was going like 160 [miles per hour] down Locust Street and it's a 30 mile per hour zone. He killed himself. I don't know what happened, there was blood all over the street.

Mark's narrative suggests displaying and racing cars on a particular weekday night in the summer is more than a source of entertainment for many teenage males. The polished super-charged cars become an emblem of difference, a resistance to authorities and a public performance of masculinity. For Mark, 'showing off' his vehicles to an appreciative peer, and often uncomplimentary public audience, is how he chooses to be noticed. The risks associated with driving them at illegal speeds are an important practice of self-definition as masculine, as well as actively reclaiming this main street. Public gatherings, display and the racing of cars in the constitution of a youthful masculinity are not unique to Townsville. Such public performances of perils are an important part of constituting teenage masculine identity (see Dunn and Hartig 1998; Kraack and Kenway 2002).

Commercial venues are also special places to be with friends, particularly shopping malls and nightclubs. Musical venues have a strong affective meaning in the everyday world of young people, with nightclubs particularly valued within narratives of self-identity (Malbon 1999). In Townsville, the nightclubs of the inner city are highly differentiated through the intersections of design, music and the ethnicity and age of the clientele. Emma recounts how she spends weekends at one particular nightclub because entry offers the illusion of travel to her former homeland:

> There is a place on Buckthorn street. It's a club called 'Kral'. It's actually a Bosnian place. It's more older, I really like the place, just because they go like rock concerts. We dance, we hang out, we get together. Like, I don't know I go, I can't say every night but usually the weekend … I know the people that go there. And, the place is as I remember back in Bosnia, even though I was younger. I didn't go that often to the club, but I would go during the day. My family used to own a nightclub back in Bosnia. So the way I remember ours, the way it looked like, these clubs are very similar to it and they make me feel as if I am back home.

This Bosnian-styled nightclub appears to offer a feeling of familiarity and older patterns of life. A sense of self for this respondent is found in the certainties of a collective familial and national identity. Conventional boundaries of self-definition for this young women are reconstituted through the safe place of this Bosnian club, in contrast to the uncertainties of a new life in the United States.

Another young woman provides a further example of the value of nightclubs to how young people negotiate their identities:

> We'll like go out, like Friday night is Club Coyote, that's on Mulberry Street. A lot of people go there on a weekend. I feel like I belong there … like when I go to the mall I feel a little bit out of place. But when I go dancing, it is like, everyone is going there to have fun, you know they're are going there to meet people, the social atmosphere … It [the club] makes me feel like I am not a minority to tell you the truth. It is also based on race, ethnics. Me, I am Hispanic. I'm Dominican. There is a lot of Hispanics around but it is not the same. It is like I am Dominican and they are Puerto Rican, it is not the same. I do not fit in with them, like I try to talk to them but I just don't fit in with them. I fit in more with, I grew up with black people and I fit in with them. They are really nice. Like people seen them really bad. And everyone, like you try not to but everyone does.

Jan highlights the highly racialized social relations in Townsville. Racism seemingly prevents opportunities for the formation of creative potential of new subjectivities. Even on the commercial dancefloors of Townsville the ephemeral groups that form through clubbing apparently remain differentiated by ethnicity. Yet, through the commercial space opened up by the dancefloor this respondent is actively redefining her boundaries, she no longer feels the social processes of marginalization but instead becomes the 'centre of attention'. The dancefloor still has creative potential. Yet, simultaneously, whilst defying layers of social-material borders that hem in her subjectivity and those of African Americans, she is also actively maintaining her identity as Hispanic. Certainties seemingly remain for young people in Townsville through their ethnic identity.

The public places of the mall are also viewed as 'cool' places in which to hang out. The 'magic of the mall' that Anthony (1985) identified, continues to cast its spell over teenagers, particularly young women. Echoing the results of McRobbie (1991), lacking alternative places that young women feel safe to gather and talk, the mall becomes one of the few places where young women can explore and establish their femininity in a highly visible way in the public realm. Lisa highlights the importance of the mall as a safe place in the lives of young women:

> All we really have is the mall. 'Cause half the things I do, I go out of town, 'cause Townsville doesn't really have [places] where a lot of young people can get together and do stuff. Somehow we always mess it up with fighting and you know violence stuff like that … I feel safe at the mall 'cause there is security and stuff around. Like if something bad is going to happen somebody

> will stop it. I guess for instance you are walking on the street, it may take a while for the cops to get there.

According to Lisa, groups of young women feel safe to gather and walk in the mall only because of the presence of security guards and their ability immediately to contain violence.

Clare discusses how gathering, shopping and walking in the mall remains important in relation to establishing and maintaining contemporary femininities in Townsville:

> It's definitely away from your house, so that is good, um, you get to shop and that is what every girl does … pretty much I either shop or just hang around, but I don't like to walk around too much, 'cause you just keep on walking in the same circle.

The mall continues to be constituted as a vantage point where these young women form and re-form styles through scrutinizing not only what each other buys and wears but also through seeking admiring comments from friends or fleeting looks from strangers. Whilst the mall remains central to the constitution of feminine identities away from adult supervision, as noted by Ruddick (1996, 1998), Clare points out a key limitation: the walking circuit or cultural practice of the promenade in Townsville is regarded as too short. In other words, smallness of the mall appears to restrict the pleasures derived from being seen and seeing others. Consequently, Clare describes how walking through the mall quickly becomes 'boring'.

## CONCLUSIONS: SMALLNESS AND A HEIGHTENED SENSE FOR AND OF PLACE

In this chapter, we have drawn upon Massey's theoretical work on progressive sense of place to argue that youthful identities are performed in and through the spaces of everyday lives. Through the themes under which this chapter is divided – changing localities, markers of difference and 'cool' or special places – we have attempted to explore how youth identities are fashioned through a sense of spatial opposition through the production of boundary markers constituted by ideas of social difference.

Townsville, a small multicultural city, is becoming smaller and more ethnically diverse. The lives of young people in this city share many commonalities with findings in other cities, both large and small. The formation of youth identities is played out in urban spaces through the intersection of age, gender, ethnicity and class in a constant process of negotiation between parental and policing controls and youth resistance. Like young people elsewhere, they have fewer choices of places to go; gatherings of young people in public places are positioned as threatening and laws have been introduced to remove young people from public places.

However, smallness appears to heighten young people's sense *for* and *of* place. In part, this amplified sensitivity to place in the everyday lives of young people operates through a heightened awareness of tensions presented by feelings of inclusion and exclusion. For young people born in Townsville, there is the familiar tension between the sense of belonging and alienation surrounding the family home. Yet, smallness only appears to compound these tensions for many young people through a sense of always 'living on the margins'. The sense of belonging sustained through the family home is countered by the marginalization often derived from an understanding of 'small as lacking'. The idea of lack is only confirmed through lived experience of employment prospects, educational opportunities and relevant everyday facilities where young people can be themselves. Permanent or temporary migration to larger cities is one strategy young people often discuss as a remedy for their sense of lack and powerlessness. Respondents equate bigger cities with better opportunities for young people (shopping, dancing, employment and education). Yet, the work conducted in other Western cities by Katz (1998), Matthews *et al.* (1999), Vanderbeck and Johnson (2000) suggests that young people in all metropolitan areas, particularly those with low levels of education, are increasingly limited in their choices of places to go and be with their friends.

As in all Western inner-city areas, the young people of Townsville are also actively involved in a diverse range of practices that sustain a sense of belonging and self-definition through making claims to public and private spaces as their own. These practices include walking, talking, gathering, shopping, clubbing, partying, drinking, smoking, sex, driving, wearing particular clothes, playing certain music and sports. As with other large and small Western cities, the favoured spaces to gather during daylight hours are in either streets or public parks of the inner-city or shopping malls. These places are perceived as 'safe' and offer public venues where young people are clearly visible, yet free from adult supervision.

On the one hand, these claims on city spaces are interpreted as an act of resistance to adult authority. On the other hand, they also operate to exclude between groups of young people. As these practices operate in and through place, they act not only to help establish and maintain particular youth subjectivities but also to help delineate and embed territorial ties or allegiances to place. Consequently, those young people whose identities rely upon different styles, ethnicity, gender, friendship networks, morals or laws (as social practices) think carefully about how to negotiate inner-city streets. Actions and ideas of a teenager's subjectivity are often aligned along border-making spatial practices that hem their movements into particular places. Smallness again seems to increase young people's awareness of how their own self-definition is premised around how particular public places are territorialized through association with particular youth groups. Smallness also heightens young people's sense of and for place through the restrictions placed on audience size. Young people living in Townsville often seek larger audiences, comprising of their peers and strangers, and increased visibility, through travelling to large cities. These young people's heightened sense of and for place would appear to be a valuable resource for policy initiatives seeking innovative urban futures that appeal to young people.

## NOTES

1 The names of the city, streets, commercial venues and informants have been made anonymous to fulfil an ethical requirement of this research.
2 The methodology was designed to investigate how youth identities are produced in iterative relationships through the meanings, experiences and actions that are embedded in and through place conceived as social practice. The principal ethnographic tools are observation, semi-structured interviews and content analysis. Ongoing observations of the social dynamics of the inner city are permitted through the research activities of Kraly. In February 2002, a total of 12 in-depth interviews were conducted with 17-year-old high-school students. Overall, a project positioned as investigating youth cultures was received enthusiastically. This interest was, in part, sustained by the social capital of the interviewer: a tall, blond, twenty-one-year-old Australian male surfer from a former industrial city. The character of the researcher–informant relationship was always conditional on Tim's positionality as a researcher and an Australian male. However, because of his shared youth and experience of growing up in a marginalized and stigmatized town, everyone with whom he spoke shared certain common social ground. Content analysis was used to interpret how participants constituted their identities through blurring, maintaining or creating boundaries of difference constituted in and through the inner-city spaces. Whilst not adhering to a particular formal procedure, a number of strategies were implemented to assist the analysis including familiarization and coding. The quotations used in the text are verbatim. The city, the streets, the venues and participants' names have been changed to ensure anonymity and confidentiality.

# 16 Small city – big ideas

## Culture-led regeneration and the consumption of place

Steven Miles

On the surface, NewcastleGateshead, in the north-east of England, provides a shining example of how cultural investment can revitalize the economic and social future of a small de-industrialized city. And yet NewcastleGateshead is not a city at all, but the fabrication of a political alliance. The very name NewcastleGateshead is symptomatic of a climate in which small cities throughout the UK are desperately competing with each other to establish their own twenty-first-century credentials, and culture provides a primary vehicle through which this can apparently be achieved. This chapter will critically discuss the degree to which, by embracing a so-called post-industrial culturally-regenerated future, cities such as NewcastleGateshead run the risk of compromising the very smallness that makes them what they are in the first place. Lorente (2002) argues that arts-led regeneration can prosper even in the most adverse urban conditions where traditional economies have become obsolete. But what really constitutes 'prosperity' in this context and can culture-led regeneration really do everything we ask of it?

When the announcement for the UK nomination for (European) Capital of Culture 2008 was made, I was resident in Newcastle and leading a ten-year longitudinal project on the social, economic and cultural impact of cultural investment on NewcastleGateshead Quayside (Bailey *et al.* 2004). The moment Tessa Jowell made the announcement in June 2003 that Liverpool was the victorious city, the profound implications of the decision hit home and they hit home hard. Not only would the decision have a potentially serious knock-on effect for the future of NewcastleGateshead and the optimism that had surrounded it during the bidding process but, at a more personal level, the future suddenly looked less bright than it had done before. In reality, the fact that NewcastleGateshead did not receive the UK nomination for Capital of Culture was devastating news all round, but it also served to highlight the fact that the cultural regeneration of small cities is a profoundly uncertain business (see Eckardt this volume).

### NewcastleGateshead

The redevelopment and re-imagining of NewcastleGateshead over the past few years has been nothing short of astounding and the magnitude of such change cannot be understood without reference to the position from which the north-east has come. Newcastle and Gateshead are, above all, the products of industrialization and its aftermath. For this reason, the River Tyne was always a regional focal point. Immigrants from Ireland, Scotland, Cumberland, Yorkshire and Scandinavia were attracted to the area by the prospect of the high wages that shipbuilding, chemical works, coalmining and other heavy industry brought with them (MacPherson 1993). The fruits of industrialization were, however, relatively short-lived. As Hudson (2000) suggests, the period between 1951 and 1958 can be described as something of a 'Golden Age' for the region in terms of high employment and rising wages. However, although the period of modernization between 1963 and 1970 saw a temporarily effective process of economic restructuring, in the longer term such developments had a negative impact upon the working classes in that, 'While to some extent policy objectives were realized (the employment structure, for example, being 'diversified'), state intervention failed to achieve its principal aim of lowering (male) unemployment and was associated with a pattern of changes which exacerbated the problems it was supposed to solve' (Hudson 2000: 53). Indeed, from the 1970s onwards, the north-east experienced a deepening and arguably irreparable depression in which the local economy continued to decline and the market forces associated with the global economy emerged as a growing economic steering mechanism (Hudson 2000).

The process of de-industrialization in the north-east brought with it urban decay, as Power and Mumford (1999) and Tomaney and Ward (2001) point out. In discussing Newcastle, Power and Mumford describe a situation in which the city had entered a cycle of escalating physical decay in which houses were progressively being abandoned and boarded up. The causes of such a development, as Hall (2002) points out, were complex but were characterized by the long-term structural decline of the economy, long-term unemployment, poorly performing schools and the perpetuation of social disorder and even gang warfare. This was a particular problem in west Newcastle, despite signs that the city centre itself was, by the turn of the millennium, beginning to show signs of something of an urban renaissance with a thriving city centre 'attracting not merely tourists and night-time visitors but also now residents who were colonizing converted warehouses and new apartment blocks: urban renaissance and urban collapse were standing side by side, sometimes as little as a mile apart' (Hall 2002: 418).

Global economic change and the de-industrialization of the north-east created a set of circumstances in which regional particularity had to be transferred from production to consumption, and this was essentially a divisive process (Vall 1999). But in the knowledge that the region could no longer depend on its industrial past it was a process that Newcastle and Gateshead felt obliged to pursue at both a symbolic and a material level (Chatterton and Hollands 2001). Nowadays, Newcastle upon Tyne (population 259,000) in particular, has an increasingly high profile, not least as a party city. Newcastle was recently voted the eighth best party city in the world by American travel experts, Weissman Travel. Combine that with the cultural provision available on Gateshead's (population 200,000) side of the River Tyne and you have a convincing tourist offer that puts NewcastleGateshead in a strong position as a weekend and conference destination, the north-east's links with the Baltic region being particularly significant in this regard. But these relatively positive developments did not occur overnight.

## CULTURE-LED REGENERATION

For many years, the city and the region had been at the bottom of any league tables on cultural attendance and participation (Bailey *et al.* 2004). But the tide was about to change. There were a number of key catalysts that ensured that culture came to play a pivotal role in realigning NewcastleGateshead's socio-economic trajectory. First, the Year of the Visual Arts in 1996 (preceded incidentally by the National Garden Festival in Gateshead in 1990 which was also a significant platform for public art) was the fifth in the series of year-long art form celebrations sponsored nationally by the Arts Council of England under the general title of 'Arts 2000'. Adopting the strap-line 'The region is the gallery', the 'year' was intended to lead to a greater awareness of the visual arts amongst the population, as well as increased attendance and participation in events. The Arts Council of England's (1997) research indicates that the Year of the Visual Arts succeeded in raising awareness, but that longer-term changes in attitude would require a more sustained programme (see Bailey *et al.* 2004).

Antony Gormley's public sculpture, The Angel of the North, was erected in the year following the Year of the Visual Arts and represented a pivotal political moment, insofar as Gateshead Borough Council managed to realize this project despite considerable local opposition. Initially, the Angel was received with considerable negativity, not least through the local press which questioned a total investment approaching £800,000. But the important thing here was the fact that the Angel was constructed at all, which proved what Gateshead City Council could achieve and gave funders renewed confidence that the Council could indeed make good on similarly large-scale capital projects; not least in terms of the potential success of subsequent large-scale developments, such as those on NewcastleGateshead Quayside. Gateshead Quays subsequently became the focal point for this apparent transition from production to consumption. At the heart of the development lay three projects: the BALTIC centre for contemporary art, costing £46 million, the Sage Gateshead Music Centre, costing £70 million and the Millennium Bridge, costing £22 million, which linked the two cultural developments (that sit on the Gateshead shore) with Newcastle city centre.

The above developments provided a platform upon which the regional capital of the north-east could launch its bid for the UK nomination for European Capital of Culture 2008. The NewcastleGateshead bid appeared to be particularly well received by local people as the longitudinal project undertaken by the Centre for Cultural Policy and Management at Northumbria University appears to indicate. Bailey *et al.* (2004) argue that the apparent success of developments on NewcastleGateshead Quayside reflects the fact that such landmarks were able to engage with the history of the region and, more pertinently, the history of the *people* of the region. But the jury is still out and the degree to which the culture associated with NewcastleGateshead is, in fact, a culture of the people remains unproven; as does the longer-term impact of cultural investment on this scale.

In order to begin to get to grips with this issue, I want to consider some of the early data emerging from the Cultural Investments and Strategic Impact Research (CISIR) programme conducted by the Centre for Cultural Policy and Management at Northumbria University, a unique ten-year project that looks at the social, cultural and economic impact of cultural investment totalling around £138 million. In order to begin to comprehend what cultural investment means for NewcastleGateshead, it is worth looking specifically at a set of interviews that were undertaken with some of the key people and organizations involved in bringing new life to the Quayside.

## UNDERSTANDING THE PROCESS

The first point to make is that although the social and economic benefits of the Quayside development soon became a key priority for those organizations behind the development, initially the driving force was a determination to provide the north-east with world-class cultural facilities. As a representative of Gateshead Borough Council indicates:

> Now we have some of the flagship icon buildings I suppose any city that takes itself seriously would expect to have and in some ways it was as simple as that: the feeling that we cannot be taken seriously unless we have that strong infrastructure and there was an opportunity, so we pushed that opportunity.

For a long time, the north-east's record for cultural provision and participation had been appalling and there was strong recognition that this had to change (Bailey *et al.* 2004). The people of the north-east should at least have the opportunity to partake in quality cultural provision. It was in this respect that a representative from Arts Council England, North East, pointed out that economic regeneration was not the initial driving force behind plans for the Quayside. It was only later that the Millennium Bridge, the BALTIC and the Sage Gateshead came together to provide the necessary magnets for inward investment.

Policy-wise, then, it seems that the economic benefits emerged from a cultural intention. The benefits of regenerating the Quayside cannot themselves be understood through purely economic lenses as one local policy maker suggests:

> De-industrialisation leaves a number of problems. The first is worklessness, unemployment and the second is communities that become abandoned … abandoned may be too strong a word but certainly they become denuded of character and purpose … Culture was seen as having three particular roles in that process. The first is potentially in being an economic sector which could bring jobs and prosperity to the area … Second, that cultural activity could play a role in providing both character and community development opportunities in local communities … The third is a more over-arching point which is

> about reputational value, image and self-confidence … places that have been de-industrialised all seem to have the characterisitics of low aspiration, low self-confidence, poor image and culture I think is generally felt can be seen to play a role in all those things.

A key theme of these interviews was pride and confidence. A representative from the Sage Gateshead felt that the regeneration of the Quayside was giving people a palpable sense of pride in their borough and their town and that this feeling was also taking root on a regional basis. As the respondent from Gateshead Borough Council also said, 'As well as the region being recognized internationally for being an exciting place to be I think in more localized ways the sub-region will have a confident identity and a future so that there is less internal competition in the region and a clearer sense of being part of a whole'. The regeneration of the Quayside can certainly be said to have fuelled a confidence amongst the people of the region, but also a confidence in the business potential of the north-east. The Quayside has traditionally been at the heart of the region and its revitalization has from this point of view also revitalized people's sense of pride in where they live.

But a key question here is the relationship between this sense of confidence and broader aspects of social change (and not least social inclusion). As one policy maker put it:

> I think the whole sense of regional identity and pride is a bit illusive to be honest … You see people going on all the time, 'oh it is beautiful; they have done a brilliant job, this is really something for us to be proud of'. Would you go and see [the] Gormley [exhibit]? 'Well no, it's not for the likes of me'. So I think that sense of pride is very strong, but that is not the same as inclusion. Pride is a different thing from inclusion.

The danger here, then, is that small cities put their faith in flagship developments that may just as likely be forces for social exclusion as they are for inclusion. The BALTIC, for example, has been criticized for endorsing an elitist version of contemporary art and one that visitors might only reasonably relate to if they have access to the appropriate cultural capital.

The relationship between Newcastle and Gateshead has clearly been an important after-effect of NewcastleGateshead's focus on culture-led regeneration. As the representative from Gateshead Borough Council points out, the relationship between the two councils and hence the two places would not have materialized without what has happened on the Quayside. NewcastleGateshead's bid for Capital of Culture 2008 played a further role in solidifying this relationship. As a representative of the destination-marketing agency, Newcastle Gateshead Initiative suggests, the key to understanding the Capital of Culture bid was in its role as a form of marketing. As a respondent from One North East, the Regional Development Agency, says, the bid is part of a process in which a much more contemporary and dynamic image of what the north-east is all about is being portrayed; the aim being, indeed, to make the north-east the most creative European region by 2008.

The rise of culture on the policy agenda has been notable and the Capital of Culture bid certainly played a key role in this process, as the following quotation from the One North East employee indicates:

> People felt good about belonging to a region that was bidding to be Capital of Culture. I think the bid has raised the profile of culture as a regenerative driver a hundredfold so much so as I say the first regional economic strategy four years ago had very little reference to culture and the new regional economic strategy has culture embedded right across it. In terms of raising the profile of the impact of culture the legacy of the bid has been enormous.

But regardless of the impact of culture on the policy landscape, what does this apparent sea-change mean for the cities of Newcastle, Gateshead and others competing on the same territory? Perhaps the most important thing to say is that a cultural investment strategy that is designed to achieve particular social and economic outputs is in grave danger of failing. The representative from Sage Gateshead compared culture-led economic impact to acupuncture. Arguing that economic impacts were (at least originally) well down the ladder of Gateshead Borough Council's objectives, it was pointed out that acupuncture works in a very perplexing way by addressing one symptom by attacking the body in a different place. So, typically, if you have a problem in one part of your body, acupuncture will involve the insertion of needles somewhere apparently quite different. This analogy was applied to explain how culture-led economic impact works: 'If you aim at say, an economic impact and you construct a cultural investment strategy with that output it will fail. All over the world it has failed. The ones that have succeeded are the ones where the targets they were aiming at were cultural success and through achieving the cultural success they generated the economic output'.

A cynic might suggest that NewcastleGateshead was somehow getting ideas above its cultural station, but in actual fact the ambition that underpinned the move towards culture-led regeneration was one based on a genuine recognition that the city simply did not have the facilities a place of its size would normally expect. As the representative from the Newcastle Gateshead Initiative pointed out, there is a hierarchy of cities, and places like NewcastleGateshead are almost obliged to place themselves within that hierarchy. First of all there are mega-cities such as London, New York and Hong Kong that are cultural and economic centres, almost countries in their own right. There is then a second tier of cities that are not quite so influential on the world stage but still have a powerful and well-defined cultural identity such as Paris, Frankfurt, Delhi, Mumbai or Singapore: the sorts of cities that bring with them a particular cultural imagery; Barcelona being a further example of a city that has moved into this category in recent years. The point here, as the representative of Newcastle Gateshead Initiative suggests, is that culture has a major part to play in the development of cities in this tier. But the key question centres on the degree to which cities such as NewcastleGateshead can or even should sensibly aspire to joining this tier (or even the tier below it). Not all cities can be cultural hot spots. Cultural investment is not a panacea, and yet the buzz that currently surrounds culture-led regeneration continues to imply otherwise.

One example of a piece of research that highlights this issue particularly effectively is Jayne's (2004b) broader work on creative industries development in Stoke-on-Trent in the Midlands of England. Jayne presents a rather more sober analysis of attempts to use the arts and cultural reproduction for urban regeneration in Stoke-on-Trent. Critiquing the suggestion that cultural investment, notably in the creative industries, can attract post-industrial jobs whilst also encouraging people to live in city centres and generally improve the urban quality of life, Jayne argues that it is important that critical rigour is applied to the ways in which creative industries development has become aligned with regeneration in our cities:

> Unlike many other Western cities, Stoke-on-Trent remains overly dominated by working-class production and consumption cultures. The city is thus, in a sense, rendered illegible to post-industrial businesses, tourists and to the many young people who leave the city for each of the more dynamic economic and cultural opportunities offered in other cities.
>
> (Jayne 2004b: 208)

The concern is that work such as that of Richard Florida (2002) which, intentionally or otherwise, flies the cultural flag, is simply too intuitively appealing to policy makers for the good of our cities. Florida's work presents culture as a 'way out' for cities. He suggests that the way forward is for cities to attract creative people through its cultural offer. As such, his approach makes all the noises that policy makers want to hear. It allows them to think big, and offers the earth in return. Florida's approach therefore tends to imply by default that culture-led regeneration is any easy option, when it is, of course, nothing of the sort, as even Florida himself would surely recognize. It is certainly true that place provides an increasingly important dimension of our identity, but culture takes many forms and many of those forms do not fit neatly into Florida's model. He does not, of course, claim that all cities can be archetypal creative cities. But he does identify a pecking order and by doing so he presents an aspirational model; an alternative that by its very nature encourages policy makers to see their city as the next potential success story, because to deny the cultural creativity of your own city is tantamount to denying your very own identity.

Perhaps the best way of contextualizing this process is by referring to Williams' (2004) work on the 'Anxious City'. Williams argues that the extraordinary degree of change that British cities have undergone since the 1990s illustrates the extent to which the idea of the city remains contested. He contends that the way English cities, in particular, are inhabited and consumed is divided along class lines, and that the current development of cities is in fact a revolution of bourgeois taste, which actively excludes that of the working class, as traditionally defined. From this point of view, our cities are becoming theatrical spectacles of consumption, impervious to the needs of the majority of 'flawed' consumers (Bauman 1998). This may shed some light upon the drift towards culture-led regeneration amongst policy makers. Developments such as NewcastleGateshead Quayside could be interpreted as staging peculiarly English uncertainties or anxieties about city life. In short, in adapting Williams' argument, we might describe the Quayside as a kind of idealization of urban life which is itself anxiety-inducing. In particular, as a public space the Quayside could be compared to the new public spaces of the regenerated Manchester which, although being worthy of commendation in the way they encourage public gatherings, according to Williams, can only do so under the umbrella of surveillance, thereby limiting the experience of the city to easily monitored practices of consumption (see Edensor this volume). In the case of NewcastleGateshead Quayside, these spaces are dominated by two bourgeois centres of entertainment, the BALTIC and the Sage Gateshead.

It could therefore be argued that small cities are grasping hold of a model of culture-led regeneration in the hope of creating a new future when the end result of this process is a model of a predictably picturesque city with consumption at its somewhat empty heart:

> What we get is a series of façades, of stage sets, in which the performers are inevitably those with time on their hands – tourists, the young, the leisured. To limit the experience of the city to these groups is a mistake. What I am arguing for, finally, is not an end to these spectacular projects – an end to them will come soon enough, as sure as they will be in favour once again in the future – but a more realistic, less neurotic way of conceptualising the city. The spectacularisation of city life has been a refreshing, and perhaps necessary response to the technocratic and functionalist approach to the city of the 1970s, but it has its limits … Let us think about margins as much as centres, of work as much as play, of ways of being in the city that do not correspond to bourgeois forms of entertainment – and let us find ways of imagining the city in these terms as well.
>
> (Williams 2004: 241)

But as far as NewcastleGateshead is concerned, the margins of city life are equally anxiety-inducing. Authors such as Chatterton and Hollands (2003) have noted the difficulties Newcastle has had in struggling with a post-industrial transition in terms of capturing new investment and encouraging new avenues for service employment. The city has had particular difficulty in balancing the cultural nirvana evident on the Quayside with the apparently less savoury night-life culture a little further west down the Quayside. As such, Chatterton and Hollands (2001: 121–122) quote a cultural industry sector spokesperson:

> There is a resistance to either run with the party city tag or completely oppose it but they have not decided what is the best image for the City. And that is part and parcel of the mechanism which is being developed within the local cultural strategy as to what the vision for the city is. And the Head of Arts and Culture obviously will lead on that. But also through the Capital of Culture for instance, as how they define culture and obviously youth culture and party city

> type activities will play a part, or whether that will undermine what other parts of the image they want to project. I do not think that the authorities are sure whether it is to the advantage of the city or disadvantage of the city in the overall scheme of things to have this party city image.

The very notion of 'NewcastleGateshead' is in itself a further illustration of the apparent anxiety that is attached to notions of the contemporary city. The tag is the construction of the destination-marketing agency, the Newcastle Gateshead Initiative, intent on cashing-in on both the reputation of Newcastle upon Tyne as a regional capital and party city, alongside the cultural iconicity on the Gateshead side of the Tyne. The marketing of NewcastleGateshead is in this sense characterized by unresolved tensions. And these tensions were perhaps best expressed in the marketing opportunity that was NewcastleGateshead's Capital of Culture bid. In discussing this issue, Lancaster (2004) laments the fact that the NewcastleGateshead bid was less of a bid of the people than it should have been: dependent upon expert contributions, it lacked a Geordie voice. Lancaster describes a city and a region dependent upon the big gesture and not least the NewcastleGateshead Quayside:

> Unable to represent ourselves, we seem to be all too willing to let the outside consultant do it for us … we seem to be incapable of shaping our destiny, big gestures in the long term only serve to confirm our backwardness and dependence … The truth is that the Newcastle/Gateshead bid offered an empty sandwich: a top slice of *grand projets* and a bottom piece of Party City monoculture. We urgently need a 'tasty filling', or we will increasingly be seen as 'the Far north', a region, disconnected not just from the rest of the nation, but also from the lifeblood of the new creative economy (Lancaster 2004: 9–10).

## CONCLUSION

Perhaps our small cities have become preoccupied with questions of image and self-confidence, and perhaps the self-confidence derived from culture-led regeneration and the prosperity that we mentally attach to that regeneration is in a sense illusory, perhaps even inauthentic. Perhaps our cities (or at least the rhetoric that surrounds our cities) are in this sense unreal: more the creation of an idealized bourgeois image of what a 'cultural city' might be than a genuine expression of what it means to live in a place. Cities simply *have* to be seen to compete in the cultural marketplace in order be taken seriously. However, by selling yourself to the cultural 'devil' there is always a danger that you lose sight of what your city was all about in the first place. In danger of being seduced by the potential for economic outputs, small cities are in particular danger of being compromised by the corporatization of culture that this apparently implies. But they are also more susceptible to the social inequalities that such an approach is likely to encourage. The dangers of regeneration of this kind are no better expressed than in Broudehoux's (2004) work on the redevelopment of Beijing:

> The image of the city that is constructed and promoted in the process of selling places is often based not on the local reality but on stereotyped notions and exaggerated representations, which seek to enhance the marketability of the locale. The ready-made identities assigned by city boosters and disseminated through the mass media often reduce several different visions of local culture into a single vision that reflects the aspirations of a powerful elite and the values, lifestyles, and expectations of potential investors and tourists. These practices are thus highly elitist and exclusionary, and often signify to more disadvantaged segments of the population that they have no place in this revitalized and gentrified urban spectacle (Broudehoux 2004: 26).

The irony here is that the immediate appeal of culture-led regeneration to the development of small and large cities alike is very likely to encourage a situation in which such cities are almost indistinguishable variations of each other. As Zukin (1982: 837) puts it, 'Competition between corporations and cities has led to a multiplicity of standardized attractions that reduce the uniqueness of urban identities even while claims of uniqueness grow more intense'. Our small cities are already dominated by predictably uniform retail opportunities. By throwing a further prefabricated form of cultural ingredient into the mix, the danger is that place identity is sacrificed on the altar of culture-led regeneration.

I do not intend to suggest for one minute that developments in NewcastleGateshead are having a negative impact on the north-east. At the very least, the Quayside experience hints at the potential of this type of regeneration. However, NewcastleGateshead has been fortunate enough to be the beneficiary of significant public and private funds of which other small cities could only dream. These funds appear, as the CISIR research seems to imply, to have had some success in tapping into the past of the region and offering it some hope for the future (Bailey *et al.* 2004). However, there is a question mark over whether or not, as far as the rest of our small cities are concerned (and not least given recent reduction in UK government funding for the arts), this hope is a profoundly false one.

As one local government official put it during the CISIR research, 'NewcastleGateshead's bid was instrumental in *expressing* the region's ambitions'. But the degree to which those ambitions can be fulfilled by NewcastleGateshead and by other small cities pursuing a culture-led regeneration route is, by the very nature of cultural investment, destined to remain uncertain, not least given how notoriously difficult it is to actually measure the impact of culture-led regeneration (Evans 2004). Although it is undoubtedly true to say that waterfront redevelopment projects in particular, such as that on NewcastleGateshead Quayside, are effective in simultaneously speaking to our future and to our past, in fact such developments are more indicative of the degree of uncertainty smaller cities have about their futures than anything else. Such developments can be described as 'attempts' to re-centre activity in urban space, 'to shift the focus from the old to the new' (Marshall 2001b). However, such an attempt may say more about the people behind the attempt than it does about the people of the city as a whole. Hunt's (2004: 337) thoughts on the Capital of Culture are also pertinent here. He describes the bidding process as a 'national narrative of reawakened civic pride'. However, from his point of view, 'Whereas nineteenth century middle-class civic culture was more often than not the indigenous product of a Nonconformist conscience, the culture of today's cities appears more of a branding and a marketing tool than a reflection of civic identity. It is frequently the work of quangos and urban regeneration consultants rather than the organic outcome of any home-grown civic sentiment' (Hunt 2004: 346).

This chapter is not intended to dismiss out of hand the undoubted civic potential of culture-led regeneration. Rather, the suggestion here is that such a model of regeneration constitutes a bandwagon that simply may not provide that one-size-fits-all solution that many policy makers may have in mind. Lancaster (1995) argues that cities are never static but are always consciously or unconsciously trying to be something else. They are obliged to deal with their particular history and to construct meaning out of that history. This much is surely true, but it is important to also remember that the struggle to be something else can also, by implication, compromise that history. NewcastleGateshead is an example of a small city that has been very ambitious in its efforts to mark out a future for its residents. But it is not enough to cite the mere existence of the redeveloped Quayside as indicative of a success story. Only time will tell if such a success is the success of NewcastleGateshead and its people or simply the success of, for and by the cultural elite.

## ACKNOWLEDGEMENT

The author wishes to acknowledge colleagues at the Centre of Cultural Policy and Management at Northumbria Univeristy who are continuing the CISIR programme.

# 17 Afterword

## Sizing up small cities

David Bell and Mark Jayne

> Cities have something more than simply 'largeness'.
>
> (Pile 1999: 5)

In September 2005, the UK daily newspaper the *Guardian* launched a new format, midway between the usual British newspaper sizes, the broadsheet and the tabloid. Among the self-referential pieces launching this new 'Berliner' format, the paper ran a satirical trend-spotting feature on the 'rise of the moderately-sized object'. The article told us that 'medium was the size of the future', adding that 'in this bold new era, you'll need to be middle-sized to thrive. … Medium will be the new big. And the new small. Both at the same time' (Burkeman 2005: 2). While the article is a joke, its sense of the in-between-ness of the middle-sized seemed resonant with the small city; the *Guardian* may have been more optimistic about the future being middle-sized than many prognoses about small cities, but we hope that the chapters in this book have shown how small cities are responding to that in-between-ness, through a variety of both strategic and happenstance responses. The Berliner format was chosen so as to differentiate the *Guardian* from its competitors, of course, and the chapters in *Small Cities* have explored how smallness may also be used in a competitive urban hierarchy, to buck the trend towards going large or going global as the only path to a bright urban future. At the same time, our chosen focus on these small places is aimed at highlighting the *ordinariness* of the small city, offering a corrective to urban theory's (and much urban policy's) obsession with world cities (Robinson 2002).

But there are clear dilemmas facing small cities the world over. Research has suggested the particularity of the small cities experience. We conceived this book in the English city of Stoke-on-Trent, where smallness was ambivalently figured both as a hindrance and as an asset. It was hard for city managers not to be aspirational, not to be envious of their larger neighbours, even as they wanted to relish some of the distinctivenesses that made Stoke the (small) city it was (and still is). But, as Dürrschmidt (2005) so vividly described in Guben-Gubin, the smallness of the city was manifest in its habitus, in the 'mental shrinkage' that cast a fatalistic pall over the place. The city has an identity crisis and, while this is in part to do with its unique history as a city cobbled together from six towns (Parker 2000), it is also to do with its perceived place on what Dürrschmidt calls the 'urban escalator' and the direction in which that escalator is felt to be moving. And while Robinson (2002: 545) is dead right to say that to 'aim to be a "global city" in the formulaic sense may well be the ruin of most cities', there is a bit of a policy void (not to mention a theory void) when it comes to pointing up alternative trajectories.

Unfortunately, both editors have recently had to move (reluctantly) from our very own small city. Dwindling student admissions were a key factor – students appeared to be increasingly ambivalent about committing to the 'unglamorous' city of Stoke-on-Trent, perhaps attracted instead to the 'buzz' and increased job opportunities of larger metropolitan centres. The difficulty of working at an institution struggling to compete in a competitive university hierarchy added to our sense of frustration with city politics and the trajectory of Stoke-on-Trent. Nonetheless, our (and others') failed attempts to overcome fatalism and inertia further strengthened our resolve to complete this book. Hence, one of our key aims in compiling *Small Cities* has been to contribute to that alternative agenda through a consideration of four dimensions of 'small urbanity': political economy, the urban hierarchy and competitive advantage, cultural economy, and identity, lifestyle and forms of sociability. Nested within each of these organizing sections are the individual chapters, each bringing to light the ways that particular small cities are dealing with smallness. Issues raised range from the articulation of entrepreneurial governance in the small-city context, to the selling of small cities, to the idea of the 'creative small city' and the ways that particular social groups mobilize identifications based on smallness (or on reactions to it).

In pondering smallness, we cannot help but cast around for ways in which the small has been figured positively, as in the *Guardian*'s spoof on medium. For example, one of the principal economic responses to de-industrialization was to place special emphasis on small businesses as the saviours of the ravaged economy. Small firms were celebrated not only for their potential role in staying the economic downturn that devastated large-scale Fordist manufacturing, but also for their culture: small businesses were personal, family-like, as well as more flexible and entrepreneurial than their leviathan predecessors (Holliday 1995). Then there are variants on the anti-global rhetoric, in which small is beautiful and the local is the proper site for action. And in the technological domain, miniaturization is the premium, making devices ever smaller because shrinkage confers 'technologicalness' (Bell 2005). Clearly, the cultural values assigned to size are case-dependent, but there may be some imaginative crossing possible. At least there may be a way to query the impulse towards growth and the fetish of bigness that overshadows the city. As Pile (1999) says, cities aren't just about largeness: could they, in fact, not be about largeness at all?

Robinson's (2002) call to end the clumsy categorizing and hierarchizing of cities is clearly pointing in this kind of direction, too: the very idea of an urban hierarchy topped by global or world cities already sets the agenda, with upward mobility the only desirable path. But the costs associated with this imperative are too much for many cities to bear, either those aspiring to move up, or those straining to maintain their ranking. Yet it is hard to think cities *without* thinking size. In an era of neo-liberal governance, inter-urban competitiveness is never going to go away and size is always likely to be one factor used in league tables and rankings. Growth will always be positive, shrinkage negative (cities being the obverse of things like mobile phones in this regard). Perhaps the task is to rethink how size and growth are used as part of this calculus. The small city that thinks big, to tweak Steven Miles' formulation, might be more competitive than the small city that tried to grow big, if growth sucked up resources and accentuated inequalities. Might it be possible to reconceive staying small as something other than stagnation or lack of amibition? And could shrinkage ever be thought of as different from loss?

The nascent 'small cities agenda', to which we see this book making a contribution, has some way to go in thinking and rethinking size, in finding ways to work with, around or against prevailing notions of urban hierarchy, to articulate the paradoxical extraordinariness of the ordinary (and at once its profound ordinariness), to assess the ways that being small need not be shouldered with reluctance, to borrow T.C. Chang's phrase, but might, in fact, go some way towards resetting the terms by which cities and their futures are imagined.

# References


Abbot, C. (1981) *Boosters and Businessmen: Popular Economic Thought and Urban Growth in the Antebellum Middle West*, Westport, CT: Greenwood Press.

African Development Bank (2003) *Mauritius: African Economic Outlook 2003.*

Agnotti, T. (1993) *Metropolis 2000: Planning, Poverty and Politics*, London: Routledge.

Aiesha, R. and Evans, G. L. (2006) 'VivaCity: mixed-use and urban tourism', in Smith, M. (ed.) *Tourism, Culture and Regeneration*, Wallingford: CAB International.

Aitchinson, C. and Evans, T. (2003) 'The cultural industries and a model of sustainable regeneration: manufacturing "pop" in the Rhondda Valleys of South Wales', *Managing Leisure*, 8, 133–44.

Amin, A. (1999) 'An institutionalist perspective on regional economic development', *International Journal of Urban and Regional Research*, 23(2), 365–78.

Amin, A. (2004) 'Regions unbound: towards a new politics of place', *Geografiska Annaler,* 86(B), 33–44.

Amin, A. and Thrift, N. (1995) 'Globalization, institutional "thickness" and the local economy', in Healey, P., Cameron, S., Madanipour, A. with Graham, S. and Davoudi, S. (eds.) *Managing Cities: The New Urban Context*, Chichester: Wiley, 91–108.

Amin, A. and Graham, S. (1997) 'The ordinary city', *Transactions of the Institute of British Geographers*, 22, 411–29.

Anonymous 1 (1999) 'Pick of the week', *The Times*, 4 July, 15.

Anonymous 2 (1999) 'Recommended today', *The Times*, 14 July, 39.

Anonymous 3 (1999) 'Cheltenham celebrates 50 years of success', *Ireland's Sunday Independent*, 17 Oct., 10.

Anonymous 4 (1999) 'The Loafer', *Guardian*, 16 Oct., 11.

Anthony, K. H. (1985) 'The shopping mall: a teenage hangout', *Adolescence*, 20, 307–12.

Armytage, M. (2000) 'Morning after pill proves a best seller', *Daily Telegraph*, 16 Mar., 42.

Arts Council of England (1997) *Visual Arts UK: Public Attitudes Towards and Awareness of the Year of the Visual Arts of England* (summary), ACE Report No. 15 (Dec.).

Ashworth, G. J. (1994) 'From history to heritage – from heritage to identity – in search of concepts and models', in Ashworth, G. J. and Larkham, P. J. (eds.), *Building a New Heritage: Tourism, Culture, and Identity in the New Europe*, London: Routledge, 13–30.

Ashworth, G. J. and Voogd, H. (1990) *Selling the City,* London: Belhaven Press.

Atkinson, D. (2006) 'Kitsch geographies: ersatz docklands and the everyday spaces of popular memory', in *Environment and Planning A*, special issue on 'mundane urban geographies' (forthcoming).

Atkinson, R. and Bridge, G. (eds.) (2005) *Gentrification in a Global Context: The New Urban Colonialism*, London and New York: Routledge.

Augé, M. (1995) *Non-Places: Introduction to an Anthropology of Supermodernity*, London: Verso.

Australia Street Company (1998a) *Point of Take Off: Cultural Policy Framework and Cultural Plan 1998–2003*, Vol. 1, Queensland: Australia Street Company.

Australia Street Company (1998b) *Point of Take Off: Cultural Policy Framework and Cultural Plan 1998–2003*, Vol. 2, Queensland: Australia Street Company.

Australian Bureau of Statistics (2001) *2001 Census of Population and Housing*, Australian Bureau of Statistics, Canberra: Australian Government Publishing Service.

Australian Council of Trade Unions (1983) *Art and Working Life: Cultural Activities in the Australian Trade Union Movement*, Carlton, Vic.: Australian Council of Trade Unions and Community Arts Board.

B+B (Sarah Carrington and Sophie Hope) (2005) *Brand New Letchworth*, unpublished project briefing.

Bagnasco, A. and Le Gales, P. (eds.) (2000) *Cities in Contemporary Europe*, Cambridge: Cambridge University Press.

Bahr, E. (2001) 'The silver age of Weimar: Franz Liszt as Goethe's successor; a study in cultural archaeology', in *Goethe Yearbook*, 10, 191–202.

Bailey, C., Miles, S. and Stark, P. (2004) 'Culture-led urban regeneration and the revitalisation of identities in Newcastle, Gateshead and the North East of England', *International Journal of Cultural Policy*, 10(1), 47–66.

Bailey, N., Flint, J., Goodlad, R., Shucksmith, M., Fitzpatrick, S. and Pryce, G. (2003) *Measuring Deprivation in Scotland: Developing a Long-Term Strategy Final Report*, Edinburgh: Scottish Executive.

Ban, K. C. (1992) 'Narrating imagination', in Ban, K. C., Pakir, A. and Tong, C. K. (eds.), *Imagining Singapore*, Singapore: Times Academic Press, 9–25.

Barringer, R., Colgan, C., DeNatale, D., Hutchins, J., Smith, D. and Wassall, G. (2004) *The Creative Economy in Maine: Measurement and Analysis*, Maine Center for Business and Economic Research, New England Environmental Finance Center, University of Southern Maine, July. (www.MaineArts.com).

Bartky, S. (1990) *Femininity and Domination: Studies in the Phenomenology of Oppression*, London: Routledge.

Barton, H. (2000) *Sustainable Communities: the Potential for Eco-Neighbourhoods*, London: Earthscan.

Bates, C. (2000) 'Communalism and identity among South Asians in Diaspora', *Heidelberg Papers in South Asian and Comparative Politics*, 2, 36–57.

Bauman, Z. (1998) *Work, Consumerism and the New Poor*, Buckingham: Open University Press.

Bauman, Z. (2004) *Modernity and Ambivalence*, Cambridge: Polity.

Bazley K (1992) 'Urban regeneration and economic development', in *100 Years: Town Planning in Dundee*, Dundee: Duncan of Jordanstone College of Art and Design (unpublished Centenary Celebration Papers).

Beecher J. and Bienvenu, R. (eds.) (1983) *The Utopian Vision of Charles Fourier*, Columbia: University of Missouri Press (previously published, 1971, Boston: Beacon Press).

Belcher, W. W. (1947) *The Economic Rivalry Between St Louis and Chicago,* New York: Columbia University Press.

Bell, D. (2005) *Science, Technology and Culture*. Maidenhead: Open University Press.

Bell, D. (2006) 'Fade to grey: reflections on policy and mundanity', *Environment & Planning A*, in press.
Bell, D. and Jayne, M. (2003a) 'Design-led regeneration: a critical review', *Local Economy*, 18(2), 121–34.
Bell, D. and Jayne, M. (2003b) 'Assessing the role of design in local and regional economies', *International Journal of Cultural Policy*, 9(3), 285–304.
Bell, D. and Jayne, M. (2004a) 'Afterword: thinking in quarters', in Bell, D. and Jayne, M. (eds.) *City of Quarters: Urban Villages in the Contemporary City*, Aldershot: Ashgate, 249–55.
Bell, D. and Jayne, M. (eds.) (2004b) *City of Quarters: Urban Villages in the Contemporary City*, Aldershot: Ashgate.
Bertels, L. (1995) 'Von der grauen zur bunten Stadt – Folgen des Umbruchs in Gotha', *Aus Politik und Zeitgeschichte*, 12, 27–35.
Best, M. (1990) *The New Competition,* London: Polity Press.
Blom, E. (2000) 'Portland's Arts District approaches cross-roads: on the brink of a golden age or doomed by its own success', *Portland Press Herald*, City Edition, 2 Apr.
Blommaert, K., De Brabander, G., Gille, A., Lezy, L., Maes, T., Verhestel, A. and Witlan, F. (1992) *Woonvoorkevren en Verhuismotiven Voor Personen*, Antwerp: Seso.
Blount, T. (1999) 'Heart of High Point still beats downtown', *High Point Enterprise*, 13 Apr., C 4.
Bogdanor, V. (1999) *Devolution in the United Kingdom*, Oxford: Oxford University Press.
Bond, R. and Rosie, M. (2002) 'National identities in post-devolution Scotland', *Scottish Affairs*, 40, 34–53.
Booth, P. and Boyle, R. (1993) 'See Glasgow, see culture' in Bianchini, F. and Parkinson, M. (eds.) *Cultural Policy and Urban Regeneration: The Western European Experience,* Manchester: Manchester University Press, 21–47.
Borsay, (1989) *The English Urban Renaissance: Culture and Society in the Provincial Town 1660–1700*. Oxford: Clarendon.
Böttner, J. (1996) *Kleine Stadt was Nun?: Weimar auf dem Weg zur Kulturstadt Europas*, Weimar: Bauhaus-Universität.
Bovaird, T. (1995) 'Urban governance and "quality of life" marketing strategies for competition between sustainable cities' (paper presented at the Regional Studies Association Conference, Gothenberg, Sweden).
Brabander, G. D. and Gijsbrechts, E. (1994) 'Cultural policy and urban marketing: a general framework and some Antwerp experiences', in Corisco, G. A. F. (ed) *Urban Marketing in Europe*, Turin: Incontra, 814–41.
Bradley, A., Hall, T. and Harrison, M. (2002) 'Selling cities: promoting new images for meetings tourism', *Cities: The International Journal of Urban Policy and Planning*, 19(1), 61–70.
Bratton, J. and Garrett-Petts, W. (2005) 'Art at work: culture-based learning and the economic development of Canadian small cities', in Garrett-Petts, W. (ed.) *The Small Cities Book: on the Cultural Future of Small Cities*, Vancouver: New Star Books.
Breen, A. and Rigby, D. (1996) *The New Waterfront: A Worldwide Urban Success Story*, London: Thames and Hudson.
Brennan, C. and Hoene, C. (2003) 'Demographic change in small cities, 1990–2000', *Research Brief on America's Cities*, 1–8.
Bridge, G. (2003) 'Time-space trajectories in provincial gentrification', *Urban Studies*, 40, 2545–56.

Briebart, M. (1998) 'Dana's mystical tunnel: young people's designs for survival and change in the city', in Skelton, T. and Valentine, G. (eds.) *Cool Places: Geographies of Youth Cultures*, London: Routledge, 100–29.

Broudehoux, A.-M. (2004) *The Making and Selling of Post-Mao Beijing*, London: Routledge.

Brown, D. and Laville, S. (2001) 'Cheltenham races cancelled over case near course', http://www.telegraph.co.uk/sport/main.jhtml?xml=%2Fsport%2F2001%2F04%2F01%2FFuhchel01.xml (last accessed 2 Apr. 2005).

Brown, J. (1978) 'Market welcome', *High Point Enterprise*, 19 Oct., A 4.

Bruttomesso, R. (2001) 'Complexity on the urban waterfront', in Marshall, R. (ed.) *Waterfronts in Post-Industrial Cities*, London: Spon Press, 39–49.

Bryson, J. R., Daniels, P. W. and Rusten, G. (2004) 'The design workshop of the world? The production of design services in the United Kingdom', XIV RESER Conference, Castres.

Buckland, A. (2002) *Brokering Cultural Changes in Creative Clusters, Conference Handbook*, First International Summit Conference on Creative Industries Regeneration, Cultural Industries Quarter, Sheffield, 20–23 Nov.

Bunnell, T., Muzaini, H. and Sidaway. J. (2006, forthcoming) 'Global city frontiers: Singapore's hinterland and the contested socio-political geographies of Bintan, Indonesia', *International Journal of Urban and Regional Research*.

Bunwaree, S. (2002) 'Economics, conflicts and interculturality in a small island state: the case of Mauritius', *Polis*, 9, 1–19.

Burayidi, M. (2001) 'Introduction: downtowns and small city development', in Burayidi, M. (ed.) *Downtowns: Revitalizing the Centers of Small Urban Communities*, New York: Garland.

Burkeman, O. (2005) 'Medium is the message', *Guardian* G2, 9 Sept., 2–3.

Callaghan, M. (1982) *Greeting From Wollongong*, produced by Nina Saunders, Steel City Pictures, Sydney Filmmakers Co-operative.

Carey, J. (ed.) (1999) *The Faber Book of Utopias*, London: Faber and Faber.

Carley, M. and Kirk, K. (1998) *Sustainable by 2020? A Strategic Approach to Urban Regeneration for Britain's Cities,* York: Joseph Rowntree Foundation.

Cashwell, W. (1987a) 'Huge new showrooms will enhance market', *High Point Enterprise*, 14 Oct., A 1–2.

Cashwell, W. (1987a) 'No skywalk access perplexes exhibitors', *High Point Enterprise*, 22 Oct., A 1,11.

Cassidy, S. (1983) *Art and Working Life – A Report Prepared for the Australia Council*, North Sydney: Australia Council.

Castells, M. (1983) *The City and the Grassroots*, Berkeley: University of California Press.

Castles, R. (1997) 'Steel city: the economy, 1945–1995', in Hagan, J. and Wells, A. (eds.) *A History of Wollongong*, Wollongong: University of Wollongong Press, 71–80.

Castree, N. (2005) 'The epistemology of particulars: human geography, case studies and "context"', *Geoforum*, 36, 541–4.

Chang, T.C. (1997) 'From "Instant Asia" to "Multi-faceted Jewel": urban imaging strategies and tourism development in Singapore', *Urban Geography*, 18(6), 542–62.

Chang, T.C. (2000) 'Renaissance revisited: Singapore as a "global city" for the arts', *International Journal of Urban and Regional Research*, 24(4), 818–32.

Chang T.C. (2001) '"Configuring new tourism space": exploring Singapore's regional tourism forays', *Environment and Planning A*, 33, 1597–619.

Chang, T.C. and Raguraman, K. (2001) 'Singapore tourism: capital ambitions and regional connections', in Teo, P., Chang, T.C. and Ho, K.C. (eds.) *Interconnected Worlds: South-east Asian Tourism in the 21st Century*, Oxford: Elsevier Science and Technology, 47–63.

Chang, T.C. and Lim, S.Y. (2004) 'Geographical imaginations of "New Asia-Singapore"', *Geografiska Annaler*, 86B, 165–85.

Chang, T.C., Raguraman, K. and Low, L. (1998) *Tourism UnLimited: Total Business Environment in Singapore*, Final Report, Singapore: Centre for Advanced Studies and Singapore Tourism Board.

Chatterton, P. and Hollands, R. (2001 ) *Changing our 'Toon': Youth, Nightlife and Urban Change in Newcastle*, Newcastle: University of Newcastle upon Tyne.

Chatterton, P. and Hollands, R. (2003) *Urban Nightscapes: Youth Cultures, Pleasure Spaces and Corporate Power*, London: Routledge.

Chatterton, P. and Unsworth, R. (2004) 'Making space for culture(s) in Boomtown: some alternative futures for the development, ownership and participation in Leeds city centre', *Local Economy*, 19(4), 361–79.

Cheltenham Borough Council (2001) 'Cheltenham: a splendid Regency town', http://www.cheltenham.gov.uk/libraries/templates/cheltenham.asp?FolderID=4 (last accessed August 2006).

Cheltenham Borough Council (2002) *Draft Cultural Strategy*, Cheltenham: Cheltenham Borough Council.

Cherry, G. (1996) *Town Planning in Britain since 1900*, Oxford: Blackwell.

Christopherson, S. (2003) 'The limits to "new regionalism": (re)learning from the media industries', *Geoforum*, 34, 413–15.

Christopherson, S. (2004) *Creative Economy Strategies for Small and Medium Size Cities: Options for New York State,* Quality Communities Marketing and Economics Workshop, Albany, New York.

City of Portland (1991) *Downtown Vision: A Celebration of Urban Living and a Plan for the Future of Portland – Maine's Centre for Commerce and Culture*, Portland, OR: City of Portland.

City of Portland (2000) 5-Year Consolidated Housing & Community Development Plan. http://www.ci.portland.me.us/planning/commdevfiveyearconsolidatedhousing.htm (last accessed 19 Mar. 2006).

City of Portland (2002) *Portland's Comprehensive Plan*, (www.portlandmaine.gov/planning/complanbook.html).

Clark, J. (2000) 'The Old Port reinvents itself again', *Down East*, Apr., 64–69, 114–15.

Clements, A. (1999) 'Textured delight: turnage's silent cities', *Guardian* 6 July, 21.

Clontz, M. (1977) 'Market's impact hard to measure', *High Point Enterprise*, 20 Oct., A 3,5.

Coates, C. (2001) *Utopia Britannica: British Utopian Experiments: 1325–1945*, London: Diggers & Dreamers.

Coleman, A. (1984) *Utopia on Trial*, London: Hilary Shipman.

Coleman, J. (1988) 'Social capital in the creation of human capital', *American Journal of Sociology,* 94, 95–120.

Colombijn, F. (2003) 'Chicks and chicken: Singapore's expansion to Riau', *IIAS Newsletter*, 31(July), 10.

Commission for Architecture and the Built Environment (CABE) (2004) *Green Space Strategies: A Good Practice Guide*, London: CABE.

Conrady, K. O. (1994) *Goethe: Leben und Werk*, München: Artemis & Winkler.

Cooke, P. (1995) 'Keeping to the high road: learning, reflexivity and associative governance in regional economic development', in Cooke, P. (ed.), *The Rise of the Rustbelt*, London: UCL Press, 231–45.

Cooke, P. and Morgan, K. (1998) *The Associational Economy: Firms, Regions and Innovation,* Oxford: Oxford University Press.

Cooper, G. (2004) 'A festival of ideas', *Local Economy* 19(4), 428–31.

Core Cities Working Group (2004) *Our Cities are Back: Competitive Cities Make Prosperous Regions and Sustainable Communities*, Third Report of the Core Cities Working Group, London: Office of the Deputy Prime Minister.

Cortright, J. (2001) 'New growth theory, technology and learning: a practitioner's guide', *Reviews of Economic Development Literature Practice*, No. 4, US Economic Development Administration.

Craver, R. (2000) 'Buyers suggest market improvements', *High Point Enterprise*, 14 Oct., B 1.

Creswell, T. (2003) 'Landscape and the obliteration of practice', in Anderson, K., Domosh, M., Pile, S. and Thrift, N. (eds.) *Handbook of Cultural Geography*, London: Sage, 269–89.

Cutler, T. (2003) 'Foreword', in Florida, R. *The Rise of the Creative Class*, North Melbourne: Pluto Press, vii-xi.

Cwi, D. (1992) 'Baltimore's promotion', in Green, K. W. (ed.) *The City as a Stage: Strategies for the Arts in Urban Economics*, Washington: Partners for Liveable Places, 100–103.

Dabinett, G. (2004) 'Creative Sheffield: creating value and changing values?' *Local Economy*, 19(4), 414–19.

Davis, M. (1990) *City of Quartz: Excavating the Future of Los Angeles*, London: Vintage.

Davis, M. (1998) *Ecology of Fear – Los Angeles and the Imagination of Disaster*, New York: Vintage.

De Lannoy, W. (1987) 'Naar een herstel van de stad?', in Knops, G. (ed.) *De Vitgeholde Stad*, Brussels: Konig, 31–47.

Department for Culture, Media and Sport (2000a) *Culture and Creativity: The Next Ten Years*, London: DCMS.

Department for Culture, Media and Sport (2000b) *Creating Opportunities*, London: DCMS.

Department for Culture, Media and Sport (2004a) *Culture at the Heart of Regeneration*, London: DCMS.

Department for Culture, Media and Sport (2004b) *Creative Cultures*, London: DCMS.

Department of the Environment, Transport and the Regions (1999) *Towards an Urban Renaissance*, London: E. & F. N. Spon.

Department of Trade and Industry (2004) *Creative People, Openness and Productivity. An Exploration of Regional Differences Inspired by "The Rise of the Creative Class"*. London: Department for Trade and Industry.

Department of Transport and the Regions (2000) *Our Towns and Cities. The Future: Delivering the Urban Renaissance*. London: Office of the Deputy Prime Minister.

Dobson Chapman, W. (1952) *The City and Royal Burgh of Dundee: Survey and Plan 1952: Part Two: The City Plan*, Macclesfield: Dobson Chapman Partners.

Dolowitz, D. and Marsh, D. (2000) 'Learning from abroad: the role of policy transfer in contemporary policy making', *Governance: An International Journal of Policy and Administration*, 13, 5–24.

Dove, G. (1999) 'The festivals fight it out', *The Times*, 25 July, 22.

DPC (Downtown Portland Corporation) (1997) *Annual Report*, Portland, OR: DPC.

Drake, G. (2003) '"This place gives me space": place and creativity in the creative industries, *Geoforum*, 34, 511–24.

du Gay, P. and Pryke, M. (2002) *Cultural Economy: Cultural Analysis and Commercial Life*, London: Sage.

Duchy of Cornwall (n.d.) *Poundbury*, Dorchester: Duchy of Cornwall.

Dundee City Council (1997) *Arts Action Plan 1998–2000*, Dundee: Dundee City Council.

Dundee City Council (2001) *Economic Development Plan 2001–2004*, Dundee: Dundee City Council.

Dundee City Council (2002) *The Council Plan 2003–2007*, Dundee: Dundee City Council.

Dundee City Council (2003a) *Dundee – A City Vision*, Dundee: Dundee City Council.

Dundee City Council (2003b) *Finalised Dundee Local Plan Review*, Dundee: Dundee City Council.

Dundee City Council (2004) *About Dundee*, Dundee: Dundee City Council.

Dunn, K. and Hartig, K. (1998) 'Roadside memorials: interpreting new deathscapes in Newcastle, New South Wales', *Australian Geographical Studies*, 36(1): 5–20.

Dürrschmidt, J. (2005) 'Model Town' and 'dying city': notes on local identity and (non)attachment to place from the European periphery, Manchester Institute for Popular Culture research seminar, Manchester, Dec.

Dutton, P. (2003) 'Leeds calling: the influence of London on the gentrification of regional cities', *Urban Studies*, 40(12), 2557–72.

Dutton, P. (2005) 'Outside the metropole: gentrification in provincial cities or provincial gentrification?', in Atkinson, R. and Bridge, G. (eds.) (2005) *Gentrification in a Global Context: The New Urban Colonialism*, London and New York: Routledge, 209–24.

Eade, J. and Sellnow, M. J. (eds.) (2000) *Contesting the Sacred: the Anthropology of Pilgrimage*, Urbana: University of Illinois Press.

Ebels, H. J. and Ostendorf, W. J. M. (1991) 'Niewe huishoudens in Amsterdam en Den Haag: De bewoners von deze woningen in central stadselen anderzacht', *Geografisch Tijdschrift*, (5), 510–20.

Eckardt, F. (2003) *Eine Periphere Gesellschaft. Regionalentwicklung zwischen Erfurt und Weimar*, Marburg: Tectum.

Eckardt, F. (2005) *Tourismus in Weimar*, Weimar: Bauhaus-Universität.

Eckardt, F. and Karwinska, A. (eds.) (2005) *Urban Magic. Cultural Life in Weimar and Krakow*, Marburg: Tectum.

Economic Development Board (1995a) *Singapore UnLimited. Singapore in the 21st Century*, Singapore: Economic Development Board.

Economic Development Board (1995b) *International Business Hub 2000*, Singapore: Economic Development Board.

EDAW and Urban Cultures (2000) *Dundee Cultural Quarter Development and Investment Strategy: Final Report*, Dundee: EDAW and Urban Cultures.

Edensor, T. (ed.) (2000) *Re-claiming Stoke-on-Trent: Leisure, Space and Identity in the Potteries*, Stoke-on-Trent: Staffordshire University Press.

Edensor, T. (2005a) *Industrial Ruins: Space, Aesthetics and Materiality*. Oxford: Berg.

Edensor, T. (2005b) 'Sensing tourism', in Minca, C. and Oakes, T. (eds.), *Travels in Paradox,* Minneapolis: University of Minnesota Press.

Edensor, T. and Augustin, F. (2001) 'Football, ethnicity and identity in Mauritius: soccer in a rainbow nation', in Armstrong, G. and Giulianotti, R. (eds.) *Fear and Loathing in World Football*, Oxford: Berg.

Edensor, T. and Kooduruth, I. (2004) 'Creolising football', in Giulianotti, R. and Armstrong, G. (eds.) *Football in Africa*, Oxford: Berg.

Edensor, T. and Kothari, U. (2004) 'Sweetening colonialism: a Mauritian themed resort', in Lasansky, M. and McClaren, B. (eds.) *Architecture and Tourism*, Oxford: Berg.

Edmonson, R. (2000) 'Business first for Fitzgerald despite the pain', *Independent*, 11 Mar., 26.

Eklund, E. (2002) *'Steel Town': The making and breaking of Port Kembla*, Carlton, Vic.: Melbourne University Press.

El-Dahdah, F. (2004), 'Brasilia – the project of Brasilia', in Robbins, E. and el-Khoury, R., *Shaping the City: Studies in History, Theory and Urban Design*, London: Routledge, 41–56.

Epperson, J. (2003) 'Viewpoint', *Furniture Supplier Spotlight*, Spring.

Erickcek, G. and McKinney, H. (2004) *'Small Cities Blues': Looking for Growth Factors in Small and Medium-Sized Cities*, Upjohn Institute Staff Working Paper 04–100, Upjohn Institute for Employment Research.

Eriksen, T. (1998) *Common Denominators: Ethnicity, Nation-Building and Compromise in Mauritius*, Oxford: Berg.

European Task Force on Culture and Development (1997) *In from the Margins: A Contribution to the Debate on Culture and Development in Europe*, Strasbourg: Council of Europe.

Evans, G. L. (1998) 'Urban leisure: edge city and the new pleasure periphery', in Collins, M. and Cooper, I. (eds.) *Leisure Management: Issues and Applications*, Wallingford: CAB International, 113–38.

Evans, G. L. (2001) *Cultural Planning: An Urban Renaissance?*, London: Routledge.

Evans, G. L. (2003) 'Hard branding the cultural city – from Prado to Prada', *International Journal of Urban and Regional Research*, 27(2), 417–40.

Evans, G. L. (2004) 'Culture and regeneration: measuring impact', *Arts Professional*, 84 (4 Oct.), 5–6.

Evans, G. L. (2004) 'Cultural industry quarters: from pre-industrial to post-industrial production', in Bell, D. and Jayne, M. (eds.), *City of Quarters: Urban Villages in the Contemporary City*, Aldershot: Ashgate Press, 71–92.

Evans, G. L. (2005) 'Measure for measure: evaluating the evidence of culture's contribution to regeneration', *Urban Studies*, 42(5/6), 959–83.

Evans, G. L. and Shaw, P. (2004) *The Contribution of Culture to Regeneration in the UK: A Review of Evidence*, Report to the Department for Culture, Media and Sport, London: London Metropolitan University.

Evans, G. L., Foord, J. and Shaw, P. (2005) *Creative Spaces: Strategies for Creative Cities*. London: London Development Agency (www.creativelondon.org).

Fairman, R. (1999) 'Musical show of variety makes impact', *Financial Times*, 14 July, 34.

Fascher, E. (1996) 'Politische Parteien und Stadtparlamente im heutigen Weimar', *Zeitschrift für Parlamentsfragen*, 1, 37–60.

Ferguson, F. (1999) 'A brief history of grassroots greening on the Lower East Side', in Wilson, P. L. and Weinberg, B. (eds.) *Avant Gardening: Ecological Struggle in the City and the World*, New York: Autonomedia, 80–90.

Fernandez, E. and Varley, A. (eds.) (1998) *Illegal Cities – Law and Urban Change in Developing Countries*, London: Zed Books.

Fernie, K. and McCarthy, J. P. (2001) 'Partnership and community involvement: institutional morphing in Dundee', *Local Economy*, 16(4), 299–311.

Filion, P. and Rutherford, T. (1996) 'Manufacturing miracles? The restructuring of Waterloo Region's manufacturing', in Filion, P., Bunting, T. and Curtis, K. (eds.) *The Dynamics of the Dispersed City*, Waterloo, Ont.: University of Waterloo, Department of Geography Publication Series, 239–71.

Fisher, M. and Owen, U. (1991) *Whose Cities?*. London: Penguin.

Fishman, R. (1987) *Bourgeois Utopias: the Rise and Fall of Suburbia*, New York: Basic Books.

Florida, R. (2002) *The Creative Class* (www.creativeclass.org).

Florida, R. (2002) *The Rise of the Creative Class*, New York: Basic Books.

Florida, R. (2005) *Cities and the Creative Class*, New York: Routledge.

Foucault, M. (1980) *Power/Knowledge*, Brighton: Harvester.

Frank, S. (2003) 'Festivalization, image politics and local identity: the Rollplatz debate in Weimar, European City of Culture 1999', in Daly, P. M. and Frischkopf, H. W. (eds.) *Why Weimar? Questioning the Legacy of Weimar from Goethe to 1999*, McGill European Studies, Vol. 5, New York: Lang, 49–61.

Fraser, N. (1993) 'Rethinking the public sphere: a contribution to the critique of actually existing democracy', in Robbins, B. (ed.) *The Phantom Public Sphere*, Minneapolis: University of Minnesota Press, 1–32.

Frey, W. and Liaw, K. (1998) 'Immigrant concentration and domestic migration dispersal: is movement to nonmetropolitan areas "white flight"?', *Professional Geographer*, 50, 215–32.

Frisby, D. (1988) *Fragments of Modernity – Theories of Modernity in the Work of Simmel, Kracauer and Benjamin*, Cambridge, MA: MIT Press.

Fukuyama, F. (1995) *Trust: The Social Virtues and the Creation of Prosperity,* London: Free Association Press.

Garcia, B. (2004a) 'Cultural policy and urban regeneration in Western European cities: lessons from experience, prospects for the future', *Local Economy*, 19(4), 312–26.

Garcia, B. (2004b) 'Urban regeneration, arts programming and major events Glasgow 1990, Sydney 2000 and Barcelona 2004', *International Journal of Cultural Policy*, 10(1), 103–18.

Garreau, J. (1991) *Edge City: Life on the New Frontier*, New York: Doubleday.

Garrett-Petts, W. and Dubinsky, L. (2005) '"Working well, together": an introduction to the cultural future of small cities', in Garrett-Petts, W. (ed.) *The Small Cities Book: On the Cultural Future of Small Cities*, Vancouver (WA): New Star Books.

Gergen, K. J., Gloger-Tippelt, G. and Berkowitz, P. (1990) 'The cultural construction of the developing child', in Semin, G. R. and Gergen, K. J. (eds.) *Everyday Understandings: Social and Scientific Understanding*, London: Sage, 108–29.

Gertler, M. S. (1997) 'The invention of regional culture', in Lee, R. and Wills, J. (eds.), *Geographies of Economies*, London: Edward Arnold, 47–58.

Gertler, M. S. (2001) 'Technology, culture and social learning: regional and national institutions of governance', in Gertler, M. S. and Wolfe, D. A. (eds.) *Innovation and Social Learning,* Basingstoke: Macmillan.

Gertler, M. S. and Wolfe, D. (1998) 'The dynamics of regional innovation in Ontario', in De La Moth, J. and Paquet, G. (eds.) *Local and Regional Systems of Innovation,* Amsterdam: Kluwer, 211–38.

Ghimrie, K. (2001) 'Regional tourism and south–south economic cooperation', *The Geographical Journal*, 167, 99–110.

Gibson, C. (2003) 'Cultures at work: why "culture" matters in research on the "cultural" industries', *Social and Cultural Geography*, 4(2), 2001–215.

Gibson, C. and Connell, J. (2004) 'Culture industry production in remote places: indigenous popular music in Australia' in Power, D. and Scott, A. J. (eds.) *Cultural Industries and the Production of Culture*, London: Routledge, 243–58.

Gibson, C. and Klocker, N. (2004) 'Academic publishing as "creative" industry, and recent discourses of "creative economies": some critical reflections', *Area*, 36(4), 423–34.

Gibson, C. and Klocker, N. (2005) 'The "Cultural Turn" in Australian regional economic development discourse: neoliberalising creativity', *Geographical Research*, 43(1), 93–102.

Gibson, C., Murphy, P. and Freestone, R. (2002) 'Employment and socio-spatial relations in Australia's cultural economy', *Australian Geographer*, 22, 173–89.

Gilmore, A. (2004) 'Local cultural strategies: a strategic review', *Cultural Trends*, 13(3), 3–32.

Girardet, H. (1990) *The Gaia Atlas of Cities: New Directions for Sustainable Urban Living*, London: Gaia Books.

Giraud, M. (1998) *Caudan Development*, Port Louis: Maurice Giraud and Zac Associates.

Goffman, E. (1959) *The Presentation of Self in Everyday Life*. New York: Doubleday.

Government of Mauritius (2003) *National Development Strategy* 2003, Ministry of Housing and Lands.

Grabher, G. (2001) 'Ecologies of creativity: the village, the group and the heterarchic organisation of the British advertising industry', *Environment and Planning*, A 33, 351–74.

Graham, S. (ed.) (2003) *The Cybercities Reader*, London: Routledge.

Gray, M. and Markusen, A. (1999) 'Colorado Springs: a military-anchored city in transition', in Markusen, A., Lee, Y. and DiGiovanni, S. (eds.) *Second Tier Cities: Rapid Growth Beyond the Metropolis*, Minneapolis: University of Minnesota Press.

Greater London Authority (2004) *The London Plan: Spatial Development Strategy*, London: Greater London Authority.

Grundy-Warr, C., Peachey, K. and Perry, M. (1999) 'Fragmented integration in the Singapore-Indonesian border zone: Southeast Asia's "growth triangle" against the global economy', *International Journal of Urban and Regional Research*, 23, 304–28.

Guppy and Associates and National Economics (2000) *Wollongong Cultural Industries Audit*, London: Guppy and Associates and National Economics.

Hackworth, J. and Smith, N. (2001) 'The changing state of gentrification', *Tijdschrift voor Economische en Sociale Geografie*, 92(4), 464–77.

Hagan, J. and Wells, A. (eds.) (1997) *A History of Wollongong*, Wollongong: University of Wollongong Press.

Haislip, B. (1970) 'High Point is Fashion Center', *High Point Enterprise*, 23 Oct., A 5.

Hall, P. (1998) *Cities in Civilisation*, London: Weidenfeld & Nicolson.

Hall, P. (2002) *Cities of Tomorrow*, Oxford: Blackwell.

Hall, P. (2004) 'European cities in a global world', in Eckhardt, F. and Hassenpflug, D. (eds.) *Urbanism and Globalization*, Frankfurt: Peter Lang, 31–46.

Hall, P. and Ward, C. (1998) *Sociable Cities: The Legacy of Ebenezer Howard*, Chichester: Wiley.

Hall, T. (1998) *Urban Geography*, London: Routledge.

Hammerthaler, R. (1998) *Die Weimarer Lähmung: Kulturstadt Europas 1999 – szenisches Handeln in der Politik*, Berlin: Lukas.

Hannemann, C. (1996) *Die Platte: Industrialisierter Wohnungsbau in der DDR*, Wiesbaden: Vieweg.

Hannigan, J. (1998) *Fantasy City: Pleasure and Profit in the Postmodern Metropolis*, London and New York: Routledge.

Haque, A. (2001) 'Does size matter? Successful economic development strategies of small cities', in Burayidi, M. (ed.) *Downtowns: Revitalizing the Centers of Small Urban Communities*, New York: Garland.

Hardoy, J. E. and Satterthwaite, D. (1986a) 'Some tentative conclusions', in Hardoy, J. E. and Satterthwaite, D. (eds.) *Small and Intermediate Urban Centres: Their Role in Regional and National Development within the Third World*, London: Hodder & Stoughton.

Hardoy, J. E. and Satterthwaite, D. (1986b) 'Why small and intermediate urban centres?' in Hardoy, J. E. and Satterthwaite, D. (eds.) *Small and Intermediate Urban Centres: Their Role in Regional and National Development within the Third World*, London: Hodder & Stoughton.

Harr, J. (1985) 'Imagine a city too good to be true: the rise of Portland', *New England Monthly*, Mar., 33–43.

Harvey, D. (1987) 'Flexible accumulation through urbanisation: reflections on postmodernism in the American city', *Antipode*, 19, 260–86

Harvey, D. (1989a) 'From managerialism to entrepreneurialism: the transformation of urban governance in late capitalism', *Geografiska Annaler*, 71 B, 3–17.

Harvey, D. (1989b) *The Urban Experience*, Baltimore, MD: Johns Hopkins University Press.

Haughton, G. (1989) 'Community and industrial restructuring: responses to the recession and its aftermath in the Illawarra region of Australia', *Environment and Planning A*, 21, 233–47.

Haughton, G. (1990) 'Manufacturing recession? BHP and the recession in Wollongong International', *Journal of Urban and Regional Research*, 14(1) 70–88.

Häußermann, H. (1996) 'Von der Stadt im Sozialismus zur Stadt im Kapitalismus', in Häußermann, H. and Neef, R. (eds.) *Stadtentwicklung in Ostdeutschland*, Opladen: Westdeutscher Verlag, 5–47.

Häußermann, H. and Siebel, W. (1987) *Neue Urbanität*, Frankfurt: Suhrkamp.

Hawkins, J. (1964) 'New exhibition building ready for fall market', *High Point Enterprise*, 18 Oct., D 1.

Hawkins, J. (1972) 'Furniture Row Goes North', *High Point Enterprise*, 20 Oct., A 5.

Hebdige, D. (1997) 'Posing ... threats, striking ... poses: youth surveillance and display', in Gelder, K. and Thorton, D. (eds.) *The Subcultures Reader*, London: Routledge, 393–405.

Henry, N. and Pinch, S. (2000) 'Spatialising knowledge: placing the knowledge community of Motor Sport Valley', *Geoforum*, 31, 191–208.

Hepworth, M. and Spencer, G. (2004) *A Regional Perspective on the Knowledge Economy in Great Britain: Report for the Department of Trade and Industry*, London: The Local Futures Group.

Herbert Sprouse Consulting (1995) *A Plan for Portland's Arts District*, prepared for the Arts and Cultural Steering Commission.

Herk, A. van (2005) 'The city small and smaller', in Garrett-Petts, W. (ed.) *The Small Cities Book: On the Cultural Future of Small Cities*, Vancouver: New Star Books.

Hertmans, S. (2001) *Intercities*, London: Reaktion.

Hesmondhalgh, D. (2002) *The Cultural Industries*, London: Sage.

Hiebert, M., (1996) 'It's a jungle out there', *Far Eastern Economic Review*, 25 Apr., 58–62.

Hills, M (2002) 'The formal and informal management of diversity in the Republic of Mauritius', *Social Identities*, 8, 287–300.

Hinrichs, W. (1992) *Wohnungsversorgung in der ehemaligen DDR – Verteilungskriterien und Zugangswege*, Diskussionspaper, Berlin: WZB.

Hirst, P. (1994), *Associative Democracy*, Cambridge: Polity Press.

Hodgson, G. (2002) 'Darwinism in economics from analogy to ontology', *Journal of Evolutionary Economics*, 12: 259–81.

Holcomb, B. (1994) 'City make-overs: marketing the post industrial city', in Gold, J. R. and Ward, S. V. (eds.) *Place Promotion: The Use of Publicity and Marketing to Sell Towns and Regions*, Chichester: John Wiley, 115–31.

Holland, J., Ramazanoglu, C., Sharpe, S. and Thomson, R. (1994) 'Power and desire: the embodiment of female sexuality', *Feminist Review*, 46, 21–38.

Holliday, R. (1995) *Investigating Small Firms: Nice Work?*, London: Routledge.

Holloway, S. L. and Valentine, G. (eds.) (2000) *Children's Geographies: Playing, Living, Learning*, London: Routledge.

Howard, E. (1898) *Garden Cities for Tomorrow*, London: Books for Business.

Hoyle, B. (2001) 'Urban renewal in East African port cities: Mombassa's Old Town waterfront', *Geojournal*, 53(2), 183–97.

Hoyle B. (2002) 'Urban waterfront revitalization in developing countries: the example of Zanzibar "Stone Town"', *Geographical Journal,* 168(2), 141–62.

Huang, S. (2001) 'Planning for a tropical city of excellence: urban development challenges for Singapore in the 21st century', *Built Environment*, 27, 112–28.

Hudson, R. (2000) *Production, Places and the Environment: Changing Perspectives in Economic Geography*, London: Pearson Education.

Hunt, T. (2004) *Building Jerusalem: The Rise and Fall of the Victorian City*, London: Weidenfeld and Nicolson.

Hüter, K.-H. (1982) *Das Bauhaus in Weimar: Studie zur gesellschaftspolitischen Geschichte einer deutschen Kunstschule*, Berlin: Akademie-Verl.

Ibrahim, A. (1996) *Asian Renaissance*, Singapore: Times Books International.

*Illawarra Mercury* (1979) 'Gallery head sees Nike in world role', 15 Aug., 2.

Ingram, S. S. (1988) 'Blocked out: High Point Cleaners and Hatters to close after 69 years', *High Point Enterprise*, 30 Oct., D 4,7.

Inman, T. (1989a), 'IHFC plans expansion of showroom complex', *High Point Enterprise*, 18 Oct., B 1–2.

Inman, T. (1989b) 'Market Square tower to satisfy triple need', *High Point Enterprise*, 19 Oct., B 1,6.

Jacobs, J. (1969) *The Death and Life of Great American Cities*. Harmondsworth: Penguin.

Jayne, M. (1999) 'Imag(in)ing a post-industrial Potteries', in Bell, D. and Haddour, A. (eds.) *City Visions*, London: Prentice Hall, 8–19.

Jayne, M. (2000) 'The cultural quarter: (re)Locating urban regeneration in Stoke-on-Trent – a "city" in name only', in Edensor, T. (ed.) *Re-claiming Stoke-on-Trent: Leisure, Space and Identity in the Potteries*, Stoke-on-Trent: Staffordshire University Press, 19–42.

Jayne, M. (2003) 'Too many voices, too problematic to be plausible: representing multiple responses to local economic development strategies', *Environment & Planning A*, 35, 959–81.

Jayne, M. (2004a) 'Culture that works? Creative industries development in a working-class city', *Capital & Class*, 84, 199–210.

Jayne, M. (2004b) 'Globalization and third-tier cities: the European experience', in Eckhardt, F. and Hassenpflug, D. (eds.) *Urbanism and Globalization*, Frankfurt: Peter Lang, 65–86.

Jayne, M. (2005) 'Creative industries: the regional dimension?', *Environment and Planning C: Government and Policy*, 23, 537–56.

Johnson, J. (1978) 'Merchants rap downtown construction', *High Point Enterprise*, 17 Oct., B 1–2.

Johnson, J. D. and Rasker, R. (1993) 'Local government, local business climate and quality of life', *Montana Policy Review*, 3(2), 19–33.

Johnson, J. D. and Rasker, R. (1995) 'The role of economic and quality of life values in rural business location', *Journal of Rural Studies*, 11(4), 405–16.

Johnson, P. (2005) 'Locals won't bow down to Vegas', *High Point Enterprise*, 11 Aug., B 1.

Jordison, S. and Kieran, D. (eds) (2003) *Crap Towns: The 50 Worst Places to Live in the UK*, London: Boxtree.

Jordison, S. and Kieran, D. (eds.) (2004) *Crap Towns II: The Nation Decides*, London: Boxtree.

Kanai, T. (1993) 'Singapore's new focus on regional business expansion', *NRI Quarterly*, 2, 18–41.

Kao, J. (1996) *Jamming: The Art and Discipline of Business Creativity,* London: Harper Collins.

Katz, C. (1998) 'Disintegrating developments: global economic restructuring and eroding ecologies of youth', in Skelton, T. and Valentine, G. (eds.) *Cool Places: Geographies of Youth Cultures*, London: Routledge, 130–44.

Kearns, G. and Philo, C. (eds.) (1993) *Selling Places: The City as Cultural Capital, Past and Present*, Oxford: Pergamon Press.

Keating, M. (2001) 'Rethinking the region: culture, institutions and economic development in Catalonia and Galicia', *European Urban and Regional Studies,* 8, 217–34.

Kelly, D. (1997) 'Work and Leisure 1828–1997', in Hagan, J. and Wells, A. (eds.) *A History of Wollongong, Wollongong*, Wollongong: University of Wollongong Press, 177–88.

Kelly, K. (1998) *New Rules for the New Economy: 10 Radical Strategies for a Connected World,* London: Fourth Estate.

Kenny, J. T. (1995) 'Making Milwaukee famous: cultural capital, urban image, and the politics of place', *Urban Geography*, 16(5), 440–58.

Knopp, L. and Kujawa, R. (1993) 'Ideology and urban landscapes: conceptions of the market in Portland, Maine', *Antipode*, 25(2), 114–39.

Knox, P. (1987) 'The social production of the built environment: architects, architecture and the postmodern city', *Progress in Human Geography*, 12(3), 154–377.

Kotler, P., Haidder, D. H. and Rein, I. (1993) *Marketing Places: Attracting Investment and Tourism to Cities, States and Nations*, New York: Free Association Press.

Kraack, A. and Kenway, J. (2002) 'Place, time and stigmatised youthful identities: bad boys in paradise', *Journal of Rural Studies*, 18, 145–55.

Kropotkin, P. (1976) *Mutual Aid, A Factor of Evolution*, Manchester (NH): Porter Sargent (first published in 1902).

Kwok, K. W. (2001) 'Singapore as cultural crucible: releasing Singapore's creative energy in the new century', in Lee, G. B. (ed.) *Singaporeans Exposed: Navigating the Ins and Outs of Globalisation*, Singapore: Landmark Books, 21–32.

Lancaster, B. (1995) 'City cultures and the "Parliaments of Birds": a letter from Newcastle', *Northern Review*, 2, 1–11.

Lancaster, B. (2004) 'As seen on TV or why Newcastle/Gateshead didn't win', *Northern Review*, 13, 5–10.

Landry, C. (2000) *The Creative City: A Toolkit for Urban Innovators*, London: Earthscan Publications.

Landry, C. (2002) *Togetherness in Difference: Culture at the Crossroads in Bosnia Herzegovina*, Strasbourg: Council of Europe.

Langendijk, A. (1999) 'The emergence of knowledge-oriented forms of regional policy in Europe', *Tijdschrift voor Economische en Sociale Geografie,* 90(1), 110–16.

Larner, G. (1999) 'Review of London Mozart players/Chillingworth Vellinger', *The Times*, 20 July, 46.

Lawless, P. (1994) 'Partnership in urban regeneration in the UK: the Sheffield Central Area Study', *Urban Studies*, 31(8), 1303–24.

Lawless, P. and Gore. T. (1999) 'Urban regeneration and transport investment: a case study of Sheffield 1992–96', *Urban Studies*, 36(3), 527–45.

Lawson, C. and Lorenz, E. (1999) 'Collective learning, tacit knowledge and regional innovative capacity', *Regional Studies,* 33, 305–17.

Lee, P. (2004) *Singapore, Tourism & Me*, Singapore: Pamelia Lee Pte. Ltd.

Lee, T. Y. (1993) 'Sub-regional economic zones in the Asia-Pacific: an overview', in Toh, M. H. and Low, L. (eds.) *Regional Cooperation and Growth Triangles in ASEAN*, Singapore: Times Academic Press, 9–58.

Lees, L. (2000) 'A re-appraisal of gentrification: towards a "geography of gentrification", *Progress in Human Geography*, 24(3), 389–408.

Lees, L. (2003a) 'Visions of "Urban Renaissance": the Urban Task Force Report and the Urban White Paper', in Imrie, R. and Raco, M. (eds.) *Urban Renaissance? New Labour, Community and Urban Policy*, Bristol: Policy Press, 61–82.

Lees, L. (2003b) 'The ambivalence of diversity and the politics of urban renaissance: the case of youth in downtown Portland, Maine, USA', *International Journal of Urban and Regional Research*, 27(3), 613–34.

Lees, L. (ed.) (2004) *The Emancipatory City: Paradoxes and Possibilities*, London: Sage.

Lefebvre, H. (1991) *The Production of Space*, Cambridge, MA: Blackwell.

LeGates, F. and Stout, R. (eds.) (2003) *The City Reader*, 3rd edn., London, Routledge.

Leibovitz, J. (2004) 'Institutional barriers to associative city-region governance: the politics of institution-building and economic governance in Canada's Technology Triangle', *Urban Studies*, 40, 2613–42.

Leidberg, M. (1995) 'Teenagers and public space', *Communications Research*, 22, 206–31.

Leitner, H., and Sheppard, E. (2002) '"The City is Dead, Long Live the Net": Harnessing European International Networks for a neo-liberal agenda', *Antipode,* 34, 495–518.

Ley, D. (1996) *The New Middle Class and the Remaking of the Central City*, Oxford: Oxford University Press.

Leyshon, A. , Matless, D. and Reville, G. (1995) 'The place of music', *Transactions of the Institute of British Geographers*, 20, 423–33.

Lilley, W. and DeFranco, L. J. (2000) *The Economic Impact of the Network Q Rally of Great Britain*, Washington DC: InContext.

Lim, L. H. (1999) *A Triangle Love Affair? Tourism in the Indonesia-Malaysia-Singapore Growth Triangle*, Singapore: Department of Geography unpublished honours thesis, National University of Singapore.

Lloyd, M. G. and McCarthy, J. (2003) 'Urban regeneration in Dundee', in Couch, C., Fraser, C. and Percy, S. (eds.) *Urban Regeneration in Europe*, Oxford: Blackwell, 56–68.

Lloyd, M. G. and Peel, D. (2003) 'Shaping national space in Scotland', *Town and Country Planning*, 72(7), 224–5.

Loader, I. (1996) *Youth, Policing and Democracy*, Basingstoke: Macmillan.

Logan, J. R. and Molotch, H. (1987) *Urban Fortunes: The Political Economy of Place*. Berkeley: University of California Press.

Lorente, J. P. 2002. 'Urban cultural policy and urban regeneration: the special case of declining port cities in Liverpool, Marseilles, Bilbao', in Crane, D., Kawashima, N. and Kawasaki, K. (eds.) *Global Culture: Media, Arts, Policy, and Globalization*, New York: Routledge, 93–104.

Lovering, J. (1999) 'Theory led by policy: the inadequacies of the "New Regionalism"', *International Journal of Urban and Regional Studies*, 23, 379–95.

Low, L. and Toh, M. H. (1997) 'Singapore: development of gateway tourism', in Go, F. M. and Jenkins, C. L. (eds.), *Tourism and Economic Development in Asia and Australasia*, London: Cassell, 237–54.

Lundvall, B. A. (ed.) (1992) *National Systems of Innovation: Towards a Theory of Innovation and Interactive Learning,* London: Pinter.

Lyle, R. and Payne, G. (1950) *The Tay Valley Plan*, Dundee: Burns and Harris Ltd.

MacCannell, D. (1999), 'New Urbanism and its discontents', in Copjec, J. and Sorkin, M. (eds.) *Giving Ground: The Politics of Propinquity*, London: Verso, 106–28.

MacLeod, G. (2001) 'Beyond soft institutionalism: accumulation, regulation, and their geographical fixes', *Environment and Planning A,* 33, 1145–67.

MacLeod, G. and Goodwin, M. (1999) 'Space, scale and state strategy: rethinking urban and regional governance', *Progress in Human Geography*, 23(4), 503–27.

MacLeod, G., Raco, M. and Ward, K. (2003) 'Negotiating the contemporary city', *Urban Studies*, 40, 1655–71.

MacPherson, T. (1993) 'Regenerating industrial riversides in the north east of England', in White, K. N., Bellinger, E. G., Saul. A. J., Symes, M. and Hendry, K. (eds.) *Urban Waterside Regeneration: Problems and Prospects*, London: Ellis Harwood, 31–42.

Maddocks, F. (1999) 'And an angel came unto him, as did several sparrows', *Observer*, 11 July, 8.

Maheras, N. (2001) 'High Point passages: the rules are new, different', *High Point Enterprise*, 11 Sept., 69–82.

Maine Arts Commission (2004) *Proceedings from the Blaine House Conference on Maine's Creative Economy*, Maine, Aug.

Malbon, B. (1999) *Clubbing: Dancing, Ecstasy and Vitality*, London: Routledge.

Mandelkow, K. R. (2001) 'Weimarer Klassik: Gegenwart und Vergangenheit eines Deutschen Mythos', in Gutjahr, O. and Segeberg, H. (eds.) *Klassik und Anti-Klassik: Goethe und seine Epoche*, Würzburg: Königshausen & Neumann, 5–17.

Marcuse, P. (1989) '"Dual City": a muddy metaphor for a quartered city', *International Journal of Urban and Regional Research*, 13(4), 309–44.

Marcuse, P. (1990) *A German Way of Revolution*, DDR-Tagebuch eines Amerikaners, Berlin: Dietz.

Marks, R. (1967) 'Size of market worrying manufacturers', *High Point Enterprise*, 20 Oct., A 15.

Marks, R. (1975) 'Theater opening begins "New Era"', *High Point Enterprise*, 6 Oct., B 1–2.

Markusen, A., Lee, Y.-S. and DiGiovanna, S. (eds.) (1999) *Second Tier Cities: Rapid Growth Beyond the Metropolis*, Minneapolis: University of Minnesota Press.

Marshall, A. (1919) *Principles of Economics*, London: Faber and Faber.

Marshall, R. (2001a) 'Contemporary urban space-making at the water's edge', in Marshall, R. (ed.) *Waterfronts in Post-Industrial Cities*, London: Spon Press, 3–15.

Marshall, R. (2001b) 'Connection to the waterfront: Vancouver and Sydney', in Marshall, R. (ed.) *Waterfronts in Post-Industrial Cities*, London: Spon Press, 17–38.

Martin, B. (1988) 'Bobbi Martin', *High Point Enterprise*, 20 Oct., B 1.

Massey, D. (1993) '"Power-geometry" and a progressive sense of place', in Bird, J., Curtis, B., Robertson, G. and Tricker, L. (eds.) *Mapping the Futures: Local Culture, Global Change*. London: Routledge, 59–69.

Massey, D. (1998) 'The spatial construction of youth cultures', in Skelton, T. and Valentine, G. (eds.) *Cool Places: Geographies of Youth Cultures*. London: Routledge.

Massey, D. (1999) 'Cities in the world', in Massey, D., Allen, J. and Pile, S. (eds.) *City Worlds*, London: Routledge.

Massey, D., Quintas, P. and Wield, D. (1992) *High Tech Fantasies,* London: Routledge.

Matarrasso, F. (1997) *Use or Ornament? The Social Impact of Participation in the Arts*, Stroud: Demos.

Matthews, H. (2003) 'Inaugural editorial: coming of age of children's geographies', *Children's Geographies*, 1(1), 3–5.

Matthews, H., Limb, M. and Taylor, M. (1999) 'Young people's participation and representation in society', *Geoforum*, 30, 135–44.

Mauersberger, V. (1999) *Hitler in Weimar: der Fall einer Deutschen Kulturstadt*, Berlin: Rowohlt.

McCann, E. (2004) '"Best places": interurban competition, quality of life and popular media discourse', *Urban Studies*, 41(10), 1909–29.

McCarthy, J. (2005) 'Promoting image and identity in "cultural quarters": the case of Dundee', *Local Planning*, 20(3), 280–93.

McCarthy, J. P. (2006) *From Property to People*, Aldershot: Ashgate.

McCarthy, J. P. and Pollock, S. H. A. (1997) 'Urban regeneration in Glasgow and Dundee: a comparative evaluation of practice', *Land Use Policy*, 14(2), 137–49.

McCrone, D. (1994) *Understanding Scotland: The Sociology of a Stateless Nation*, London: Routledge.

McKay, G. (1996) *Senseless Acts of Beauty: Cultures of Resistance since the Sixties*, London: Verso.

McKie, R. (1972) *Singapore*, Sydney: Angus and Robertson.

McLaughlin, K. (1990) *Waterloo: An Illustrated History*, Windsor, Ont.: Windsor Publications.

McNaughton, N. (1998) *Local and Regional Government in Britain*, London: Hodder and Stoughton.

McRobbie, A. (1991) *Feminism and Youth Culture: From 'Jackie' to 'Just Seventeen'*, London: Macmillan.

Meller, H. E. (ed.) (1979) *The Ideal City, including Barnett's The Ideal City (1893) and Geddes' Civics: as Applied Sociology (1905)*, Leicester: Leicester University Press.

Meller, H. E. (2001) *European Cities 1890–1930s: History, Culture and the Built Environment*, Chichester: Wiley.

Mercer, C. (2003) 'From indicators to governance to the mainstream: tools for cultural policy and citizenship', prepared for *Accounting for Culture: Examining the Building Blocks of Cultural Citizenship Colloquium*, Canadian Cultural Research Network.

Merseburger, P. (1999) *Mythos Weimar. Zwischen Geist und Macht*, Frankfurt: DTV.

Miles, M. (2000) *The Uses of Decoration: Essays on the Architectural Everyday*, Chichester: Wiley.

Miles, M. (2004) 'Drawn and quartered: El Raval and the Haussmannisation of Barcelona', Bell, D. and Jayne, M. (eds.) *City of Quarters: Urban Villages in the Contemporary City*, Aldershot: Ashgate, 37–55.

Miller , S. (2004) 'In New England, a city revival built on creativity', *The Christian Science Monitor*, 28 Sept.

Millington, S. (2002) *Place Marketing as a Entrepreneurial Strategy in Local Governance*, Unpublished PhD Thesis, Manchester: Manchester Metropolitan University.

Mills, C. (1993) 'Myths and meanings of gentrification', in Duncan, J. and Ley, D. (eds.) *Place/Culture/Representation*, New York: Routledge, 149–70.

Mills, D. and Brown, P. (2004) *Art and Wellbeing*. Australian Council for the Arts, New South Wales: Australia Council.

Ministry of Trade and Industry (1986) *Report of the Economic Committee. The Singapore Economy: New Directions*, Singapore: Ministry of Trade and Industry.

Ministry of Trade and Industry (1998) *Committee on Singapore's Competitiveness*, Singapore: Ministry of Trade and Industry.

Mommaas, H. (2004) 'Cultural clusters and the post-industrial city: towards the remapping of urban cultural policy', *Urban Studies*, 41(3), 507–32.

Morales, C. (2005) 'Student Life in Weimar', in Eckardt, F. and Karwinska, A. (eds.) *Urban Magic. Cultural Life in Weimar and Krakow*, Marburg: Tectum.

Moss, L. (2002) 'Sheffield's Cultural Industries Quarter 20 years on: what can be learned from a pioneering example', *International Journal of Cultural Policy*, 8(2), 211–19.

Mumford, L. (1968) *The City in History*, Fort Washington, PA: Harvest Books.

Murray, P. (2004) 'Brighton and Hove: a natural festival', *Local Economy*, 19(4), 420–22.

Myers-Jones, H. J. and Brooker-Gross, S. R. (1994) 'Newspapers as promotional strategists for regional definition', in Gold, J. R. and Ward, S. V. (eds.) *Place Promotion: The Use of Publicity and Marketing to Sell Towns and Regions*, Chichester: John Wiley, 195–212.

Nash, P. H. and Carney, G. O. (1996) 'The seven themes of music geography', *The Canadian Geographer*, 40, 69–74.

Nathan, M. (2005) *The Wrong Stuff: Creative Class Theory, Diversity And City Performance*. London: Institute for Public Policy Research.

National Economics (2002) *State of the Regions Report 2002. Summary and Regional Indicators*. Clifton Hill, Vic.: National Economic and the Australian Local Government Acossiation.

Nelson, R. (ed.) (1993) *National Innovation Systems,* Oxford: Oxford University Press.

Nelson, R. (2005) 'A cultural hinterland? Searching for the creative class in the small Canadian city', in Garrett-Petts, W. (ed.) *The Small Cities Book: On The Cultural Future of Small Cities*, Vancouver: New Star Books.

Newman, M. (2005) 'Developers in Las Vegas put money on furniture', *New York Times*, 23 Feb., C 8.

Niethammer, L. (1990) *Die Volkseigene Erfahrung*, Berlin: Rowohlt.

Noteboom, C. (1991) *Berliner Notizen*, Frankfurt: Suhrkamp.

Oaksey, J. (2000) 'Johnson makes the most of his second chance', *Daily Telegraph*, 17 Mar., 40.

Office of Federal Housing Enterprise Oversight (2005) *House Price Index*, last accessed 19 Mar. 2006 from http://www.ofheo.gov/HPI.asp

Office of National Statistics (2001) *The 2001 Census*, London: Office of National Statistics.

Office of the Deputy Prime Minister (ODPM) (2003) *Sustainable Communities: Building for the Future*. London: Office of the Deputy Prime Minister.

Olwig, K. (2002) *Landscape, Nature, and the Body Politic: From Britain's Renaissance to America's New World*, Madison: University of Wisconsin Press.

Ooi, C. S. (2002) *Cultural Tourism and Tourism Cultures: The Business of Mediating Experiences in Copenhagen and Singapore*, Copenhagen: Copenhagen Business School Press.

Overesch, M. (1995) *Buchenwald und die DDR oder die Suche nach Selbstlegitimation*, Göttingen: Vandenhoeck & Ruprecht.

Page, S. (1995) *Urban Tourism*, London: Routledge.
Paradis, T. (2000a) 'Conceptualizing small towns as urban places: the process of downtown redevelopment in Galena, Illinois', *Urban Geography*, 21, 61–82.
Paradis, T. (2000b) 'Main Street transformed: community sense of place for non-metropolitan tourism business districts', *Urban Geography*, 21, 609–39.
Paradis, T. (2002) 'The political economy of theme development in small urban places: the case of Roswell, New Mexico', *Tourism Geographies*, 4, 22–43.
Parker, M. (2000) 'Identifying Stoke: contesting North Staffordshire', in Edensor, T. (ed.) *Reclaiming Stoke-on-Trent: Leisure, Space and Identity in The Potteries*, Stoke-on-Trent: Staffordshire University Press.
Parkinson, M. and Bianchini, F. (eds.) (1993) *Cultural Policy and Urban Regneration: The Western European Experience*. Manchester: Manchester University Press.
Parr, J. B. (2005) 'Perspectives on the city-region', *Regional Studies,* 39, 555–66.
Peel, D. (2006) 'Varsity real estate in Scotland: new visions for Town and Gown?', in Perry, D. C. and Wiewel, W. (eds.) *Universities as Developers: International Experiences*, Boston, MA: The Lincoln Institute.
Peel, D. and Lloyd, M. G. (2005) 'City-Visions: visioning and delivering Scotland's economic future?' *Local Economy*, 20(1), 40–52.
Phelps, N. A. and Tewdwr-Jones, M. (2000) 'Scratching the surface of collaborative and associative governance: identifying the diversity of social action in institutional capacity building', *Environment and Planning A,* 32, 111–30.
Phillips, A. D. M. (ed.) (1993) *The Potteries: Continuity and Change in a Staffordshire Conurbation*, Stroud: Allan Sutton.
Pierre, J. (1999) 'Models of urban governance: the institutional dimension of urban politics', *Urban Affairs Review*, 34(3), 372–96.
Pile, S. (1999) 'What is a city?', in Massey, D., Allen, J. and Pile, S. (eds.) *City Worlds*, London: Routledge.
Pine, J. and Gilmore, J. H. (1999) *The Experience Economy: Work is Theatre and Every Business a Stage*, Boston: Harvard Business School Press.
Pippi, L. (2004) 'Weimar West', in Eckardt, F. (ed.) *Die soziale Stadt in Thüringen*, Marburg: Tectum, 129–52.
Ploch, L. A. (1984) *Recent Immigration to Maine*, Department of Agricultural and Resource Economics Staff Paper, ARE 369, Orono: University of Maine.
Ploch , L. A. (1985) *Economic and Occupational Aspects of Recent Immigration to Maine*, Department of Agricultural and Resource Economics Staff Paper, ARE 371, Orono: University of Maine.
Podmore, J. (1998) '(Re)reading the "loft living" habitus in Montreal's inner city', *International Journal of Urban and Regional Research*, 22(2), 283–302.
Poon, A. (1989) 'Competitive strategies for a "new tourism"', in Cooper, C. P. (ed.), *Progress in Tourism, Recreation and Hospitality Management*, London: Belhaven Press, 91–102.
Poon, A. (1996) *Tourism, Technology and Competitive Strategies*, Wallingford: CAB International.
Porter, M. E. (1993) 'The competitive advantage of the inner city', *Harvard Business Review*, May–June.
Porter, M. E. (1990) *The Competitive Advantage of Nations*, New York: The Free Association Press.
Portland City Planning Board (1946) *Urban Redevelopment for Portland*, Portland: City Planning Board.

Portland Market (2001) www.portlandmarket.com/main/aboutmkt.html (last accessed 19 Mar. 2006).

Poundbury (2003) *Poundbury, 10th Anniversary Exhibition Review*, Dorchester: Poundbury Publishing and *Dorset Magazine*.

Power, A. and Mumford, K. (1999) *The Slow Death of Great Cities? Urban Abandonment or Urban Renaissance*?, York: Joseph Rowntree Trust.

Prakash, O. (ed.) (1982) *Small Cities and National Development*, Nagoya: United Nations Centre for Regional Development.

Pratt, A. (1999) *Technological and Organisational Change in the European Audio-Visual Industries: An Exploratory Analysis of the Consequences for Employment*, London: European AudioVisual Observatory/London School of Economics.

Pratt, A. (2006) 'The new economy or the emperor's new clothes', in Daniels, P.W. and Beaverstock, J. (eds.) *Geographies of the New Economy*, London: Routledge.

Putnam, R. (1993) *Making Democracy Work*, Princeton, NJ: Princeton University Press.

Putnam, R. (1995) 'Bowling alone: America's declining social capital', *Journal of Democracy,* 6(1), 65–78.

Putnam, R. D. (2000) *Bowling Alone: The Collapse and Revival of American Community*, New York: Simon & Schuster.

Rasche, R. (1992) 'Südviertel 1990', in Schriften der HAB Weimar (ed.) *Europäische Provienz. Weimar*, Weimar: HAB, 107–123.

Ravetz, A. (1980) *Remaking Cities*, London: Croom Helm.

Reed, J. (2000) 'Jumping to the summit', *Guardian*, 13 Mar., 6.

Regnier, P. (1993) 'Spreading Singapore's wings worldwide: a review of traditional and new investment strategies', *The Pacific Review*, 6, 305–12.

Reis, E. (1986) 'Battle lines drawn in dispute over Downtown plans'. *High Point Enterprise*, 26 Oct., A 1–2.

Riedel, M. (2000) *Nietzsche in Weimar: ein Deutsches Drama*, Leipzig: Reclam.

Rifkin, J. (2000) *The Age of Access: How the Shift from Ownership to Access is Transforming Capitalism*, London: Penguin.

Robbins, B. (ed.) (1993) *The Phantom Public Sphere*, Minneapolis, University of Minnesota Press.

Robbins, E. (1996), 'Thinking space/seeing space: Thamesmead revisited', *Urban Design International*, 1(3), 283–91.

Robertson, K. (2001) 'Downtown development principles for small cities', in Burayidi, M. (ed.) *Downtowns: Revitalizing the Centers of Small Urban Communities*, New York: Garland.

Robertson, R. (1992) *Globalisation: Social Theory and Global Culture*, London: Sage.

Robinson, J. (2002) 'Global and world cities: a view from off the map', *International Journal of Urban and Regional Research*, 26, 531–54.

Rodger, R. (1996) 'Urbanisation in Twentieth Century Scotland', in Devine, T. M. and Finlay, R. J. (eds.) *Scotland in the Twentieth Century*, Edinburgh: Edinburgh University Press, 122–52.

Rodgers, P. (1989) *A Summary of the Economic Importance of the Arts in Britain: A Research Volume in Four Parts by John Myerscough*, London: Policy Studies Institute.

Rofe, M. (2003) '"I want to be global": theorising the gentrifying class as an emergent elite global community', *Urban Studies*, 40(12), 2511–26.

Rogers, R. (1997) *Cities for a Small Planet*, London: Faber and Faber.

Rogers, R. and Power, A. (2000) *Cities for a Small Country*, London: Faber and Faber.

Romer, P., (1993) 'Ideas and things: the concept of production is being retooled', *The Economist*, 150(suppl.), 11 Sept., 64–8.

Rondinelli, D. (1983) 'Towns and small cities in developing countries', *The Geographical Review*, 73, 379–95.

Rossi, U. (2004) 'New regionalism contested: some remarks in light of the case of the Mezzogiorno of Italy', *International Journal of Urban and Regional Research*, 28, 466–76.

Roth, S. (2003) 'Goethe and Buchenwald: Re-constructing German national identity in the Weimar Year 1999', in Daly, P. M. and Frischkopf, H. W. (eds.) *Why Weimar?: Questioning the Legacy of Weimar from Goethe to 1999*, McGill European Studies, Vol. 5, New York: Lang, 93–106.

Ruddick, D. (1996) *Young and Homeless in Hollywood*, London: Routledge.

Ruddick, S. (1998) 'Modernism and resistance: how "homeless" youth sub-cultures make a difference', in Skelton, T. and Valentine, G. (eds.) *Cool Places: Geographies of Youth Cultures*, London: Routledge, 343–60.

Ryan, B. (1992) *Making Capital from Culture: The Corporate Form of Capitalist Cultural Production*, Berlin: Walter de Gruyter.

Safranski, R. (2004) *Friedrich Schiller oder die Erfindung des deutschen Idealismus*, München: Carl Hanser.

Sandercock, L. (1998) *Towards Cosmopolis*, Chichester: Wiley.

Sanders, I. and Helfgo, J. (1977) *Portland, Maine: Upbeat Downeast – A Community Social Profile*, Community Sociology Monograph Series, Vol. V, Community Sociology Training Program, Department of Sociology, Boston: Boston University.

Santagata, W. (2002) 'Cultural districts, property rights and sustainable economic growth', *International Journal of Urban and Regional Research*, 26(1), 9–23.

Savill, R. (2001) 'Cheltenham's hotels saddled with a loser', http://news.telegraph.co.uk/news/main.jhtml?xml=/news/2001/03/09/nfnm309.xml (last accessed 9 Mar., 2005)

Schätzke, A. (1991) *Zwischen Bauhaus und Stalinallee: Architekturdiskussion im Östlichen Deutschland 1945–1955*, Braunschweig: Vieweg.

Schley, J. (1999) *Nachbar Buchenwald: die Stadt Weimar und ihr Konzentrationslager, 1937–1945*, Köln: Böhlau.

Schlosberg, J. (1988) 'Portland pulls it off', *American Demographics* (Mar.), 45–8.

Schoon, N. (2001) *The Chosen City*, London: Spon.

Schultz, J. (1985) *Steel City Blues*, Ringwood, Vic: Penguin Books.

Schulz, M. S. (2003) 'Cultural politics after Buchenwald: imagining Weimar', in Daly, P. M. and Frischkopf, H. W. (eds.) *Why Weimar?: Questioning the Legacy of Weimar from Goethe to 1999*, McGill European Studies, Vol. 5, New York: Lang, 39–47.

Schwartz, W. and Schwartz, D. (1998) *Living Lightly: Travels in Post-Consumer Society*, Charlbury: Jon Carpenter.

Scott, A. J. (1996) 'The collective order of flexible production agglomerations: lessons for local economic development policy and strategic choice', *Economic Geography*, 72(3), 219–313.

Scott, A. J. (2000) *The Cultural Economy of Cities: Essays on the Geography of Image-Producing Industries*, London: Sage.

Scottish Executive (2000a) *Scotland's National Cultural Strategy*, Edinburgh: Scottish Executive.

Scottish Executive (2000b) *The Way Forward: Framework for Economic Development in Scotland*, Glasgow: Scottish Executive.

Scottish Executive (2002a) *Better Cities, New Challenges: A Review of Scotland's Cities*, Edinburgh: The Stationery Office.

Scottish Executive (2002b) *Meeting the Needs ... Priorities, Actions and Targets for Sustainable Development in Scotland* (Paper 2002/14), Edinburgh: Scottish Executive.

Scottish Executive (2002c) *Better Communities in Scotland: Closing the Gap,* Edinburgh: Scottish Executive.

Scottish Executive (2002d) *Building Better Cities: Guidance and Next Steps*, Edinburgh: Scottish Executive.

Scottish Executive (2003a) *Partnership Agreement*, Edinburgh: Scottish Executive.

Scottish Executive (2003b) *Building Better Cities – A Policy Statement*, Edinburgh: Scottish Executive.

Scottish Executive (2003c) *The Cities Review*, Edinburgh: Scottish Executive.

Scottish Executive (2004a) *National Planning Framework*, Edinburgh: Scottish Executive.

Scottish Executive (2004b) *A Framework for Economic Development*, Edinburgh: Scottish Executive.

Scudamore, P. (2000) 'The fear of failure that drives McCoy', *Daily Mail*, 13 Mar., 69.

Sharpe, D. (2004) '*Film Illawarra 05–08*. A proposal for a continuing regional film office for the Illawarra', Wollongong: Film Illawarra.

Shaw, S., Bagwell, S. and Karmowska, J. (2004) 'Ethnoscapes as spectacle: reimaging multicultural districts as new destinations for leisure and tourism', *Urban Studies*, 41(10), 1983–2000.

Sheehan, S. M. (2003) *Anarchism*, London: Reaktion.

Sheffield City Council (2005) *Closing the Gap*, Cabinet Report: Sheffield.

Shields, R. (1991) *Places on the Margins: Alternative Geographies of Modernity*, London and New York: Routledge.

Shipley, R., Feick, R., Hall, B. and Earley, R. (2004) 'Evaluating municipal visioning', *Planning Practice and Research*, 19(2), 195–210.

Shires, W. (1963) 'Furniture bellwether of economy', *High Point Enterprise*, 23 Oct., A 1.

Sibley, D. (1988) 'Survey 13: purification of space', *Environment and Planning D: Society and Space*, 6, 409–21.

Simmonds, N. (1997) 'A beach for a backyard', *Business Traveller Asia Pacific*, Oct., 60–67.

Simmonds, R. and Hack, G. (eds.) (2000) *Global City Regions: Their Emerging Forms*, London: Spon.

Singapore Tourist Board (undated) *STB Publicity Material for Travel Agents*, Singapore: Singapore Tourism Board.

Singapore Tourist Promotion Board (1966) *VIP, The Prestige Magazine*, Singapore: Singapore Tourist Promotion Board.

Singapore Tourist Board (1996) *Tourism 21: Vision of a Tourism Capital*, Singapore: Singapore Tourism Board and National Tourism Plan Committees.

Singapore Tourist Promotion Board (1993a) *A Handy Guide to Singapore*, Singapore: Singapore Tourist Promotion Board.

Singapore Tourist Promotion Board (1993b) *Strategic Plan for Growth 1993–1995*, Singapore: Singapore Tourist Promotion Board.

Sklair, L. (1995) *Sociology of the Global System*, London: Prentice-Hall.

Smeenk, B. (1992) 'Lokalisatie na 1992', *Economisch en Social Tijdschrift* (2), 299–317.

Smith, D. (2004) '"Studentification": the gentrification factory?', in Atkinson, R. and Bridge, G. (eds.) *The New Urban Colonialism*, London: Routledge.

Smith, L. (1996) *Dynamic Profile of the Economy of the Waterloo Region*, Kitchener, ON: Regional Municipality of Waterloo.

Smith, N. (1996) *The New Urban Frontier: Gentrification and the Revanchist City*, London: Routledge.

SOCDS (2005) *State of the Cities Data Systems (SOCDS)*, US Department of Housing and Urban Development, (last accessed 19 Mar. 2006 from http://socds.huduser.org/index.html).

Soja, E. (1989) *Postmodern Geographies: The Reassertion of Space in Critical Social Theory*, London: Verso.

Sparke, M., Sidaway, J., Bunnell, T. and Grundy-Warr, C. (2004) 'Triangulating the borderless world: geographies of power in the Indonesia-Malaysia-Singapore growth triangle', *Transactions of the Institute of British Geographers*, 29, 485–98.

Stephens, C. (1986) 'Jury still out on success of market', *High Point Enterprise*, 22 Oct., A 1,11.

Stevens, Q. and Dovey, K. (2004) 'Appropriating the spectacle: play and politics in a leisure landscape', *Journal of Urban Design*, 9(3), 351–65.

Stevenson, D. (2004) '"Civic Gold" rush: cultural planning and the politics of the Third Way', *International Journal of Cultural Policy*, 10(1), 119–31.

Storper, M. (1995) 'The resurgence of regional economies, ten years later', *European Urban and Regional Studies*, 2(3), 191–221.

Storper, M. (1997) *The Regional World: Territorial Development in a Global Economy,* New York: Guilford Press.

Sutherland, A. (1995) 'Birth of a cultural arts district', *Portland Press Herald,* 17 Dec., 1B.

Talen, E. (1999) 'Sense of community and neighbourhood form: an assessment of the social doctrine of new urbanism', *Urban Studies*, 36, 1361–79.

Taylor, P. (2001) 'Sports facility development and the role of forecasting: a retrospective on swimming in Sheffield', in Gratton, C. and Henry, I. (eds.) *Sport in the City. The Role of Sport in Economic and Social Regeneration*, London: Routledge, 214–26.

Teo, C. C. (1982) *The Mental Images of Package Tourists: A Study of Singapore's Tourist Attractions*, Singapore: Department of Geography unpublished Master's dissertation, University of Singapore.

Teo, P. and Chang, T. C. (2001) 'From rhetoric to reality: cultural heritage and tourism in Singapore', in Low, L. and Johnston, D. (eds.) *Singapore Inc. Public Policy Options in the Third Millennium*, Singapore: Asia Pacific Press, 273–303.

Tewdwr-Jones, M. and McNeill, D. (2000) 'The politics of city-region planning and governance – reconciling the national, regional and urban in the competing voices of institutional restructuring', *European Urban and Regional Studies*, 7(2), 119–34.

Tham, E. (2001) 'Regionalisation as a strategy for Singapore's tourism development', in Tan, E. S., Yeoh, B. S. A. and Wang, J. (eds.) *Tourism Managements and Policy. Perspectives from Singapore*, Singapore: World Scientific, 50–54.

The Course Inspector (2000) 'Party invitation extends to all', *The Times*, 15 Mar., 37.

*The Straits Times* and *The Straits Times Weekly Edition*, various issues, Singapore: Singapore Press Holdings.

Theile, G. (2000), 'The Weimar myth: from city of the arts to Gobal village', in Henke, B. and Kott, S. (eds.) *Unwrapping Goethe's Weimar: Essays in Cultural Studies and Local Knowledge*, Rochester: Camden House, 310–28.

Thrift, N. (2000) 'Not a straight line but a curve, or, cities are not mirrors of modernity', in Bell, D. and Haddour, A. (eds.) *City Visions*, Harlow: Prentice Hall.

Timothy, D. (2003) 'Tourism planning in Southeast Asia: bringing down borders through cooperation', in Chon, K. (ed.), *Tourism in Southeast Asia: A New Direction*, New York: Haworth Hospitality Press, 2–38.

Tomaney, J. and Ward, N. (2001) *A Region in Transition: The North East at the Millennium*, London: Ashgate.

Tomlinson, J. (1999) *Globalisation and Culture*, Cambridge: Polity Press.

Toon, I. (2000) '"Finding a place in the street": CCTV surveillance and young people's use of urban public space', in Bell, D. and Haddour, A. (eds.) *City Visions*, London: Prentice Hall, 141–65.

Toronto Culture (2003) *Culture Plan: For the Creative City*, Toronto: City of Toronto.

Turner, J. F. C. (1976) *Housing By People*, London: Marion Boyars.

United Nations Development Program (2003): *UNDP Human Development Report 2003: Mauritius 2003*, New York: UNDP.

Urban and Economic Development Group (URBED) (1988) *Developing the Cultural Industries Quarter in Sheffield*, Sheffield: Sheffield City Council.

Urry, J. (1995) *Consuming Places*, London: Routledge.

Valentine, G. (1996) 'Angels and devils: moral landscapes of childhood', *Environment and Planning D: Society and Space*, 14, 581–99.

Vall, N. (1999) 'Where are you from?' *Northern Review*, 8, 27–34.

Vanderbeck, R. and Johnson, J. H. (2000) '"That's the only place where you can hang out": urban young people and the space of the mall', *Urban Geography*, 21, 5–25.

Varma, R. (2004) 'Provincializing the global city: from Bombay to Mumbai', *Social Text*, 81, 65–89.

Vasagar, J., Kelso, P. and Hall, S. (2001) 'Cost of outbreak soars as crisis goes on' http://www.guardian.co.uk/footandmouth/story/0,7369,452133.00.html (last accessed 19 Mar. 2006).

Vasil, R. (1992) *Governing Singapore*, Singapore: Mandarin Paperbacks.

Vegh, S. G. (1996) 'Downtown lures a lot of people with its new mix', *Portland Press Herald*, 8 Dec., 1B, 8B.

Verwijnen, J. and Lehtovuori, P. (eds.), (1999) *Creative Cities: Cultural Industries, Urban Development and the Information Society*, Helsinki: University of Art and Design.

Violette, Z. (2005) *Aesthetics of Class and Ethnicity in the Renewal of Portland's Vine-Deer-Chatham Neighborhood, 1920–1960*, Undergraduate honors thesis, Department of History, University of Southern Maine.

Vlandys, V. (1999) *Survey of Outdoor Public Art*, Cultural Services Division, Wollongong: Wollongong City Council.

Voye, L. (1985) 'Le retour à la ville: Qui revient où et pourquoi?' *Revue Belge de Géographie*, (4), 183–94.

Waitt, G. (2004) 'Pyrmont-Ultimo: the newest chic quarter of Sydney', in Bell, D. and Jayne, M. (eds.) *City of Quarters: Urban Villages in the Contemporary City*, Aldershot: Ashgate, 15–36.

Walker, A. (1999) 'Creating communities is the way forward', *Westcountry News*, 17 July, 27.

Walker, D. F. (1987) 'Expansion and adaptation in the post-war years', in Walker, D. F. (ed.) *Manufacturing in Kitchener-Waterloo: A Long-Term Perspective*, Waterloo, ON: University of Waterloo Press, 61–82.

Ward, K. and Jonas, A. (2004) 'Competitive city-regionalism as a politics of space: a critical reinterpretation of the new regionalism', *Environment and Planning A*, 36, 2119–39.

Ward, S. V. (1998) *Selling Places: The Marketing and Promotion of Towns and Cities 1850–2000*, London: E. & F. N. Spon.

Waterman, S. (1998) 'Carnivals for elites? The cultural politics of arts festivals', *Progress in Human Geography*, 22(1), 54–74.

Watson, S. (1990) 'Gilding the smokestacks: The symbolic representations of deindustrialised regions', *Environment and Planning D: Society and Space*, 9, 59–70.

Watt, P. and Stenson, L. (1998) 'The street: "It's a bit dodgy around there": safety, danger, ethnicity and young people's use of public space', in Skelton, T. and Valentine, G. (eds.) *Cool Places: Geographies of Youth Cultures*, London: Routledge.

Weaver, P. (2001) 'Counting the cost of lost festival', http://www.guardian.co.uk/Archive/Article/0,4273,4148945,00.html (last accessed 9 Mar., 2001).

Weiske, C. and Schäfer, U. (1992) 'Europäische Provinz Weimar – Deutung und Selbstdeutung', in Schriften der HAB Weimar (ed.) *Europäische Provinz*, Weimar: HAB, 1–14.

Whatley, C. A. (1991) 'The making of "Juteopolis" – and how it was', in Whatley, C. A. (ed.) *The Remaking of Juteopolis*, Proceedings of the Abertay Historical Society's Octocentenary Conference, Abertay Historical Society. Dundee: Abertay Historical Society, 7–22.

Whittington, E. (1998) 'Showrooms continue to flourish', *High Point Enterprise*, 13 Jan., A 1.

Williams, R. J. (2004) *The Anxious City*, London: Routledge.

Wilson, E. (1991) *The Sphinx in the City*, Berkeley: University of California Press.

Wilson, P. L. (1999) 'Avant Gardening', in Wilson, P. L. and Weinberg, B. (eds.) *Avant Gardening: Ecological Struggle in the City and the World*, New York: Autonomedia, 7–34.

Wilson, W. D. (2000) 'Skeletons in Goethe's closet: human rights, protest and the myth of political liberty', in Henke, B. and Kott, S. (eds.) *Unwrapping Goethe's Weimar: Essays in Cultural Studies and Local Knowledge*, Rochester: Camden House, 295–309.

Wollongong City Council (2004) *The Wollongong Image Campaign 1999–2004: Re-branding a City*, Wollongong: Wollongong City Council.

Wood, P. and Taylor, C. (2004) 'Big ideas for a small town: the Huddersfield Creative Town Initiative', *Local Economy*, 19(4), 380–95.

World Commission on Culture and Development (1995) *Our Creative Diversity: Report on the World Commission on Culture and Development*, Paris: UNESCO.

Wright, J. (1999) 'Clearcutting the East Village', in Wilson, P. L. and Weinberg, B. (eds.) *Avant Gardening: Ecological Struggle in the City and the World*, New York: Autonomedia, 127–31.

Yeung, H. W. C. (1998) 'The political economy of transnational corporations: a study of the regionalization of Singaporean firms', *Political Geography*, 17, 389–416.

Yeung, H. W. C. (2000a) 'State intervention and neoliberalism in the globalizing world economy: lessons from Singapore's regionalization programme', *The Pacific Review*, 13, 133–62.

Yeung, H. W. C. (2000b) 'Local politics and foreign ventures in China's transitional economy: the political economy of Singaporean investments in China', *Political Geography*, 19, 809–40.

Young, C. and Lever, J. (1997) 'Place promotion, economic location and the consumption of city image', *Tijdschrift voor Economische en Sociale Geografie*, 88(4), 332–41.

Zukin, S. (1982) *Loft Living: Culture and Capital in Urban Change*, Baltimore, MD: Johns Hopkins University Press.

Zukin, S. (1990) 'Socio-spatial prototypes of a new organisation of consumption: The role of cultural capital', *Sociology* (24), 35–56.

Zukin, S. (1995) *The Cultures of Cities*, Cambridge, MA: Blackwell.

Zukin, S. (1998) 'Urban lifestyles: diversity and standardisation in spaces of consumption', *Urban Studies*, 35(1), 825–39.

Zukin, S. and Smith Macquire, J. S. (2004) 'Consumers and consumption', *Annual Review of Sociology*, 30, 173–97.

# Index